Advances Research of Volatile Compounds, Composition, Stability and Thermal Behavior of Foods

Advances Research of Volatile Compounds, Composition, Stability and Thermal Behavior of Foods

Editor

Thomas Dippong

Basel • Beijing • Wuhan • Barcelona • Belgrade • Novi Sad • Cluj • Manchester

Editor
Thomas Dippong
Chemistry and Biology
Technical University of Cluj Napoca
Cluj Napoca
Romania

Editorial Office
MDPI
St. Alban-Anlage 66
4052 Basel, Switzerland

This is a reprint of articles from the Special Issue published online in the open access journal *Foods* (ISSN 2304-8158) (available at: https://www.mdpi.com/journal/foods/special_issues/2BX0KQKKMQ).

For citation purposes, cite each article independently as indicated on the article page online and as indicated below:

Lastname, A.A.; Lastname, B.B. Article Title. *Journal Name* **Year**, *Volume Number*, Page Range.

ISBN 978-3-7258-1496-1 (Hbk)
ISBN 978-3-7258-1495-4 (PDF)
doi.org/10.3390/books978-3-7258-1495-4

© 2024 by the authors. Articles in this book are Open Access and distributed under the Creative Commons Attribution (CC BY) license. The book as a whole is distributed by MDPI under the terms and conditions of the Creative Commons Attribution-NonCommercial-NoDerivs (CC BY-NC-ND) license.

Contents

About the Editor vii

Preface ix

Thomas Dippong, Lacrimioara Senila and Laura Elena Muresan
Preparation and Characterization of the Composition of Volatile Compounds, Fatty Acids and Thermal Behavior of Paprika
Reprinted from: *Foods* **2023**, *12*, 2041, doi:10.3390/foods12102041 1

Flavia Pop and Thomas Dippong
The Antioxidant Effect of Burdock Extract on the Oxidative Stability of Lard and Goose Fat during Heat Treatment
Reprinted from: *Foods* **2024**, *13*, 304, doi:10.3390/foods13020304 19

Baoxiang Zhang, Weiyu Cao, Changyu Li, Yingxue Liu, Zihao Zhao, Hongyan Qin, et al.
Study on the Effect of Different Concentrations of SO_2 on the Volatile Aroma Components of 'Beibinghong' Ice Wine
Reprinted from: *Foods* **2024**, *13*, 1247, doi:10.3390/foods13081247 36

Yanli He, Hongyan Qin, Jinli Wen, Weiyu Cao, Yiping Yan, Yining Sun, et al.
Characterization of Key Compounds of Organic Acids and Aroma Volatiles in Fruits of Different *Actinidia argute* Resources Based on High-Performance Liquid Chromatography (HPLC) and Headspace Gas Chromatography–Ion Mobility Spectrometry (HS-GC-IMS)
Reprinted from: *Foods* **2023**, *12*, 3615, doi:10.3390/foods12193615 58

Zhijian Long, Shilin Zhao, Xiaofeng Xu, Wanning Du, Qiyang Chen and Shanglian Hu
Dynamic Changes in Flavor and Microbiota in Traditionally Fermented Bamboo Shoots (*Chimonobambusa szechuanensis* (Rendle) Keng f.)
Reprinted from: *Foods* **2023**, *12*, 3035, doi:10.3390/foods12163035 86

KübraÖztürk, Zeynep Feyza Yılmaz Oral, Mükerrem Kaya and Güzin Kaban
The Effects of Sheep Tail Fat, Fat Level, and Cooking Time on the Formation of Nε-(carboxymethyl)lysine and Volatile Compounds in Beef Meatballs
Reprinted from: *Foods* **2023**, *12*, 2834, doi:10.3390/foods12152834 101

Wengang Zhang, Xijuan Yang, Jie Zhang, Yongli Lan and Bin Dang
Study on the Changes in Volatile Flavor Compounds in Whole Highland Barley Flour during Accelerated Storage after Different Processing Methods
Reprinted from: *Foods* **2023**, *12*, 2137, doi:10.3390/foods12112137 114

Huaixiang Tian, Ling Zou, Li Li, Chen Chen, Haiyan Yu, Xinxin Ma, et al.
Characterisation of the Aroma Profile and Dynamic Changes in the Flavour of Stinky Tofu during Storage
Reprinted from: *Foods* **2023**, *12*, 1410, doi:10.3390/foods12071410 132

Nur Cebi and Azime Erarslan
Determination of the Antifungal, Antibacterial Activity and Volatile Compound Composition of *Citrus bergamia* Peel Essential Oil
Reprinted from: *Foods* **2023**, *12*, 203, doi:10.3390/foods12010203 146

Agnieszka Pluta-Kubica, Dorota Najgebauer-Lejko, Jacek Domagała, Jana Štefániková and Jozef Golian
The Effect of Cow Breed and Wild Garlic Leaves (*Allium ursinum* L.) on the Sensory Quality, Volatile Compounds, and Physical Properties of Unripened Soft Rennet-Curd Cheese
Reprinted from: *Foods* **2022**, *11*, 3948, doi:10.3390/foods11243948 . . . . **158**

About the Editor

Thomas Dippong

Thomas Dippong (an associate professor at the Technical University of Cluj-Napoca) is a chemical engineer with a Ph.D. and habilitation in chemistry. His current research activities are related to the characterization of nanoparticles for various applications as part of an ongoing research projects in partnership with the Technical University of Cluj-Napoca within the field of ferrite embedded in silica matrix. He is an expert in analytical chemistry, organic and inorganic chemistry, thermal treatment, instrumental analysis, and the synthesis of nanomaterials. Dr. Dippong has published 151 peer-reviewed publications (95 papers in high-ranked scientific ISI-Thomson journals (54 Q1, 17 Q2, and 24 Q3) and 56 in other national and international journals); the cumulative impact factor of the journals in which the 95 articles were published is 400, with 2250 citations and an h-index of 40 (WoS). He has given 35 lectures at international conferences (the 14th ICTAC in Brazil, the 5th CEEC-TAC in Roma, the JTACC in Budapest, the ESTAC in Brasov, etc.). He has also published two books with international publishing houses and fifteen books with national publishing houses. In the last three consecutive years (2021 to 2023), Dr. Thomas Dippong was included in the prestigious list of the world's top 2% researchers. He has been a contract manager for many projects and is currently an active member of four projects. Dr. Dippong has reviewed 560 scientific articles for 110 ISI-Thomson journals. He has been a guest editor for 10 Special Issues published by six prestigious ISI-Thomson journals (five Q1 journals and one Q2 journal).

Preface

This Special Issue of *Foods* contains articles related to spectrophotometric techniques (AAS, ICP, UV-VIS, IR, MS, and NMR), chromatographic characterization (CG, HPLC, and TLC), and other physical–chemical techniques of food analysis, as well as research concerning the thermal behavior of solid foods.

We aimed to include new methods or procedures of food treatment to ensure the stability of food products with less sensory and compositional changes. Thermal treatment has been the most widely utilized technique in the sterilization of foodborne pathogens, inactivating enzymes, drying, thawing, and extracting bioactive compounds. Although this method is convenient and can ensure food safety, it alters, or can even destroy, the sensory, nutritional, and physicochemical properties of food. The use of ultra-high-performance liquid chromatography coupled with mass spectrometry is becoming a valid method/platform in food science for assessing food quality, safety, and traceability in a robust, efficient, sensitive, and cost-effective way.

Volatile organic compounds with low odor thresholds make significant olfactory contributions to solid food flavor. Advanced spectroscopic techniques and chemometric tools can have great applications in food science and technology, as well as in achieving consumer confidence.

The Special Issue topics are as follows: thermal behavior of solid food products; flavor analysis of foods and volatile profile; volatile compounds in food and their transformation during process; volatile oils used for improvements in food stability; chromatographic methods in food analysis; spectroscopic methods in food analysis; physico-chemical characterization and metal content in drinking water; metal composition of foods; and food active packaging and superior valorization of food raw materials.

Thomas Dippong
Editor

Article

Preparation and Characterization of the Composition of Volatile Compounds, Fatty Acids and Thermal Behavior of Paprika

Thomas Dippong [1,*], Lacrimioara Senila [2] and Laura Elena Muresan [3]

[1] Department of Chemistry and Biology, Technical University of Cluj-Napoca, 76 Victoriei Str., 430122 Baia Mare, Romania
[2] INCDO-INOE 2000, Research Institute for Analytical Instrumentation, 67 Donath Street, 400293 Cluj-Napoca, Romania; lacri.senila@icia.ro
[3] Raluca Ripan' Institute for Research in Chemistry, Babes Bolyai University, Fantanele, 30, 400294 Cluj-Napoca, Romania; laura_muresan2003@yahoo.com
* Correspondence: dippong.thomas@yahoo.ro

Abstract: This study aimed to investigate the thermal behavior and composition of volatile compounds, fatty acids and polyphenols in paprika obtained from peppers of different countries. The thermal analysis revealed various transformations in the paprika composition, namely drying, water loss and decomposition of volatile compounds, fatty acids, amino acids, cellulose, hemicellulose and lignin. The main fatty acids found in all paprika oils were linoleic (20.3–64.8%), palmitic (10.6–16.0%) and oleic acid (10.4–18.1%). A notable amount of omega-3 was found in spicy paprika powder varieties. The volatile compounds were classified into six odor classes (citrus (29%), woody (28%), green (18%), fruity (11%), gasoline (10%) and floral (4%)). The total polyphenol content ranged between 5.11 and 10.9 g GA/kg.

Keywords: paprika; thermal behavior; VOCs; sensorial profile; fatty acids

1. Introduction

Paprika is widely used within the food industry as a natural colorant (i.e., in soups, sausages, cheeses and snacks) due to its ability to improve upon the flavor of food through its characteristic taste and pungency [1]. Paprika is obtained by the dehydration of some pepper fruit varieties (*Capsicum annuum* L.) followed by milling of the dried pepper to obtain a fine powder [1]. Drying conditions affect the rehydration capacity of dehydrated paprika quite significantly [2]. The intensity of its characteristic red color is the main quality criterion of paprika powder, although this parameter depends on the variety of pepper used as well as the employed preparation method [1]. Most manufacturers, however, lack the knowledge to produce a safe and standardized food product to prevent contamination with any foreign matter, molds or toxins [3]. In paprika, the most important compounds are carotenoids and capsaicinoids, as well as vitamins E and C [2]. Carotenoids are responsible for the color of paprika, their content within the product being connected to the variety and ripeness of peppers alongside their growing condition (i.e., cool and rainy seasons tend to yield fruit with more β-carotene and technological factors) [2]. The types of pigments found confer paprika its particular color (yellow, green or red). The red pigments are specific to Capsicum species and reveal the presence of pungent capsaicinoids capsaicin and dihydrocapsaicin (dominant constituents), nordihydrocapsaicin and homocapsaicin (minor constituents) [2]. In addition, paprika spice has advantageous health properties, such as analgesic, anti-obesity, cardio-protective or neurologic properties, among others. Therefore, this spice is readily used in the pharmaceutical and cosmetic industries [4]. Recently, the adulteration of condiment powders, such as paprika, pepper, curry, chili and saffron, has increased. In the case of paprika, this nutritional integrity is important as it contains carotenoid pigments, neutral lipids such as tocopherols and vegetable oil [5].

The formation of volatile compounds that generate the characteristic aroma is caused by chemical conversions, such as hydrolytic reactions, amino acid conversion, oxidative degradation reactions of lipids (fatty acids) and carotenoids (lipid-soluble pigments), Maillard reactions and caramelization browning [6]. Some studies highlight the fact that the quality and quantity of aroma and flavor compounds of paprika are decisive parameters for quality control [1]. More than 125 volatile compounds have been identified in fresh and processed paprika, although the significance of these compounds for the aroma is not yet well known [1]. The main VOCs in paprika are esters and terpenoids, followed by other minor compounds, such as lipoxygenase derivatives, nitrogen and sulfur compounds, phenol derivatives, norcarotenoids, carbonyls, alcohols and other hydrocarbons [7]. Esters usually confer a fruity aroma, while many aldehydes, referred to as green leaf volatiles, can create a grassy aroma [7].

Paprika oil contains saturated (SFAs), monounsaturated (MUFAs) and polyunsaturated (PUFAs) fatty acids. SFAs are undesirable in large quantities because they can cause cardiovascular problems. Moreover, they are stable at room temperature, and the unsaturated fatty acids start to increase the fluidity and oxidation process, leading to the formation of free radicals. Replacing SFAs with PUFAs has positive benefits for cholesterol in the blood. Among all PUFAs, linoleic (C18:2) and α-linolenic acids (C18:3) are essential fatty acids for food [8]. Abbeddou and co-workers studied the fatty acid profile of paprika oleoresin. The following content was reported: linoleic acid C18:2 (55.97%), C16:0 (15.16%), C18:1 (13.81%), C18:1 (13.18%), C18:3 (5.11) and small quantities (below 2%) of C12:0, C14:0, C16:1, C18:0, C20:0, C22:0, C24:0 and C20:1 [9]. Zaki et al. (2013) analyzed red varieties of paprika and found that the fat present in paprika is in an esterified form with carotenoids [4]. The analyzed paprika presents a content of lipids of around 8%, a content of carbohydrates of approximately 55% and is rich in some metals (K, P, Ca, Mg) while being poor in Na. Linoleic (C18:2), oleic (C18:1) and palmitic (C16:0) acids are the most predominant FAs [4]. Paprika contains high quantities of PUFAs which have valuable properties in decreasing cholesterol levels and reducing the risk of obesity. Kim et al. (2017) investigated the use of red paprika on the lipid metabolism of obese male mice for eight weeks [10]. The results revealed that red paprika reduces obesity, fatty acid oxidation and lipid droplet size [10]. The fatty acid content depends on the geographical origin, meteorological conditions and production process. Paprika cultivation under traditional conditions was preferred because of its higher quality.

The methods used for oil extraction from plants are solvent extraction, maceration, cold-press extraction, steam distillation, CO_2 extraction, ultrasound, enfleurage and microwave irradiation. Oil is the fluid separated from the plant in the presence of a solvent. The most common solvent used for the extraction of oil is n-hexane. Studies show that polar solvents can extract all fatty acids and are preferred for this purpose. In addition, extraction methods influence the amounts of fatty acids. Combining solvent extraction, microwave irradiation, ultrasonic extraction and supercritical carbon dioxide techniques could significantly improve oil performance [11]. Kostrzewa et al. (2022) used supercritical CO_2 extraction of dry paprika and n-hexane, and the results showed that there is no difference between the two extraction methods. [11]. Ultrasound-assisted extraction is a simple and ecologically safe method for the extraction of oil. The ultrasonic extraction method increases the yield of active compounds such as antioxidants, phenols and fatty acids. The advantages of UAE were lower extraction time, lower volume of solvent and lower temperature of extraction, as well as higher efficiency [12].

This study focused on the determination of the key volatile compounds, the comparative analysis of fatty acids, polyphenol and thermal behavior among wild-harvested and commercial paprika samples obtained from pepper varieties from different countries. Ultrasound-assisted extraction (UAE) was applied to extract oil from paprika powder varieties. The interest of this study consists in the following: (i) the literature is enriched with the obtained extensive characterization of different paprikas, (ii) the relationship between the thermal behavior and the content of volatile compounds, cellulose, hemicellulose, lignin and fatty acids, (iii) the aroma profile and classification of volatile compounds by classes of chemical compounds and types of aromas and (iv) the statistical analysis of the content of volatile substances and fatty acids of the eight varieties of paprika.

2. Materials and Methods

2.1. Chemicals

The chemicals, including methanol (CH_3OH), chloroform ($CHCl_3$), potassium chloride (KCl), sodium sulfate (Na_2SO_4), isooctane (C_8H_{18}), potassium hydroxide (KOH), sodium hydrogen sulphate monohydrate ($NaHSO_4 \cdot H_2O$), ethanol (C_2H_6O), diethyl ether ($(C_2H_5)_2O$), phenolphthalein ($C_{20}H_{14}O_4$), sodium chlorite ($NaClO_2$), sulphuric acid (H_2SO_4) and sodium hydroxide (NaOH), all of analytical grade, were purchased from Merck (Darmstadt, Germany). The standard FAME mixture (Supelco 37 component FAME mix, CRM47885) was purchased from Sigma-Aldrich (St. Louis, MO, USA). Ultrapure water (18.2 $M\Omega \cdot cm^{-1}$ at 20 °C) was obtained from a Direct-Q3 UV Water Purification System (Millipore, Molsheim, France).

2.2. Sample Description

Paprika 1 (P1) was obtained from red Kapia peppers of Romanian origin, light reddish-brown in color, very pleasant in smell, sweet feeling with a well-defined pepper aroma.

Paprika 2 (P2) was obtained from hot peppers, with Romania as the country of origin. The uniform powder had a specific smell of pepper with a spicy taste.

Paprika 3 (P3) was obtained from golden pepper, with Romania as the country of origin—a yellow-brown powder with a specific and pleasant smell.

Paprika 4 (P4) was obtained from hot peppers, with Morocco as the country of origin—light red-orange powder, with a specific smell of peppers and a sweet spicy taste.

Paprika 5 (P5) was obtained from red Kapia peppers, with Turkey as the country of origin, and it was characterized by a light red-orange color, specific sweet pepper smell and a weaker taste compared to the others.

Paprika 6 (P6) is a ground chili originating from India, with a light red-brown color and a very spicy, aromatic and pleasant taste.

Paprika 7 (P7) is obtained from red Kapia peppers originating from China. The taste and smell of this red-orange colored paprika powder is pleasant and specific but weakly pronounced compared to the rest of the samples.

Paprika 8 (P8) is obtained from Kapia peppers originating from Hungary. It is characterized by a red-brick color, with a taste and smell specific to this particular pepper assortment.

Paprika Preparation Methods

Each variety of paprika (P1–P8) was prepared under the same working conditions. After washing the peppers, they were cut into rings and placed in the oven at a temperature of 80 °C and then left to dry until they reached a light red-brown color and acquired a crunchy texture. After cooling, they were chopped.

All the samples were freeze-dried (FreeZone 2.5 LiterBenchtop freeze dry system, Labconco, Kansas, MO, USA) at −40 °C and −25 psi for 24 h to uniformize their moisture content. The freeze-dried samples were ground using an agate mortar and pestle to obtain homogenized powders. The moisture of the samples was determined by drying the samples to a constant mass at 105 °C in a universal oven (UFE 400, Memmert, Schwabach, Germany).

2.3. Extraction of Lipids from Paprika Powder

The dried samples (1 g) were extracted with 20 mL chloroform: methanol (2:1, v/v) and introduced into an ultrasonic bath (ISOLAB, Germany, tank dimensions: 150 × 138 × 65 mm^3, tank volume 1.3 L, ultrasonic power: 60 W, frequency: 40 kHz) for 15 min (repeated for four times) at room temperature, according to Pohndorf et al. (2016) with modifications [13]. The obtained mixture was extracted with 10 mL KCl (0.74%). The extracts were centrifuged

(10 min at 5000 rpm), and the organic phase was recovered. Finally, the organic phase was filtered using Na_2SO_4 to obtain a clear solution. The solvent was evaporated using the rotary evaporator Laborota 4010 (Heidolph, Schwabach, Germany), and the oil obtained was dried at 60 °C in an oven. The yield and the lipid content were calculated using Equation (1).

$$\text{Lipid content (\%)} = \frac{m_L}{m_P} \times 100 \tag{1}$$

where m_L is the extracted lipid weight, and m_p is the mass of dried paprika powder.

2.3.1. Lipid Extraction Yield

The oils obtained using ultrasound extraction with chloroform: methanol (2:1, v/v) were converted to FAME using transesterification methods according to [14]. The fatty acids were separated on Zb-WAX, a polyethylene glycol column stationary phase suitable for separating fatty acids.

2.3.2. Fatty Acid Methyl Esters (FAMEs)

Fatty acid compositions were determined using a gas chromatography coupled with flame ionization detector technique, after lipid extraction and transesterification to fatty acid methyl esters. The obtained lipids were converted into FAMEs by transesterification with potassium hydroxide. The samples (0.06 g) were dissolved in isooctane, treated with 0.2 mL methanolic potassium hydroxide solution (CH_5KO_2) (2 mol/L) and vigorously stirred for 30 s. Lastly, the mixture was treated with 1 g of sodium hydrogen sulphate to avoid saponification of methyl esters and neutralize excess alkali. Each oil sample was trimethylated and analyzed in three replicates.

2.3.3. Free Fatty Acid (FFA) Content from Extracted Oil

The free fatty acids were determined by titrating the oil obtained with KOH (0.1 M in ethanol). An amount of m g of oil was dissolved in a solvent mixture of ethanol: diethyl ether (1:1, v/v), using phenolphthalein (2%) as the indicator. The FFA content was calculated in accordance with Equation (2):

$$\text{FFA} = \frac{56.1 * V * C}{m} \text{mg KOH/g} \tag{2}$$

where 56.1 is the molecular weight of KOH, V is the volume of KOH used for titration (mL), C is the concentration of KOH used for titration, and m is the biomass of oil used for analysis.

2.3.4. GC Analysis

The FAME content was determined using GC-FID (Agilent Technologies, Santa Clara, CA, USA, 6890 N) equipped with a Zebron ZB-WAX capillary column (30 m × 0.25 mm × 0.25 µm) and a flame ionization detector (FID, Agilent Technologies 7683). The gas carrier was helium with a constant flow rate of 1 mL min^{-1}. The injection volume was 1 µL in a 1:20 split mode. The GC oven temperature program consists of three stages: 60 °C for 1 min, 60 to 200 °C (rate 10 °C min^{-1}, 2 min), 200 to 220 °C (5 °C min^{-1}, 20 min). The temperature of the injector and detector was set to 250 °C. FAs in samples were identified by comparing their retention time with that of the Supelco FAME standard mixture.

2.4. Thermal Analysis

The thermal behavior of all samples was evaluated based on thermal analysis (TG-DTA) carried out with a Mettler-Toledo TGA/SDTA851. The measurements were performed at a heating rate of 20 °C/min in a controlled atmosphere using air or nitrogen with a flow rate of 60 mL/min.

2.5. Volatile Composition

For the HS-SPME GC-MS analysis of volatile organic compounds, 3 g of ground paprika P1–P8 was transferred to a 20 mL headspace vial, and 3 mL of NaCl saturated solution was added to enhance the volatile organic compounds in the headspace and to inhibit any enzymatic reactions. The method was developed according to Martín et al., with improvement [1]. The headspace vials were sealed with crimp-top caps with TFE-silicone headspace septa (Thermo Fischer Scientific, Waltham, MA, USA). Each vial was incubated for 20 min at 60 °C. Afterward, the SPME fiber Divinylbenene/Carboxen/Polydimethylsiloxane (50 µm DVB/30 µm CAR/30 µm PDMS) was exposed for 15 min (60 °C) at the headspace of the sample to perform the HS-SPME extraction of volatile organic compounds. Furthermore, the extracted volatile organic compounds were desorbed for 7 min from the fiber coating into the Thermo Fischer Scientific Trace 1310 GC gas chromatograph injection port set at 250 °C. The volatile organic compounds were separated using a DB-WAX capillary column (30 m × 0.25 mm i.d. × 0.25 µm film thickness, J&W Scientific Inc. (Folsom, CA, USA)). Ultrahigh purity helium was used as a carrier gas at a linear velocity of 1 mL/min. The oven temperature program was as follows: initial temperature of 35 °C, heated to 180 °C at a rate of 5 °C·min^{-1}, increased to 230 °C at a rate of 15 °C·min^{-1} and then held at a plateau for 7 min. Mass spectra were recorded in electron impact (EI) ionization mode at 70 eV using a TSQ 9000 MS, Thermo Fischer Scientific mass spectrometer. The quadrupole mass detector, ion source and transfer line temperatures were set at 150, 230 and 280 °C, respectively. Mass spectra were scanned in the range m/z 50–450 amu. VOCs were identified by comparing the mass spectra with the NIST 14 database system library and linear retention index. The criteria for compound identification required a mass spectrum matching score of $\geq$80%. The results were expressed as a percentage of the relative peak area (% RPA) of a peak in each paprika sample that was calculated by dividing the peak area by the total peak area of all identified peaks in each chromatogram. The total ion chromatogram (TIC) of each sample was used for peak area integration.

All measurements were conducted in triplicate, and data are presented as the mean ± standard deviation.

2.6. Antioxidant Characterization

Polyphenols were measured using the Folin–Ciocalteu colorimetric method using a Perkin Elmer Lambda 25 spectrophotometer to measure the blue complex at 760 nm, and gallic acid was used as a reference standard [15]. All measurements were conducted in triplicate, and data are presented as the mean ± standard deviation.

2.7. Cellulose, Hemicellulose and Lignin Content

The content of cellulose, hemicelluloses and lignin in paprika varieties was determined according to Senila et al. [16]. The content of holocellulose (cellulose and hemicelluloses) was determined by delignification of samples with $NaClO_2$ in acetic acid (10%). The content of cellulose was determined by treatment of holocellulose with NaOH (17.5%). The lignin was determined as present residue after treatment of samples with 72% H_2SO_4 solution.

2.8. Statistical Analysis

For the statistical processing of the data, OriginPro Data Analysis and Graphing Software (OriginLab Corporation, Northampton, MA, USA) was used. Descriptive data analyses, including standard deviation and Pearson correlation, explained by very strong correlation (0.9–1), strong correlation (0.70–0.89), moderate correlation (0.40–0.7), weak correlation (0.10–0.39) and negligible correlation, were realized. Two different sets of variables, including fatty acids and volatiles and polyphenols, were evaluated in order to separate the geographical provenance and types of the paprika samples. The paprika samples were grouped according to sets of variable contents by Agglomerative Hierarchical Clustering (AHC) using the squared Euclidian distance and the Ward method for combining clusters, using XLStat software version 2019.3.2 (Addinsoft, Paris, France).

3. Results and Discussion

3.1. Thermal Behavior

The decomposition stages of the paprika samples were investigated in both air (Figure 1) and nitrogen atmospheres (Figure 2) up to 1000 °C. All paprika samples have a similar thermal behavior in three or four stages under the air atmosphere. In the case of P1, P2 and P5, the DTA curve shows the first stage was characterized by an endothermic effect at 64–73 °C, accompanied by a mass loss of 3.6–4.6%, attributed to the evaporation of adsorbed water and solvent [17–19]. The second stage involved the decomposition of volatile compounds and polyphenols characterized by an exothermic effect at 206–214 °C, accompanied by a mass loss of 29.6–33.7% [17–19]. The third stage at 324–344 °C, with a mass loss of 22.1–24.5%, can be attributed to the decomposition of fatty acids and proteins [17–19]. The fourth stage of decomposition for P1 and P5 corresponds to the degradation of cellulose, hemicellulose and lignin, with three exothermic effects at 493–489 °C, 552–560 °C and 590–606 °C accompanied by a mass loss of 36.3%. In the case of P2, it was observed only through two exothermic effects at 490 °C (decomposition of hemicellulose and cellulose) and 618 °C (decomposition lignin), accompanied by a mass loss of 28.7 and 8.1% [14–16]. In the case of the P3, P4, P6, P7 and P8 samples, the DTA curve shows, for the first stage, an endothermic effect between 58 and 89% °C associated with a 2.7–3.6% mass loss on the TG curve. This can be attributed to the drying of paprika powders and the desorption of physically absorbed water molecules. For the second stage, an exothermic effect in the range of 317–346 °C associated with a mass loss of 50.4–58.2% on the TG curve is attributed to the decomposition of volatile compounds and polyphenols, fatty acids and proteins, and for third stage, an exothermic effect at 478–493 °C is associated with a mass loss of 33.7–41.5% corresponding to the degradation of cellulose, hemicellulose and lignin (only in the case of the P8 sample for the third stage a split peak in two exothermic effects at 446 and 487 °C can be observed) [17–19].

The thermal behavior is different for decompositions in a nitrogen atmosphere (Figure 2) compared to an air atmosphere (Figure 1). In all cases, five stages of decomposition can be discerned. The first stage of decomposition is observed by the endothermic effect on the DTA curve, with a mass loss of 2.4–6.7% corresponding to the evaporation of adsorbed water and solvent [17–19]. The second stage of visible decomposition through the exothermic effect from 210 to 220 °C, with a mass loss of 20.7–34.4%, can be attributed to the decomposition of volatile compounds and polyphenols [17–19]. The third stage of decomposition visible through the exothermic effect at around 243–284 °C, with a mass loss of 19.7–41.7%, is attributed to the decomposition of fatty acids, amino acids and proteins [17–19]. The fourth stage of decomposition, equivalent to a mass loss of 22.8–37.0%, was observable through exothermic effects at around 364–399 °C, 440–450 °C and 483–546 °C and can be attributed to the degradation of cellulose, hemicellulose and lignin [17–19]. In addition, for samples P1, P3, P4, P5, P7 and P8, an exothermic effect occurs at 881–981 °C, with a mass loss of around 14%, which occurs only in the nitrogen atmosphere. We infer that it could be attributed to mineral substances with nitrogen content present in the residue. The total mass losses were in the range of 95.0–98.6% in the air atmosphere compared to 80.6–97.2% in the nitrogen atmosphere. The thermal degradation stages of paprika varieties are in agreement with their compositions.

In our previous studies on coffee [20], the thermal analysis revealed various transformations in coffee composition, namely, drying, water loss and decomposition of polysaccharides, lipids, amino acids and proteins. Thermal analysis also revealed transformations in cocoa powder's composition [21]: drying and water loss; decomposition of pectic polysaccharides, lipids, amino acids and proteins; and crystalline phase transformations and carbonizations.

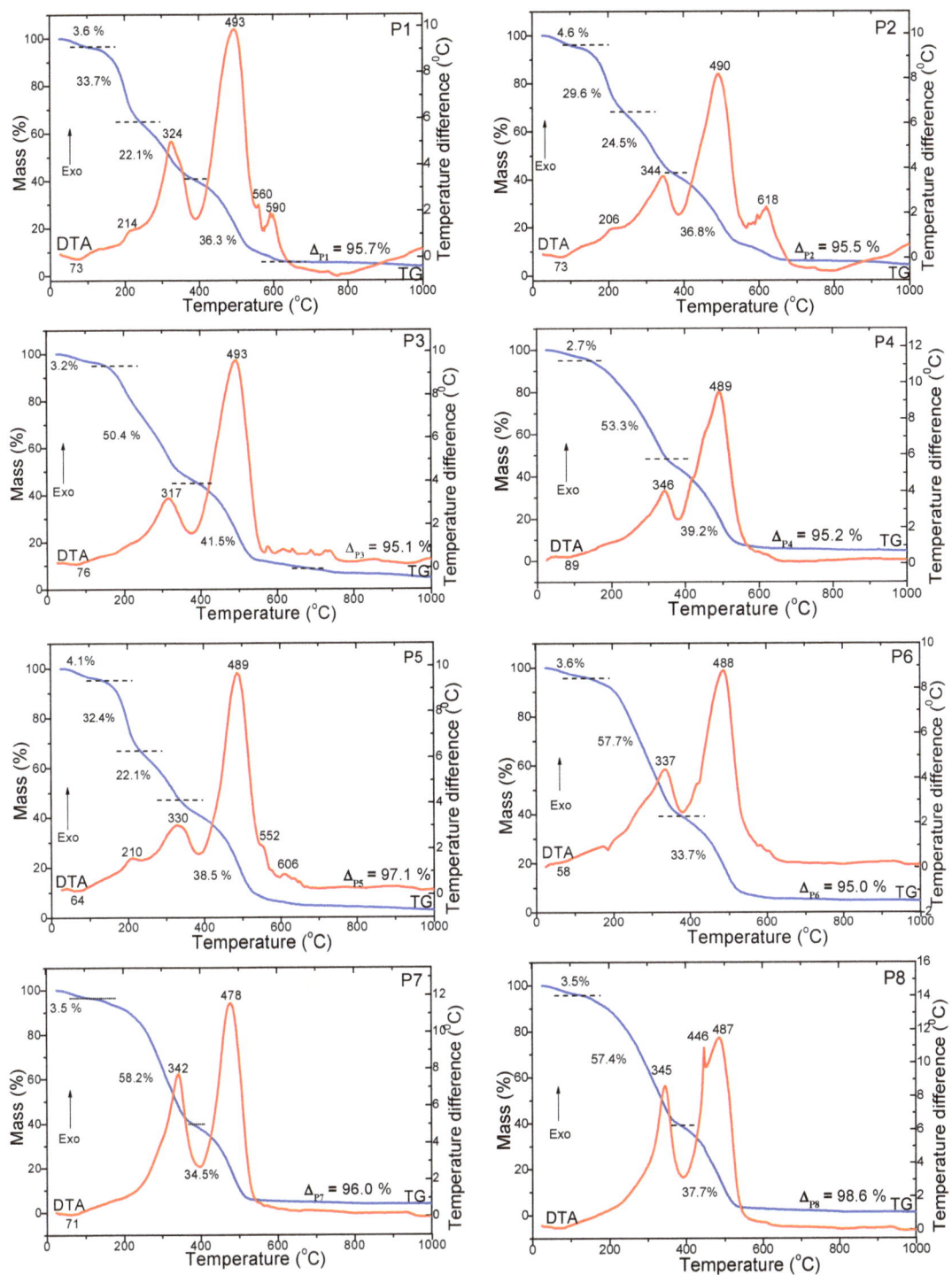

Figure 1. TG (blue line) and DTA (red line) curves of the paprika sample (P1–P8) under air atmosphere.

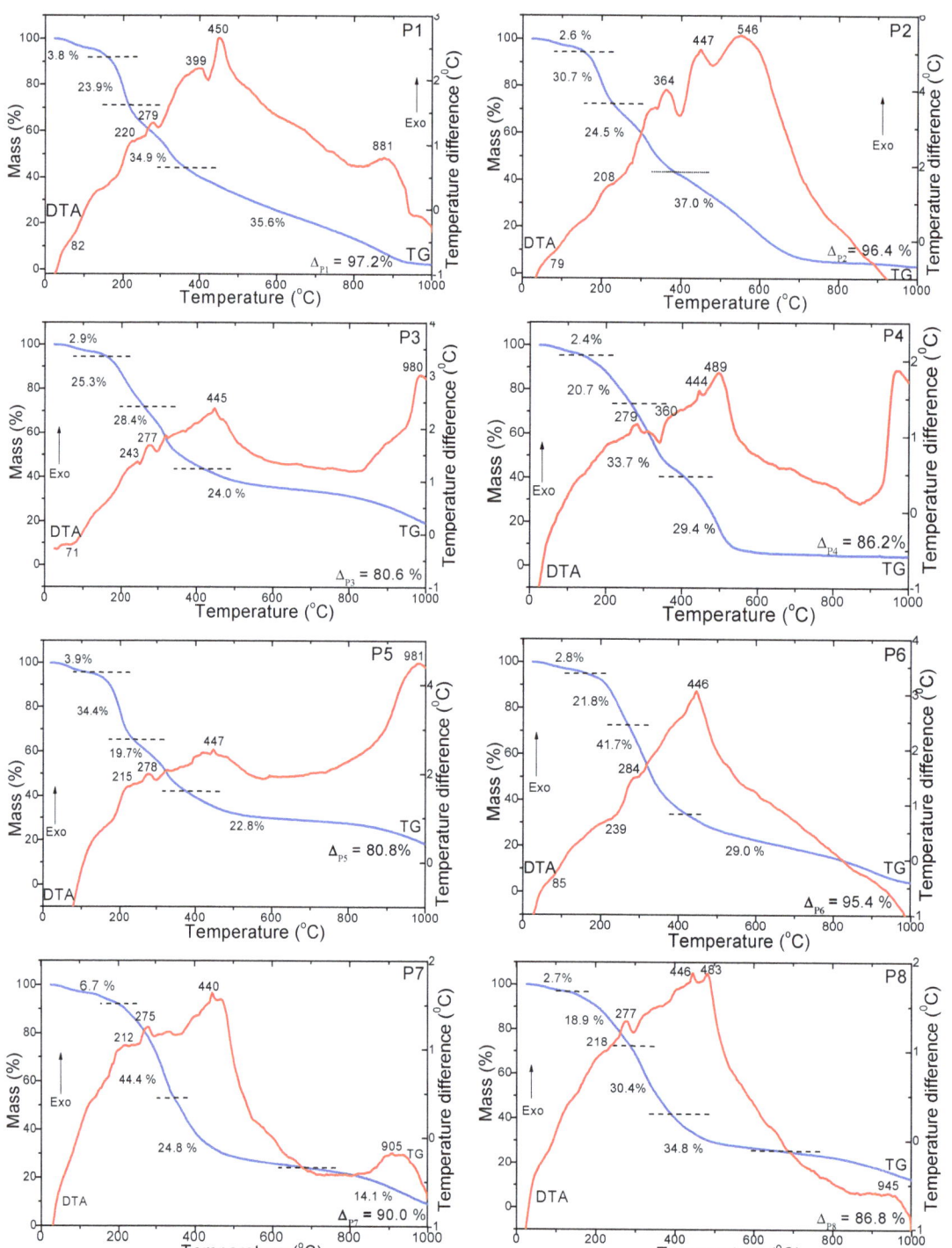

Figure 2. TG (blue line) and DTA (red line) curves of the paprika sample (P1–P8) under nitrogen atmosphere.

3.2. Cellulose, Hemicellulose and Lignin Content of Paprika Samples

Lignin is a natural phenolic polymer found in higher plant tissues and the second most abundant polymer after organic cellulose [22]. In all samples, cellulose, hemicellulose and lignin contents were identified. The lignin content varied in the following order (%): P1 (15.2 ± 1.1) > P4 (13.1 ± 1.0) > P8 (13.0 ± 0.89) > P7 (12.8 ± 0.85) > P2 (11.3 ± 1.1) > P5 (11.2 ± 1.0) > P6 (8.3 ± 0.62) > P3 (7.2 ± 0.52). A high concentration of lignin was identified in red Kapia pepper. Our results are in accordance with other researchers' results regarding lignin determination from pepper [23]. Estrada et al. [23] reported 4–9% lignin from the fruit of *Capsicum annum* pepper species and reported the variation of lignin during the maturation period. It was concluded that the maturation process caused the lignin content to decrease. The process can be explained by the rearrangement of cell structure by plant maturation and the interaction between lignin-like substances derived from phenyl propanoid precursors [23]. Celluloses and hemicelluloses are carbohydrates that provide the taste of the pepper and can create links with protein, lipids and biomolecules in the paprika powder [24].

The cellulose content varied in the following order (%): P2 (32.2 ± 2.1) > P4 (28.6 ± 1.8) > P6 (25.5 ± 2.0) > P1 (23.1 ± 1.9) > P5 (23.0 ± 1.3) > P7 (21.1 ± 1.6) > P8 (19.2 ± 1.2) > P3 (19.1 ± 1.0), whereas hemicellulose contents were (%) P6 (13.5 ± 0.8) > P4 (12.1 ± 1.0) > P2 (11.1 ± 0.9) > P8 (8.9 ± 0.5) > P1 (8.1 ± 0.5) > P5 (7.2 ± 0.6) > P7 (5.4 ± 0.3) > P3 (5.2 ± 0.4). The highest content of cellulose was found in spicy varieties. According to Mudrić et al., in Serbian paprika, the following sugars were identified: glucose, fructose, arabinose, xylose, mannose, disaccharides (trehalose and maltose), trisaccharides and sugar alcohols [24]. The results show a good agreement between the thermogravimetric analysis and the presence of cellulose, hemicelluloses and lignin in paprika samples.

3.3. HS-SPME GC-MS Analysis of Volatile Organic Compounds

Volatile compounds of Romanian paprika were analyzed using SPME followed by GC-MS, as shown in Table 1. This study identified 32 volatile compounds divided into seven classes in the samples, including 16 hydrocarbons, 5 aldehydes, 4 ketones, 3 alcohols, 2 esters, 1 sulfur compound and 1 heterocyclic compound (Table 1).

Hydrocarbons had the highest proportion (50.6%, Figure 3a) in the samples, inducing fruity, camphorous, sweet, lemony, pine-like, minty, woody, resinous, balsamic, plastic, roasted, citrus, floral, aromatic, green, herbaceous, clove, pepper or spicy odors [6,25]. The limonene was the most abundant in the paprika samples, although its amount varied between samples as follows: P7 (100%) > P8 (42.8%) > P6 (41.6%) > P3 (30.2%). M-cymene may be formed from oxidation of monoterpene hydrocarbons by isomerization and oxidation [26,27].

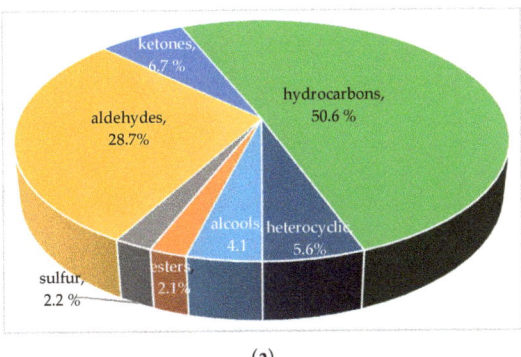
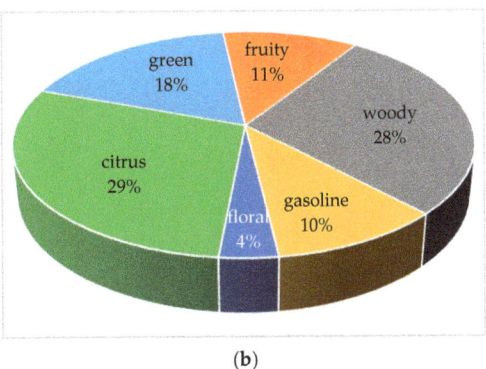

(a)   (b)

Figure 3. Classification of volatile compounds identified in paprika by chemical class (**a**) and aroma profile (**b**).

Table 1. Retention time, volatile organic compounds, group, odor type and content (%) identified using HS-SPME GC-MS for P1–P8 paprika samples expressed as averages ± standard deviation ($n = 3$).

R_t (min)	Volatile Compounds	Chemical Group	Odor Types	P1	P2	P3	P4	P5	P6	P7	P8
5.1	2-methylbutan-1-ol	alcohols	green	17.2 ± 1.9	9.5 ± 1.0	<0.02	<0.02	<0.02	<0.02	<0.02	<0.02
5.2	dimethyl disulfide	sulfur comp	green	<0.03	<0.03	<0.03	<0.03	17.2 ± 1.8	<0.03	<0.03	<0.03
6.5	2-methylisovalerate	esters	fruity	<0.05	<0.05	<0.05	<0.05	<0.05	<0.05	<0.05	<0.05
7.4	hexanal	aldehydes	green	<0.02	<0.02	<0.02	1.3 ± 0.1	<0.02	5.3 ± 0.7	<0.02	18.6 ± 2.2
8.9	furfural	aldehydes	woody	28.1 ± 3.1	52.4 ± 5.9	44.3 ± 4.6	2.7 ± 0.3	53.5 ± 6.1	<0.03	<0.03	<0.03
10.2	2-methyloctane	hydrocarbons	gasoline	54.7 ± 6.4	17.2 ± 1.8	<0.06	<0.06	11.1 ± 1.2	<0.06	<0.06	<0.06
11.2	2-heptanone	ketones	fruity	<0.04	<0.04	7.1 ± 0.8	<0.04	<0.04	<0.04	<0.04	<0.04
12.0	2-Propanone	ketones	fruity	<0.03	4.4 ± 0.5	6.7 ± 0.7	<0.03	3.6 ± 0.4	<0.03	<0.03	<0.03
12.6	4-carene	hydrocarbons	fruity	<0.05	<0.05	<0.05	2.1 ± 0.2	<0.05	1.6 ± 0.2	<0.05	<0.05
12.8	α-pinene	hydrocarbons	woody	<0.02	<0.02	<0.02	6.3 ± 0.7	<0.02	4.9 ± 0.5	<0.02	<0.02
13.0	3-(bromomethyl)cyclohexane	hydrocarbons	woody	<0.03	<0.03	2.5 ± 0.3	<0.03	<0.03	<0.03	<0.03	<0.03
13.4	camphene	hydrocarbons	green	<0.02	<0.02	<0.02	<0.02	<0.02	<0.02	<0.02	<0.02
13.8	benzaldehyde	aldehydes	woody	<0.02	3.9 ± 0.4	<0.02	<0.02	<0.02	<0.02	<0.02	<0.02
14.0	5-methyl furfural	aldehydes	woody	<0.03	5.2 ± 0.5	6.9 ± 0.7	<0.03	3.9 ± 0.4	<0.03	<0.03	<0.03
14.1	2-methyl-7-norbornanol	alcohols	fruity	<0.05	3.3 ± 0.4	<0.05	<0.05	<0.05	<0.05	<0.05	<0.05
14.3	β-phellandrene	hydrocarbons	fruity	<0.03	<0.03	<0.03	11.1 ± 1.2	<0.03	<0.03	<0.03	<0.03
14.4	β-ocimene	hydrocarbons	fruity	<0.02	<0.02	<0.02	10.3 ± 1.1	<0.02	<0.02	<0.02	<0.02
14.5	β-pinene	hydrocarbons	woody	<0.04	<0.04	<0.04	<0.04	<0.04	4.2 ± 0.5	<0.04	<0.04
15.0	(-)-cis-myrtenol	alcohols	green	<0.02	<0.02	2.8 ± 0.3	<0.02	<0.02	<0.02	<0.02	<0.02
15.0	β-myrcene	hydrocarbons	woody	<0.02	<0.02	<0.02	3.7 ± 0.4	<0.02	<0.02	<0.02	<0.02
15.5	1.3.8-p-menthathiene	hydrocarbons	woody	<0.04	<0.04	<0.04	2.5 ± 0.3	<0.04	<0.04	<0.04	<0.04
15.7	terpinyl acetate	esters	green	<0.02	<0.02	<0.02	12.1 ± 1.2	<0.02	3.2 ± 0.3	<0.02	<0.02
16.2	m-cymene	hydrocarbons	fruity	<0.05	<0.05	<0.05	8.1 ± 0.8	<0.05	10.8 ± 1.1	<0.05	<0.05
16.3	limonene	hydrocarbons	citrus	<0.04	4.1 ± 0.4	5.4 ± 0.5	30.2 ± 3.1	7.5 ± 0.8	41.7 ± 4.3	100 ± 9.9	42.8 ± 4.3
17.0	β-ocimene	hydrocarbons	fruity	<0.02	<0.02	5.3 ± 0.6	<0.02	<0.02	<0.02	<0.02	<0.02
17.4	γ-terpinene	hydrocarbons	fruity	<0.03	<0.03	<0.03	4.3 ± 0.4	<0.03	4.2 ± 0.4	<0.03	<0.03
18.3	terpinolene	hydrocarbons	floral	<0.02	<0.02	<0.02	1.5 ± 0.1	<0.02	<0.02	<0.02	<0.02
19.6	2-nonen-4-one	ketones	fruity	<0.03	<0.03	5.8 ± 0.6	<0.03	<0.03	<0.03	<0.03	<0.03
20.6	2-n-pentylthiophene	heterocyclic	green	<0.03	<0.03	6.4 ± 0.7	<0.03	<0.03	<0.03	<0.03	<0.03
22.1	phenylsulfanylacetaldehyde	aldehydes	green	<0.01	<0.01	5.1 ± 0.5	<0.01	<0.01	<0.01	<0.01	38.6 ± 3.9
23.2	d-carvone	ketones	floral	<0.02	<0.02	1.7 ± 0.2	<0.02	3.2 ± 0.3	21.6 ± 2.2	<0.02	<0.02
29.7	cedrene	hydrocarbons	woody	<0.03	<0.03	<0.03	3.8 ± 0.4	<0.03	<0.03	<0.03	<0.03

< Below limit of quantification (LQ).

Aldehydes (28.7%, Figure 3a) are formed through the oxidative degradation of amino acids during their interaction with sugars at high temperatures or the interaction of amino acids and polyphenols in the presence of polyphenol oxidase and mainly contribute to fatty and cowy flavor [6,7,25]. The formation of aldehydes indicates that the Maillard reaction occurred during the heating or roasting processes, correlated to the concentrations of free amino acids present in the samples [3,28]. The aromas given by aldehydes are fruity, fatty, green, oil, green grassy, very strong, harsh and lemon-like [6]. Furfural (highest content in P1, P2, P3 and P5 paprika samples) was the major volatile component in this group and is usually an indicator of thermal damage during roasting [3,29]. Hexanal (P4, P6 and P8), as an oxidation product of enzymatic as well as autoxidative linoleic acid oxidation, increased strongly with heating, and is the main compound found in fresh pepper and is responsible for the odor of freshly cut grass or ground leaves of green plant materials [1,3,28,30]. Benzaldehyde, the only odor-active aromatic aldehyde, only present in the B2 sample, is characterized by a caramel-like or roasted odor and commonly exists as the glycosidically bound form in paprika [6,27].

The content of ketones (6.7%, Figure 3a) in paprika samples is associated with fruity, spicy, cinnamon, banana, mushroom, camphor, cedar leaf, mint and bitter aromas [6,15]. Oxidative decomposition of unsaturated fatty acids is the main pathway for ketone formation during heating treatments. In addition, oxidative decomposition of saturated fatty acids could produce volatiles, such as alkenes and alcohols, which might be further oxidized to produce ketones under a high temperature treatment [7,27,31]. 2-Heptanone (only present in sample P2) contributed to 'stale' and 'cabbage' odors and very little to the entire aroma formation because of its high threshold and low content [6]. Carvone (only in samples P3, P5 and P6) provided 'mint', 'basil' and 'fennel' odors and was detected as one of the odor-active compounds of paprika [6].

Alcohol compounds (4.1%, Figure 3a) account for cooling, camphoraceous, fresh pine, ozone, citrus, floral, green, peppermint, woody, earth and sweet odors [6]. 2-Methyl-butanal (P1 and P2 samples) showed the highest abundance, which provided a characteristic roasted garlic odor [3].

Esters (2.1%, Figure 3a) in paprika samples induced fruity (apple, cherry, pear, etc.), floral, herbaceous, sweet, refreshing, green, grassy, bergamot, lavender and minty odors [6]. Esters were produced by oxidation or pyrolysis of unsaturated fatty acids in peppers under high temperature, especially pericarp and seeds, which could explain the increase in esters during the initial drying process [3,7,32].

Sulfur compounds are decomposition products of thiosulfinates and are derived from amino acid flavor precursors of the *Allium* family, including garlic and shallot [3,33,34]. Dimethyl disulfide (2.2% only in the P5 sample) (pungent, spicy) is a sulfur-containing volatile from garlic that is responsible for medicinal properties, presents the most abundant flavor in garlic oil and plays a major role in the formation of di- and trisulfides found as components of garlic [3,35].

The predominant aroma is citrus (29%), where the limonene and woody aroma (28%) of paprika is represented by furfural, 5-methyl furfural, α-pinene, β-pinene, 3-(bromomethyl)cyclohexane, benzaldehyde, β-myrcene, 1.3.8-p-menthathiene and cedrene [6]. Green-aroma-associated compounds (18%, Figure 3b), after the processing of peppers, showed that terpenes, sesquiterpenes and terpene derivatives are more abundant in pungent paprikas than sweet ones and 2-methylbutan-1-ol, dimethyl disulfide, hexanal, camphene, cis-myrtenol, terpinyl acetate, 2-n-pentylthiophene and phenylsulfanylacetaldehyde decreased upon maturation [6,36]. Fruity aroma (11%, Figure 3b) is represented by compounds such as 2-methylisovalerate, 2-heptanone, 2-Propanone, 4-carene, 2-methyl-7-norbornanol, β-phellandrene, β-ocimene, m-cymene, β-ocimene, γ–terpinene and 2-nonen-4-one. These compounds are sensitive to compositional alterations and variations in the metabolic pathways during ripening, harvest, post-harvest and storage and many factors related to the variety and type of technological treatment [6,36]. The gasoline flavor

(10%) is given by 2-methyloctane, and the floral aroma (4%) is given by terpinolene and d-carvone [6].

3.4. Fatty Acid Content in Paprika Oil Varieties

However, the cis (oleic acid) and trans (elaidic acid) isomers of C18:1 and cis (linoleic acid) and trans (linolaidic acid) isomers of C18:2 were not separated and were quantified together. The lipid content varied in the following order: P5 (16.0 ± 1.1%) > P2 (15.5 ± 1.0%) > P6 (13.7 ± 1.1%) > P7 (13.0 ± 1.0%) > P8 (12.4 ± 0.89%) > P4 (12.1 ± 0.90) > P1 (10.01 ± 0.90%) > P3 (4.2 ± 0.2%). The fatty acid methyl esters found in all oil from paprika powder varieties are presented in Table 2 and Figure 4. All oil samples analyzed contain SFAs, MUFAs and PUFAs in different quantities within sample types. The SFAs found in paprika oil are C12:0, C14:0, C15:0, C16:0, C17:0, C18:0, C20:0, C21:0, C22:0 and C23:0. Palmitic acid (C16:0) is the predominant SFA in all paprika samples. The highest content of C16:0 was found in the P1 sample. The identified MUFAs are C14:1, C16:1, C17:1, C18:1(c + t) and C20:1. The highest oleic and octadecenoic acid (C18:1(c + t)) content was found in P8 (11.31%) and P4 (10.38%).

Table 2. Profile of fatty acids found in paprika oils; data are expressed in % (w/w), expressed as averages ± standard deviation (n = 3).

Acid Type	P1	P2	P3	P4	P5	P6	P7	P8
C12:0	2.49 ± 0.1	3.13 ± 0.12	<0.032	0.54 ± 0.01	3.56 ± 0.12	0.91 ± 0.01	1.32 ± 0.01	0.32 ± 0.01
C14:0	3.61 ± 0.1	3.76 ± 0.11	3.16 ± 0.15	0.96 ± 0.01	4.70 ± 0.15	1.31 ± 0.01	1.76 ± 0.01	0.79 ± 0.01
C14:1	1.40 ± 0.08	1.69 ± 0.07	<0.036	<0.036	<0.036	0.64 ± 0.01	0.91 ± 0.02	2.01 ± 0.05
C15:0	1.56 ± 0.01	1.52 ± 0.02	<0.014	0.24 ± 0.01	1.84 ± 0.02	0.41 ± 0.01	0.64 ± 0.02	0.23 ± 0.01
C16:0	16.0 ± 1.1	14.6 ± 1.0	13.0 ± 1.1	13.0 ± 1.0	15.5 ± 1.2	10.6 ± 0.8	12.2 ± 1.0	11.2 ± 1.1
C16:1	1.93 ± 0.08	2.11 ± 0.08	1.65 ± 0.05	0.63 ± 0.01	2.13 ± 0.08	0.69 ± 0.02	0.93 ± 0.03	0.46 ± 0.01
C17:0	2.85 ± 0.1	1.68 ± 0.1	2.42 ± 0.08	0.31 ± 0.01	2.49 ± 0.07	1.64 ± 0.01	1.24 ± 0.03	0.65 ± 0.01
C17:1	1.66 ± 0.04	1.77 ± 0.05	<0.023	0.23 ± 0.01	<0.023	<0.023	<0.023	0.15 ± 0.01
C18:0	7.57 ± 0.2	7.82 ± 0.41	5.29 ± 0.21	3.14 ± 0.12	6.64 ± 1.3	3.40 ± 0.01	4.17 ± 0.1	3.33 ± 0.13
C18:1(c + t)(n9)	5.12 ± 0.1	6.85 ± 0.12	5.02 ± 0.31	10.4 ± 1.0	5.60 ± 1.3	14.0 ± 1.1	9.61 ± 0.2	11.3 ± 1.2
C18:2(c + t)(n6)	21.8 ± 1.8	20.3 ± 1.8	24.3 ± 1.5	64.8 ± 2.3	20.7 ± 1.8	61.4 ± 2.5	56.8 ± 2.3	61.9 ± 4.1
C18:3(n6)	4.30 ± 0.2	4.22 ± 0.13	3.75 ± 0.18	0.53 ± 0.02	3.44 ± 0.05	1.03 ± 0.02	1.43 ± 0.08	0.36 ± 0.01
C18:3(n3)	19.6 ± 1.2	16.68 ± 1.3	10.03 ± 0.9	1.85 ± 0.012	17.78 ± 1.1	1.80 ± 0.04	4.17 ± 0.12	2.33 ± 0.1
C20:0	2.82 ± 0.1	4.27 ± 0.21	3.59 ± 0.2	0.74 ± 0.02	3.42 ± 0.02	0.80 ± 0.01	1.28 ± 0.08	0.29 ± 0.01
C20:1	<0.0052	<0.0052	3.06 ± 0.2	0.34 ± 0.01	<0.0052	0.53 ± 0.02	0.71 ± 0.02	2.62 ± 0.1
C20:2	1.56 ± 0.1	<0.0096	<0.0096	0.26 ± 0.01	2.56 ± 0.05	<0.0096	1.71 ± 0.01	0.19 ± 0.01
C21:0	2.21 ± 0.08	3.04 ± 0.10	<0.0086	0.40 ± 0.01	3.18 ± 0.01	<0.0086	<0.0086	0.29 ± 0.01
C20:3(n3)	<0.0061	2.22 ± 0.11	<0.0061	0.21 ± 0.01	1.72 ± 0.01	<0.0061	<0.0061	0.13 ± 0.01
C20:5(n3)	<0.0058	1.46 ± 0.08	<0.0058	<0.0058	<0.0058	<0.0058	<0.0058	0.13 ± 0.01
C22:1	<0.0063	<0.0063	<0.0063	<0.0063	<0.0063	<0.0063	<0.0063	0.13 ± 0.01
C22:0	1.15 ± 0.02	1.42 ± 0.07	1.77 ± 0.08	0.33 ± 0.01	1.46 ± 0.07	0.47 ± 0.01	0.62 ± 0.01	0.26 ± 0.01
C22:2	1.30 ± 0.01	<0.0091	<0.0091	<0.0091	1.62 ± 0.05	<0.0091	<0.0091	0.13 ± 0.01
C23:0	<0.056	<0.056	21.62 ± 1.8	0.77 ± 0.01	<0.056	<0.056	<0.056	<0.056
C24:0	1.08 ± 0.08	1.42 ± 0.07	<0.050	0.32 ± 0.01	1.62 ± 0.1	0.43 ± 0.02	0.50 ± 0.01	0.23 ± 0.01
SFA	41.4 ± 3.1	42.7 ± 2.1	50.8 ± 4.1	20.7 ± 1.6	44.4 ± 2.6	19.9 ± 1.1	23.8 ± 2.0	18.1 ± 1.2
MUFA	10.1 ± 1.1	12.4 ± 1.0	9.7 ± 0.5	11.6 ± 1.0	7.7 ± 0.2	15.9 ± 1.4	12.2 ± 1.1	16.7 ± 1.3
PUFA	48.5 ± 3.1	44.9 ± 3.5	38.0 ± 2.3	67.6 ± 4.2	47.8 ± 3.1	64.2 ± 2.6	64.1 ± 6.0	65.2 ± 4.6
ω-6	29.0 ± 1.8	24.5 ± 1.8	28.0 ± 1.5	65.6 ± 4.3	28.3 ± 1.9	62.4 ± 5.2	59.9 ± 4.2	62.6 ± 4.2
ω-3	19.6 ± 1.1	20.4 ± 1.9	10.0 ± 0.9	2.1 ± 0.1	19.5 ± 1.2	1.80 ± 0.08	4.17 ± 0.3	2.59 ± 0.1
ω-6/ω-3	1.5 ± 0.08	1.2 ± 0.07	2.8 ± 0.1	31.8 ± 2.6	1.5 ± 0.1	34.6 ± 2.1	14.4 ± 1.2	24.2 ± 2.1

< Below limit of quantification (LQ).

The PUFAs are divided into omega-6 (ω-6) and omega-3 (ω-3). The highest PUFA (ω-6) quantities found in all samples are reported for linoleic acid (C18:2(c + t)) and varied in the following order: P4 (64.7%) > P8(61.9%) > P6(61.4%) > P7 (56.8%) > P3(24.3%) > P1, P5 and P2 (approx.20%). The highest PUFA content was found in spicy samples. According to Figure 4, the highest content of SFA was found in golden pepper (P3), due to a high content of tricosanoic acid (C23:0). Red paprika varieties were found to be free of saturated tricosanoic acid. Rutkowska and Stolyhwo (2009) reported a content of 59.4% C18:2 and 5.1% C18:3 in red paprika powder oil. The method by which the oil was extracted was Soxhlet with hexane/ethyl ether (1:1) [37]. Several studies reported that food rich in PUFAs reduces the risk of cardiovascular diseases, and they were recently shown to prevent the risk of SARS-CoV-2 infection [38]. Omega-3 was found in high quantities in samples P2 (20.6%), P1 and P3 (19.5%) and in lower amounts in samples P7, P4, P8 and P6 (lower than

4%). The conclusion is that spicy paprika has a high omega-3 content and is appropriate for consumption. It is essential that the consumption of SFAs is replaced with the consumption of PUFAs. It is recommended that the consumption of SFAs should not exceed 10%, according to the Food and Agricultural Organization of United Nations [39]. A lower ω-6/ω-3 ratio is desirable due to the reduction in the risk of cardiovascular problems. The free fatty acids were analyzed in all samples (as acid values). All the oil samples contained below 1%, demonstrating the solubility of fatty acids in used solvents in the presence of the ultrasound method. The fatty acid profile of paprika varieties is important for food chemistry due to differences in variety chemistry, assigned quality, origin, food taste and color.

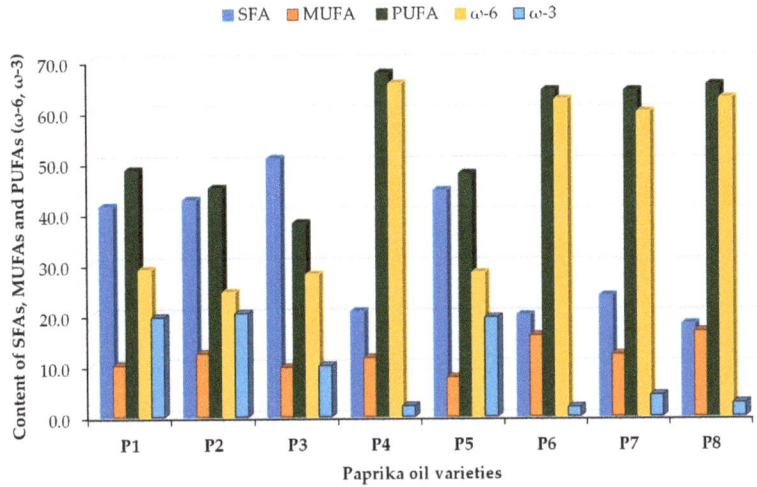

Figure 4. The content of SFAs, MUFAs and PUFAs.

3.5. Total Polyphenol Contents

Polyphenols are often responsible for the antioxidant capacity of plant products, and they could be important constituents to explain the protective effects of plant-derived foods and beverages [2]. The results of the total polyphenolic content of paprika spices, measured with Folin–Ciocalteau reagent using the spectrometric method, are given in Table 3. Polyphenols are linked to health-promoting properties as they show antioxidant, anti-inflammatory, antidiabetic and anticarcinogenic activity [40]. The antioxidant character was attributed to the higher number of polyphenols present in the paprika and to the presence of lignin. The highest content of polyphenols was measured in P3 (10.9 g GA kg^{-1}), and the lowest content was measured in P8 (5.11 g gallic acid/kg). The sample with the lowest total polyphenol content (sweet paprika spice) had only about 50% of the content of paprika delicate with the highest value; there is only a weak connection between the pungency of the spices and the polyphenolic amount [2]. The amount of polyphenols in paprika products could be influenced by the variety of pepper and could also be dependent on the time of harvesting and processing [2].

Table 3. Total polyphenol contents in the studied paprika samples expressed as averages ± standard deviation (n = 3).

Paprika	P1	P2	P3	P4	P5	P6	P7	P8
Polyphenol g GA kg^{-1}	8.35 ± 0.79	9.44 ± 0.96	10.9 ± 1.1	8.18 ± 0.82	5.61 ± 0.61	7.81 ± 0.81	7.44 ± 0.73	5.11 ± 0.52

3.6. Principal Component Analysis

Hierarchical clustering (dendrogram), AHC of the paprika varieties presented in Figure 5, was conducted to find the inter-connectivity and closeness of the studied paprika samples and individual volatile organic compounds. The dendrogram cluster is divided into two groups, the first containing P1, P2, P3 and P5 and the second which contains P4, P6, P7 and P8. Paprika species from Romania and Turkey have similar chemical compositions, whereas samples from China, India, Morocco and Hungary can be differentiated. Pearson correlation regarding fatty acids confirms a very strong correlation between P1, P5, P2 and P3 and P7, P8, P4 and P6. The cluster analysis shows the differences between samples separated in each cluster based on the different content of fatty acids and volatile compounds.

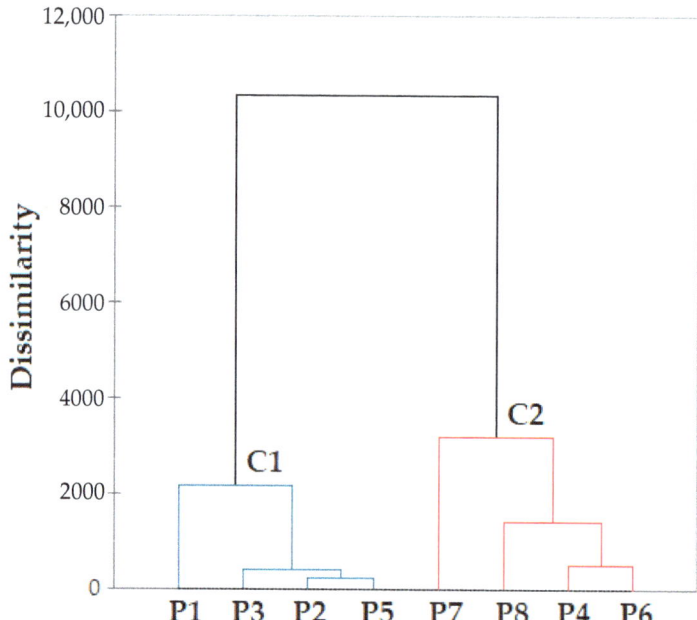

Figure 5. Hierarchical clustering (dendrogram) of different paprika varieties.

Figure 6 presents the principal component analysis regarding fatty acid, volatile compound and polyphenol correlation. The PCA is an unsupervised method to visualize the difference/similarity among sample profiles and detect significant variables contributing to these discrepancies among the eight paprika types. The data for PCA were evaluated for each class of fatty acids, volatile compounds and polyphenols as a preliminary test. Very strong positive correlations were obtained for C18:3(n6) with C16:1 (0.97), C18:0 (0.95) and C14:0 (0.92), and it was negatively correlated with C18:2(c + t)(n6) (−0.98). In addition, a strong positive correlation was obtained for C18:3(n3) with C18:0 (0.93), C16:1 (0.96) and C14:0 (0.95), and it was strongly negatively correlated with C18:2(c + t)(n6) (−0.95). A moderate correlation was obtained for C23:0 with C20:1 (0.71) (Figure 6a). Linoleic acid,

C18:2, was positively correlated with the 18:1 isomer (0.95). Finally, C14:1 was positioned alone and was not related to any other components in the PCA.

In the case of volatile compounds, four clusters can be observed. A very strong positive correlation was obtained for d-carvone with camphene and β-pinene (0.99). Four clusters of volatile compounds were identified (Figure 6b). The presence of aldehydes, alcohols, esters, hydrocarbons and ketones contributes to the flavor of paprika varieties originating from different countries.

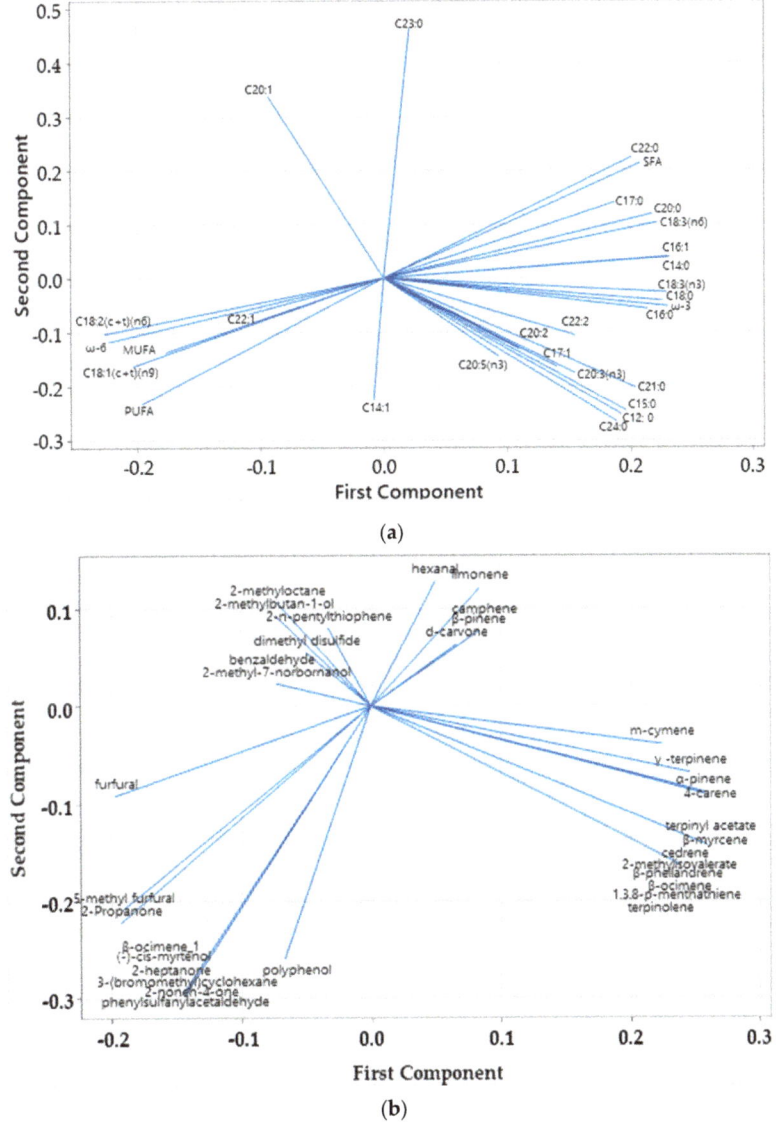

Figure 6. Principal component analysis loading plot of PC1 and PC2 for fatty acids (a) and volatiles and phenols (b) of different paprika varieties.

4. Conclusions

This study conducted a comparative analysis of volatile compounds, fatty acids, cellulose, hemicellulose, lignin and total polyphenols and evaluated the thermal behavior among paprika samples from different countries. The thermal analysis of paprika samples was in agreement with the moisture evaporation of adsorbed water and solvent (at 64–73 °C and a mass loss of 3.6–4.6%), decomposition of volatile compounds (206–214 °C and mass loss of 29.6–33.7%), decomposition of fatty acids and proteins (at 324–344 °C and mass loss of 22.1–24.5%) and degradation of cellulose, hemicellulose and lignin (at 493–489 °C, 552–560 °C and 590–606 °C and mass loss of 33.7–41.5%). In total, 32 volatile compounds divided into seven chemical classes were identified in the samples, including 16 hydrocarbons, 5 aldehydes, 4 ketones, 3 alcohols, 2 esters, 1 sulfur compound and 1 heterocyclic compound. The predominant aromas in paprika samples were citrus (29%, limonene) and woody (28%), and other aromas identified in smaller quantities were green (18%), fruity (11%), gasoline (10%) and floral (4%). The limonene was the most abundant in the paprika samples, although its amount varied between samples as follows: P7 (100%) > P8 (42.8%) > P6 (41.6%) > P3 (30.2%). The highest PUFA (ω-6) quantities were found in hot pepper from Turkey, India, China and Hungary. Linoleic acid (C18:2(c + t)) is the major PUFA fatty acid. The highest SFA content was found in golden pepper. The highest content of polyphenols was measured in P3 (10.9 g GA/kg), and the lowest content was measured in P8 (5.11 g gallic acid/kg). This could be influenced by pepper variety and also be dependent on the time of harvesting and processing. The obtained results will help create new research perspectives regarding the use of paprika powders in the food, cosmetic and pharmaceutical industries, thus creating new natural ingredients and bioactive compounds.

Author Contributions: Conceptualization T.D.; methodology, T.D., L.S. and L.E.M.; validation, L.S. and L.E.M.; formal analysis, T.D., L.S. and L.E.M.; investigation, T.D., L.S. and L.E.M.; resources, T.D., L.S. and L.E.M.; data curation, T.D.; writing—original draft preparation, T.D. and L.S.; writing—review and editing, T.D. and L.S.; visualization, L.S. and L.E.M.; supervision, T.D. All authors have read and agreed to the published version of the manuscript.

Funding: The APC was funded by the Technical University of Cluj-Napoca Grant Support CA 11/28.06.2022.

Data Availability Statement: The data presented in this study are available on request from the corresponding author.

Acknowledgments: The authors would like to express their gratitude to Melinda Haydee Kovacs for the VOC measurements.

Conflicts of Interest: The authors declare no conflict of interest.

References

1. Martín, A.; Hernández, A.; Aranda, E.; Casquete, R.; Velázquez, R.; Bartolomé, T.; Córdoba, M.G. Impact of volatile composition on the sensorial attributes of dried paprika. *Food Res. Int.* **2017**, *100*, 691–697. [CrossRef] [PubMed]
2. Škrovánková, S.; Mlček, J.; Orsavová, J.; Juríková, T.; Dřímalová, P. Polyphenols content and antioxidant activity of paprika and pepper spices. *Slovak. J. Food Sci.* **2017**, *11*, 52–57. [CrossRef] [PubMed]
3. Sukchum, N.; Surasereewong, S.; Chaethong, K. Volatile compounds and physicochemical characteristics of Thai roasted chilli seasoning. *Food Res.* **2022**, *6*, 309–318. [CrossRef] [PubMed]
4. Zaki, N.; Hakmaoui, A.; Ouatmane, A.; Fernandez-Trujillo, J.P. Quality characteristics of Moroccan sweet paprika (*Capsicum annum* L.) at different sampling times. *Food Sci. Technol.* **2013**, *33*, 577–585. [CrossRef]
5. Monago-Maraña, O.; Durán-Merás, I.; de la Peña, A.M.; Galeano-Díaz, T. Analytical techniques and chemometrics approaches in authenticating and identifying adulteration of paprika powder using fingerprints: A review. *Microchem. J.* **2022**, *178*, 107382. [CrossRef]
6. Burdock, G.A. *Fenaroli's Handbook of Flavor Ingredients*, 6th ed.; CRC Press: Boca Raton, FL, USA, 2010; pp. 134, 135, 238, 239, 253, 262, 369, 492, 799, 828, 1090, 1091, 1261, 1262, 1269, 1270, 1305, 1306, 1448, 1451, 1452, 1496, 1621, 1632, 1695, 1696, 1697, 1875, 1877.

7. Heng, Z.; Xu, X.; Xu, X.; Li, Y.; Wang, H.; Huang, W.; Yan, S.; Li, T. Integrated transcriptomic and metabolomic analysis of chili pepper fruits provides new insight into the regulation of the branched chain esters and capsaicin biosynthesis. *Food Res. Int.* **2023**, *169*, 112856. [CrossRef]
8. Tepić, A.; Zeković, Z.; Kravić, S.; Mandić, A. Pigment content and fatty acid composition of paprika oleoresins obtained by conventional and supercritical carbon dioxide extraction. *CyTA-J. Food* **2009**, *7*, 95–102. [CrossRef]
9. Abbeddou, S.; Petrakis, C.; Pérez-Gálvez, A.; Kefalas, P.; Hornero-Méndez, D. Effect of Simulated Thermo-Degradation on the Carotenoids, Tocopherols and Antioxidant Properties of Tomato and Paprikan Oleoresins. *J. Am. Oil Chem. Soc.* **2013**, *90*, 1679–1703. [CrossRef]
10. Kim, J.-S.; Ha, T.-Y.; Kim, S.; Lee, S.-J.; Ahn, J. Red paprika (*Capsicum annuum* L.) and its main carotenoid capsanthin ameliorate impaired lipid metabolism in the liver and adipose tissue of high-fat diet-induced obese mice. *J. Funct. Foods* **2017**, *31*, 131–140. [CrossRef]
11. Kostrzewa, D.; Dobrzyńska-Inger, A.; Mazurek, B.; Kostrzewa, M. Pilot-Scale Optimization of Supercritical CO_2 Extraction of Dry Paprika *Capsicum annuum*: Influence of Operational Conditions and Storage on Extract Composition. *Molecules* **2022**, *27*, 2090. [CrossRef]
12. Matei, P.L.; Deleanu, I.; Brezoiu, A.M.; Chira, N.A.; Busuioc, C.; Isopencu, G.; Ciltea-Udrescu, M.; Alexandrescu, E.; Stoica-Guzun, A. Ultrasound-Assisted Extraction of Blackberry Seed Oil: Optimization and Oil Characterization. *Molecules* **2023**, *28*, 2486. [CrossRef]
13. Pohndorf, R.S.; Camara, A.S.; Larrosa, A.P.Q.; Pinheiro, C.P.; Strieder, M.M.; Pinto, L.A.A. Production of lipids from microalgae *Spirulina* sp. sp.: Influence of drying, cell disruption and extraction methods. *Biomass Bioener.* **2016**, *93*, 25–32. [CrossRef]
14. Petrović, M.; Kezić, N.; Bolanća, V. Optimization of the GC method for routine analysis of the fatty acid profile in several food samples. *Food Chem.* **2010**, *122*, 285–291. [CrossRef]
15. Nielsen, S.D.; Crafack, M.; Jespersen, L.; Jakobsen, M. The Microbiology of Cacao Fermentation. In *Chocolate in Health and Nutrition*; Humana Press: Totowa, NJ, USA, 2012; Volume 7, pp. 39–60.
16. Senila, L.; Kovacs, E.; Scurtu, D.A.; Cadar, O.; Becze, A.; Senila, M.; Levei, E.A.; Dumitras, D.E.; Tenu, I.; Roman, C. Bioethanol production from vineyard waste by autohydrolysis pretreatment and chlorite delignification via simultaneous saccharification and fermentation. *Molecules* **2020**, *25*, 2606. [CrossRef] [PubMed]
17. Zhang, X.; Yu, Z.; Lu, X.; Ma, X. Catalytic co-pyrolysis of microwave pretreated chili straw and polypropylene to produce hydrocarbons-rich bio-oil. *Bioresour. Technol.* **2021**, *319*, 124191. [CrossRef]
18. Feraz, M.C.; Procopio, F.R.; Furtado, G.F.; Hubinger, M.D. Co-encapsulation of paprika and cinnamon oleoresin by spray drying using whey protein isolate and maltodextrin as wall material: Development, characterization and storage stability. *Food Res. Int.* **2022**, *162*, 112164. [CrossRef]
19. Qiu, J.; Zheng, Q.; Fang, L.; Wang, Y.; Min, M.; Shen, C.; Tong, Z.; Xiong, C. Preparation and characterization of casein-carrageenan conjugates and self-assembled microcapsules for encapsulation of red pigment from paprika. *Carbohydr. Polym.* **2018**, *196*, 322–331. [CrossRef]
20. Dippong, T.; Dan, M.; Kovacs, M.H.; Kovacs, E.D.; Levei, E.A.; Cadar, O. Analysis of Volatile Compounds, Composition, and Thermal Behavior of Coffee Beans According to Variety and Roasting Intensity. *Foods* **2022**, *11*, 3146. [CrossRef]
21. Dippong, T.; Mihali, C.; Vosgan, Z.; Daniel, A.; Dumuta, A. Thermal behavior of different cocoa powder varieties and their physicochemical, phytochemical and microbiological characteristics. *J. Therm. Anal. Calorim.* **2021**, *143*, 4217–4228. [CrossRef]
22. Ciolacu, D.; Oprea, A.M.; Anghel, N.; Cazacu, G.; Cazacu, M. New cellulose-lignin hydrogels and their application in controlled release of polyphenols. *Mater. Sci. Eng. C* **2012**, *32*, 452–463. [CrossRef]
23. Estrada, B.; Bernal, M.A.; Díaz, J.; Pomar, F.; Merino, F. Fruit development in *Capsicum annum*: Changes in capsaicin, lignin, free phenolics, and peroxidase patterns. *J. Agric. Food Chem.* **2000**, *48*, 6234–6239. [CrossRef]
24. Mudrić, S.Ž.; Gašić, U.M.; Dramićanin, A.M.; Ćirić, N.Ž.; Milojković-Opsenica, D.M.; Popović-Dordević, J.B.; Momirović, N.M.; Tešić, Ž.L. The polyphenolics and carbohydrates as indicators of botanical and geographical origin of Serbian autochthonous clones of red spice paprika. *Food Chem.* **2017**, *217*, 705–715. [CrossRef] [PubMed]
25. Mateo, J.; Aguirrezábal, M.; Domínguez, C.; Zumalacárregui, J.M. Volatile Compounds in Spanish Paprika. *J. Food Comp. Anal.* **1997**, *10*, 225–232. [CrossRef]
26. Korkmaz, A.; Hayaloglu, A.A.; Atasoy, A.F. Evaluation of the volatile compounds of fresh ripened *Capsicum annuum* and its spice pepper (dried red pepper flakes and isot). *LWT Food Sci. Technol.* **2017**, *84*, 842–850. [CrossRef]
27. Korkmaz, A.; Atasoy, A.F.; Hayaloglu, A.H. Changes in volatile compounds, sugars and organic acids of different spices of peppers (*Capsicum annuum* L.) during storage. *Food Chem.* **2020**, *311*, 125910. [CrossRef]
28. Cremer, D.R.; Eichner, K. Formation of volatile compounds during heating of spice paprika (*Capsicum annuum*) powder. *J. Agric. Food Chem.* **2000**, *48*, 2454–2460. [CrossRef]
29. Vanderhaegen, B.; Neven, H.; Coghe, S.; Verstrepen, K.J.; Verachtert, H.; Derdelinckx, G. Evolution of chemical and sensory properties during aging of top-fermented beer. *J. Agric. Food Chem.* **2003**, *51*, 6782–6790. [CrossRef]
30. Caporaso, N.; Paduano, A.; Nicoletti, G.; Sacchi, R. Capsaicinoids, antioxidant activity, and volatile compounds in olive oil flavored with dried chili pepper (*Capsicum annuum*). *Eur. J. Lipid Sci. Technol.* **2013**, *115*, 1434–1442. [CrossRef]
31. Yu, J.; Zhang, Y.; Wang, Q.; Yang, L.; Karrar, E.; Jin, Q.; Zhang, H.; Wu, G.; Wang, X. Capsaicinoids and volatile flavor compounds profile of Sichuan hotpot as affected by cultivar of chili peppers during processing. *Food Res. Int.* **2023**, *165*, 112476. [CrossRef]

32. Ge, S.; Chen, Y.; Ding, S.; Zhou, H.; Jiang, L.; Yi, Y.; Deng, F.; Wang, R. Changes in volatile flavor compounds of peppers during hot air drying process based on headspace-gas chromatography-ion mobility spectrometry (HS-GC-IMS). *J. Sci. Food Agri.* **2020**, *100*, 3087–3098. [CrossRef]
33. Block, E.; Ahmad, S.; Catalfamo, J.L.; Jain, M.K.; Apitz-Castro, R. Antithrombotic organosulfur compounds from garlic: Structural, mechanistic and synthetic studies. *J. Am. Chem. Soc.* **1986**, *108*, 7045–7055. [CrossRef]
34. Wu, J.L.; Chou, C.C.; Chen, M.H.; Wu, C.M. Volatile flavor compounds from shallots. *J. Food Sci.* **1982**, *47*, 606–608. [CrossRef]
35. Calvo-Gómez, O.; Morales-López, J.; López, M.G. Solid-phase microextraction-gas chromatographic-mass spectrometric analysis of garlic oil obtained by hydrodistillation. *J. Chromatogr. A* **2004**, *1036*, 91–93. [CrossRef]
36. Mazida, M.M.; Salleh, M.M.; Osman, H. Analysis of volatile aroma compounds of fresh chilli (*Capsicum annuum*) during stages of maturity using solid phase microextraction (SPME). *J. Food Comp. Anal.* **2005**, *18*, 427–437. [CrossRef]
37. Rutkowska, J.; Stolyhwo, A. Application of carbon dioxide in subcritical state (LCO_2) for extraction/fractionation of carotenoids from red paprika. *Food Chem.* **2009**, *115*, 745–752. [CrossRef]
38. Sánchez, C.A.O.; Zavaleta, E.B.; García, G.R.U.; Solano, G.L.; Díaz, M.P.R. Krill oil microencapsulation: Antioxidant activity, astaxanthin retention, encapsulation efficiency, fatty acids profile, in vitro bioaccessibility and storage stability. *LWT-Food Sci. Technol.* **2021**, *147*, 111476. [CrossRef]
39. Food and Agricultural Organization of United Nations. *Interim Summary of Conclusion and Dietary Recommendations on Total Fat & Fatty Acids*; FAO: Roma, Italy, 2008.
40. Janda, K.; Jakubczyk, K.; Baranowska-Bosiacka, I.; Kapczuk, P.; Kochman, J.; Rebacz-Maron, E.; Gutowska, I. Mineral Composition and Antioxidant Potential of Coffee Beverages Depending on the Brewing Method. *Foods* **2020**, *9*, 121. [CrossRef]

Disclaimer/Publisher's Note: The statements, opinions and data contained in all publications are solely those of the individual author(s) and contributor(s) and not of MDPI and/or the editor(s). MDPI and/or the editor(s) disclaim responsibility for any injury to people or property resulting from any ideas, methods, instructions or products referred to in the content.

Article

The Antioxidant Effect of Burdock Extract on the Oxidative Stability of Lard and Goose Fat during Heat Treatment

Flavia Pop and Thomas Dippong *

Department of Chemistry and Biology, Faculty of Science, Technical University of Cluj-Napoca, 76A Victoriei St., 430122 Baia Mare, Romania; flavia_maries@yahoo.com
* Correspondence: dippong.thomas@yahoo.ro

Abstract: Concerns regarding product quality and nutrition are raised due to the effects of high temperatures on frying fats. The aim of this research was to examine the effects of temperature and burdock extract addition in relation to quality parameters for dietary lard and goose fat exposed to heating. In order to monitor quality changes, animal fats and 0.01% additivated fats were heated at different temperatures (110, 130, 150, 170, 190, and 210 °C for 30 min). Thiobarbituric acid-reactive substances test (TBARS), peroxide value (PV), iodine value (IV), acid value (AV), saponification value (SV), total polar compounds (TPoC), total phenolic content (TPC), fatty acid (FA) content, and microscopic examination were established in order to quantify the level of oxidative rancidity. Heating temperature and additivation had a significant ($p < 0.001$) effect on peroxide value. In all fats, values of thiobarbituric acid-reactive substances significantly ($p < 0.001$) increased with heating temperature, but values decreased when burdock extract was added in a proportion of 0.01%. Positive correlations were found between AV and PV for lard ($r = 0.98$; $p < 0.001$) and goose fat ($r = 0.96$; $p < 0.001$). The heating temperature had a significant effect on total MUFAs in both lard and goose fat (mostly in non-additivated fat). Statistical analysis of the data showed that the addition of burdock extract at a concentration of 0.01% significantly ($p < 0.01$) reduced the installation of oxidation process in alimentary fats heated at different temperatures. Animal fats were well protected from oxidation by burdock extract, which demonstrated its efficacy as an antioxidant; it may be used to monitor the fats oxidation and to estimate their shelf-life stability.

Keywords: goose fat; lard; burdock extract; heating; oxidative stability

Citation: Pop, F.; Dippong, T. The Antioxidant Effect of Burdock Extract on the Oxidative Stability of Lard and Goose Fat during Heat Treatment. *Foods* **2024**, *13*, 304. https://doi.org/10.3390/foods13020304

Academic Editor: Emma Chiavaro

Received: 4 January 2024
Revised: 16 January 2024
Accepted: 17 January 2024
Published: 18 January 2024

Copyright: © 2024 by the authors. Licensee MDPI, Basel, Switzerland. This article is an open access article distributed under the terms and conditions of the Creative Commons Attribution (CC BY) license (https://creativecommons.org/licenses/by/4.0/).

1. Introduction

Since it allows for quick product preparation, frying is a widely utilized food preparation technique. The impact of high temperatures on fats and oils subjected to frying is a serious concern for both product quality and nutrition because of the widespread use of these fats. Fried foods are very popular among consumers because they have desirable sensory properties such as crunchy texture, appealing colour, and unique flavours [1]. However, fats and oils undergo a variety of chemical reactions at high frying temperatures which change the fried product's flavour and the frying fat's quality. The quality of frying fat has a major impact on the taste, flavour, consumer acceptance, and shelf life of the fried food [2].

Obtaining alimentary animal fats consists of melting fats obtained from raw materials at a temperature between 65–75 °C, and then filtering them to remove water and solid contaminants. A significant amount of these fats are used in Romanian cooking and frying of various foods. Using animal fats in food preparation has nutritional benefits. The fat-soluble vitamins A, E, D, and K, as well as important fatty acids, are transported by dietary animal fats. Additionally, fats help these vitamins to be absorbed and transport the vitamins and their precursors throughout the body. Dietary fats improve the flavour, texture, and smell of the food. Moreover, they take more time to be digested in the stomach than proteins or carbohydrates, delaying the feeling of hunger. Fats not only supply energy but also serve as the precursors of glycolipids and phospholipids, two crucial components in cell membranes [3]. Several research studies have shown a correlation between the consumption of animal fats in a balanced diet and a lower incidence of cardiovascular illnesses and related risk factors, including obesity, insulin resistance, and tumours [4–6].

Common chemical processes when frying fat include hydrolysis, oxidation, and polymerization, which result in the production of volatile or nonvolatile chemicals. These reactions speed up the decomposition of the used frying fat by altering and modifying its chemical composition [7]. Glycerides undergo partial hydrolysis, unsaturated fatty acids undergo oxidative degradation, and the glycerol becomes dehydrated and transforms into acrylic aldehyde throughout the heat process. Changes are encouraged by the faster rate of oxidation and reactivity of the hydrolysis products, free fatty acids, mono- and diglycerides, compared to the original triacylglycerides [8,9]. Fats undergo oxidation, which results in the production of aldehydes, ketones, and dimerized triglycerides through oxygen bridges, free or oxidized fatty acids, etc. Factors that increase oxidation include temperature, the type of fat, the fat/air contact, and the rate at which food absorbs fat [10]. The primary reaction resulting from high heat treatment is the cyclization of fatty acid molecules. The dimers and polymers of triacylglycerides are the most significant class of alteration products from a quantitative perspective, and the high temperatures attained throughout the process also catalyse their synthesis [11,12]. New alterations occur during the autoxidation process, including physical (an increase in viscosity and scum production), chemical (creation of polymers, volatile chemicals), and organoleptic (flavour alteration, palatability, darkening). The fried food absorbs the non-volatile substances which are kept in the fat and end up in the consumer's diet [13].

Since animal fats are primarily saturated and monounsaturated fats, they are more heat-resistant and have a longer shelf life than vegetable fats. Reduced oxidation in animal fats makes them less vulnerable to the toxins and cancer-causing substances produced when using vegetable oil. Owing to their greater stability, food cooked in animal fats absorbs less oil and fat [14]. Foods and vegetable oils which are rich in unsaturated fatty acids are more vulnerable to oxidation. The oil is increasingly prone to oxidation and thermal degradation as more highly unsaturated double bonds are present in the sample. Compared to unsaturated oils, animal fats are safer and produce less carcinogenic compounds [9].

In the food industry, antioxidants are frequently utilized, especially those derived from plants. Extracts made from the various burdock plant components have a range of pharmacological and biological properties, including antioxidant, anticancer, anti-inflammatory, antidiabetic, antiviral, and antimicrobial effects. It was stated that burdock extract contains chlorogenic, quinic, caffeic, p-coumaric, cinnamic, and gallic acids. Additionally, the functional qualities of burdock inulin were demonstrated, as it showed a better oil-holding capacity and promising swelling characteristics [15].

The antioxidant activity of phenolic compounds such us gallic, gentisic, protocatechuic, syringic, vanillic, caffeic, and ferulic acids was studied in pork lard [16–18]. It has been established that natural antioxidants are effective in stabilizing lard's oxidative process and the antioxidant activity declined with increasing temperatures. The effect of green tea extract on quality parameters and shelf life of animal fats during storage was investigated [19]. Organoleptic analyses revealed that, after a month of storage, no deterioration was seen in samples with green tea extract. The benefits of adding plant extracts such as rosemary extract, oregano, black tea, sage, and thyme to lard in order to inhibit lipid oxidation have been the subject of various reports [20–22]. Their activity was effective in the inhibition of fats oxidation at a concentration of 0.01%, and also exhibited antimicrobial efficiency.

Investigations assessing the antioxidative properties of natural antioxidants in dietary fats are still lacking, and to the best of our knowledge, no investigations have been conducted concerning the effects of burdock extract on the oxidative stability of lard and goose fat subjected to heating.

The goal of this study was to examine the impact of varying temperatures (110, 130, 150, 170, 190, and 210 °C) and the addition of burdock extract at a concentration of 0.01% on the qualitative characteristics of lard and goose fat exposed to heating. Iodine value (IV), peroxide value (PV), thiobarbituric acid-reactive substances (TBARS), acid value (AV), saponification value (SV), total polar compounds (TPoC), total phenolic content (TPC), fatty acid (FA) content, and microscopic examination are among the methods utilized for the assessment of fats stability and monitoring of their degradation when heating. The oxidation process in lard and goose fat subjected to heating was reduced by the addition of burdock extract in a proportion of 0.01%.

2. Materials and Methods

2.1. Materials

Two categories of dietary fats were the subject of the study: lard and goose fat. These fats were selected based on the differences and similarities between them. They differ in the composition of polyunsaturated fatty acids (goose fat is richer compared to lard), but they have a similar colour (white-yellowish) and similar consistency, and both can be used as spreads or frying fats. Raw materials were collected immediately after slaughtering. Mangalica pig fat from the recently obtained subcutaneous adipose tissue was melted at 75 °C on a water bath. Goose fat was extracted from male and female goslings of the Landaise breed at 18 weeks of age. Animals' diets were based on soybean and corn meal, with phosphorus, salt calcium, vitamins, and minerals. Animals' diets were not enhanced with long-chain polyunsaturated fatty acids or antioxidants. Approximately 500 g of raw fat was divided into small pieces, heated at 75 °C on a water bath, centrifuged, and filtered. The burdock was purchased from the local market. The leaves were manually separated, cleaned with active chlorine, then dried in an air oven at 50 °C to a constant weight. The solvent extraction method was used to obtain the extract. Crushed leaves were dipped into a 70:29.5:0.5 $v/v/v$ mixture of acetone, ethyl alcohol, and acetic acid, with a 1:10 ratio for dehydrated leaves and solvent. After the mixture was triturated in a homogenizer equipment, Velp OV 5 model (Velp Scientifica, Usmate Velate, Italy), it was vacuum filtered through a grade 2 paper filter. The supernatants were then mixed. A rotary evaporator was utilised to evaporate the remaining solvent at 60 °C [23].

In the melted fats, burdock extract was dissolved in the proportion of 0.01% (0.01 g/100 g of fat). One sample was used without heating for each fat under study. About 50 g of fat was heated in an electric oven HBG633NB1 model (Bosch GmbH, Gerlingen-Schillerhöhe, Germany) at 110, 130, 150, 170, 190 and 210 °C for 30 min. Glass vessels with an air-exposed surface area of 80 cm^3 were used for each fat and heating temperature. Before being analysed, the cold samples were placed in glass tubes and refrigerated. Every sample was examined using three replications. All chemicals utilised were of analytical quality and were procured from Merck (Merck, Darmstadt, Germany).

2.2. Methods

2.2.1. Peroxide Value (PV) Determination

The UV-VIS T60U spectrophotometer (Bibby Scientific, London, UK) was used to measure the peroxide value. Its operating temperature range was 5–45 °C, its field wavelength was 190–1100 nm, and its wavelength precision was 0.1 nm. The procedure was established on the spectrophotometer measurement of ferric ions (Fe^{3+}), which are produced when hydroperoxides oxidize ferrous ions (Fe^{2+}) in the presence of ammonium thiocyanate (NH_4SCN). Each solution's absorbance was measured at 500 nm. Thiocyanate ions (SCN^-) combine with Fe^{3+} ions to create a red-violet homogeny that can be measured spectrophotometrically. PV was measured by creating a calibration curve (absorbance at 500 nm vs. Fe^{3+}, reported in μg). The peroxide value was given as meq O_2 kg^{-1} fat [24].

2.2.2. Thiobarbituric Acid-Reactive Substances Test (TBARS)

The following procedure was used to determine TBARS: TBA Reagent (0.02 M 2-thiobarbituric acid in 90% glacial acetic acid) was obtained, then a glass-stoppered test tube containing 1 g of fat sample was filled with 5 mL of TBA reagent. The contents of the tube were combined after it was stopped. The tube was then immersed for 35 min in a boiling water bath. A blank for the TBA reagent in distilled water was made and handled the same as the sample. The sample was heated and then allowed to cool for ten minutes in tap water. After transferring a portion to a cuvette, the sample's optical density was read against the blank at a wavelength of 538 nm in a T60U spectrophotometer (Bibby Scientific, London, UK) model spectrophotometer. By creating suitable dilutions of the 1×10^{-3} M 1,1,3,3-tetraethoxypropane standard solution, a standard curve was created to provide amounts that vary from 1×10^{-8} to 7×10^{-8} mol of malondialdehyde in 1 mL. The optical densities of these dilutions were measured at a wavelength of 538 nm after they were treated with TBA reagent. The TBARS value was stated as mg malondialdehyde (MDA) kg^{-1} fat [25].

2.2.3. Iodine Value (IV) Determination

The Hanus method was used to determine the iodine value. To halogenate the double bonds, 0.5 g of sample (dissolved in 15 mL CCl_4) was combined with 25 mL of Hanus solution (IBr). After 30 min in the dark, the mixture was exposed to 20 mL of KI (100 g/L) and 100 mL of distilled water, which converted the excess IBr to free I_2. Titration was used to quantify free I_2 with 24.9 g/L $Na_2S_2O_3 \cdot 5H_2O$ using starch (1.0 g/100 mL) as an indicator. The iodine value was represented as g I_2 100 g^{-1} fat [24].

2.2.4. Acid Value (AV) Determination

Acid value (AV) was determined by using phenolphthalein as an indicator and sodium hydroxide solution 0.1 N to neutralize the sample's acidity. A conical flask containing 2 g of the material was filled with 50 mL of 99% ethanol that had been neutralized with 0.1 N NaOH and phenolphthalein as an indicator. After shaking the flask, two drops of the phenolphthalein indicator solution were added, then the mixture was neutralized by adding 0.1 N NaOH until a pale pink colour was formed. The acid value was reported as g oleic acid 100 g^{-1} fat [24].

2.2.5. Saponification Value (SV) Determination

About 5.0 g of fat was mixed with 50 mL of 4% KOH solution, and the combination was slowly heated until the sample was fully saponified. After that, it was titrated with 0.5 N HCl while 1% phenolphthalein was used as an indicator. The saponification value was given as mg KOH g^{-1} fat [25].

2.2.6. Total Polar Compounds (TPoC) Determination

A cooking oil tester (Testo 265, Testo SE & Co. KGaA, Lenzkirch, Germany) was used to measure the total polar compounds (TPoC). This instrument provides the content of polar materials with an accuracy of $+/-2\%$. The analysis was carried out by putting the sensor into heated oil and, after about 30 s, reading the temperature and the TPoC content in % from the display. The sensor was calibrated with the manufacturer-supplied calibration oil. Between measurements, the equipment was cleaned with warm water and a neutral detergent. According to Mlcek et al. [26], the allowed limit for TPoC has been established as being 25%.

2.2.7. Total Phenolic Content (TPC) Determination

TPC content was calculated using Folin–Ciocalteu reagent in accordance with the method of Singleton and Orthofer [27]. Folin–Ciocalteu reagent in a volume of 0.5 mL was combined with 0.1 mL of oil after being diluted 1:10 with distilled water. Before adding 1.7 mL of sodium carbonate solution (20%), the mixture was left to stand at 25 °C for five minutes. After 20 min of incubation with agitation at room temperature, 10 mL of distilled water was added to the mixture and the absorbance was measured at λ = 735 nm. The outcomes were stated as mg of gallic acid kg^{-1} of fat.

2.2.8. GC Analysis

In a round-bottomed flask, 2 mL of sample was added, along with 20 mL of sulphuric acid methanol solution and three pieces of porous porcelain. The flask was placed in a water bath and boiled for approximately 60 min with a reflux cooler attached. Clarifying the solution and observing that there are no longer any fat globules indicated that the reaction was complete. The flask's contents were allowed to cool to room temperature and then was quantitatively passed in a separatory funnel using 20 mL of water. Twenty millilitres of heptanes were used in two steps of the methyl esters' extraction. Following the addition of the extracts into a second separatory funnel, the extracts were thoroughly cleaned with 20 mL of water using methyl orange to ensure that all traces of sulphuric acid were absent.

The extracts were filtered into a flask after being dehydrated by the addition of anhydrous sodium sulphate. Solvent traces were eliminated from the sample by nitrogen blowing after the solvent was distilled out in a water bath under vacuum. Following the collection of methyl esters in 1 mL of hexane, 1 µL of the sample was put into the gas chromatograph. A Shimadzu GC-17 A gas chromatograph (Shimadzu, Tokyo, Japan) in conjunction with a flame ionization detector was used to evaluate the composition of fatty acids. Alltech AT-Wax (60 m × 0.32 mm × 0.5 µm) is the gas chromatography column and stationary phase (polyethylene). At 147 kPa of pressure, helium was used as carrier gas, and the injector and detector temperatures were set to 260 °C. The program of the oven was the following: 70 °C for 2 min, after which the temperature was increased to 150 °C with a gradient of 10 °C min^{-1}, for a 3 min level, then the temperature was raised to 235 °C with a gradient of 4 °C min^{-1}. Comparing the results with standards allowed for the identification and measurement of fatty acids. Results were expressed as g 100 g^{-1} [24].

2.2.9. Microscopic Examination

An Optika-B290 microscope with a tablet was used for the microscopic investigation (Optika, Ponteranica, Italy). Technical characteristics were the following: binocular, 360° rotating and 30° inclined; built-in 3.1 MP camera; interpupillary distance from 48 to 75 mm; vernier scale on the two axes, accuracy: 0.1 mm; double layer rackless mechanical sliding stage, 150 × 139mm, 75 × 33 mm X-Y movement range; objectives: N-PLAN 4×/0.10, N-PLAN 10×/0.25, N-PLAN 40×/0.65, N-PLAN 100×/1.25.

2.2.10. Statistical Analysis

The impact of antioxidant addition and heating temperature on the physicochemical parameters of lard and goose fat were examined using factorial ANOVA with the General Linear Model in Minitab 16.1.0 (LEAD Technologies, Inc., Charlotte, NC, USA). Each sample was examined using three replications. Tukey's honest significance test was performed at a level of 95% confidence ($p < 0.05$). The degree of connection between chemical parameters was estimated using Pearson's correlation ($\alpha = 0.05$) and two-tailed probability values.

3. Results and Discussion

The goal of this study was to determine the impact of 0.01% burdock extract added in two alimentary fats (lard and goose), which are commonly used in Romania for culinary cooking, in terms of their oxidative stability and rates of deterioration at various heating temperatures (110, 130, 150, 170, 190, and 210 °C). Chemical analysis results and fatty acid composition for fats and 0.01%-additivated fats upon heating are presented in Tables 1–4. All the measured parameters were significantly impacted by the heating temperature, with IV and SV being the least affected. PV, TBARS, and TPoC were significantly ($p < 0.01$) influenced by the additive × heat treatment interaction and significantly ($p < 0.001$) affected by the fat type × additive × heat treatment interaction. The first- and second-degree interactions had a significant ($p < 0.05$) effect on IV and a significant ($p < 0.01$) impact on AV (Table 1).

Table 1. The impact of the fat type, additivation, heat treatment, and their first- and second-degree interactions, on PV (meq O_2 kg^{-1} fat), TBARS (mg MDA kg^{-1} fat), IV (g I_2 100 g^{-1} fat), AV (g oleic acid 100 g^{-1} fat), SV (mg KOH g^{-1} fat), TPoC (%), and TPC (mg gallic acid kg^{-1} fat).

Factor	PV	TBARS	IV	AV	SV	TPoC	TPC
Fat type							
Lard	4.02 [a]	3.15 [a]	81.2 [b]	0.29 [a]	195 [b]	1.4 [a]	164 [a]
Goose fat	5.19 [b]	5.38 [b]	82.5 [a]	0.20 [b]	198 [a]	1.6 [b]	168 [b]
p	<0.01 **	<0.001 ***	<0.01 **	<0.01 **	<0.01 **	<0.05 *	<0.05 *
Additivation							
Non-additivated	5.60 [a]	4.56 [a]	82.7 [b]	0.45 [a]	198 [a]	2.7 [a]	165 [a]
Additivated with 0.01% burdock extract	4.40 [b]	3.20 [b]	83.2 [a]	0.39 [b]	195 [b]	1.5 [b]	184 [b]
p	<0.001 ***	<0.001 ***	<0.05 *	<0.01 **	<0.01 **	<0.01 **	<0.01 **
Heat treatment							
Unheated	1.95 [f]	0.96 [g]	84.7 [a]	0.21 [f]	195 [e]	1.4 [a]	164 [a]
110 °C	2.46 [e]	1.14 [f]	84.5 [a]	0.28 [e]	196 [de]	1.9	161 [b]
130 °C	3.16 [d]	1.87 [e]	84.1 [b]	0.35 [d]	197 [d]	2.7 [ab]	157 [c]
150 °C	5.47 [c]	2.98 [d]	83.2 [b]	0.41 [c]	198 [c]	3.8 [a]	149 [d]
170 °C	7.64 [b]	4.56 [c]	81.5 [c]	0.63 [b]	200 [bc]	5.3 [ab]	142 [e]
190 °C	8.84 [a]	6.48 [b]	80.4 [c]	0.78 [ab]	201 [b]	8.2 [ab]	135 [f]
210 °C	8.70 [a]	7.64 [a]	78.6 [d]	0.86 [a]	204 [a]	12.4 [b]	118 [g]
p	<0.001 ***	<0.001 ***	<0.01 **	<0.001 ***	<0.01 **	<0.001 ***	<0.001 ***
Additive * Heat treatment							
Additive * unheated	1.83 [e]	0.84 [d]	76.8 [a]	0.25 [d]	194 [b]	1.1 [a]	172 [a]
Additive * 110 °C	2.54 [d]	1.16 [cd]	76.3 [a]	0.27 [cd]	194 [b]	1.8 [ab]	170 [a]
Additive * 130 °C	2.96 [cd]	1.85 [c]	75.5 [b]	0.32 [c]	195 [ab]	2.7 [b]	162 [b]
Additive * 150 °C	3.48 [c]	2.17 [bc]	75.1 [b]	0.39 [c]	195 [ab]	3.5 [bc]	154 [bc]
Additive * 170 °C	6.53 [bc]	3.61 [b]	74.3 [c]	0.46 [bc]	195 [ab]	4.9 [c]	141 [c]
Additive * 190 °C	7.12 [b]	5.28 [ab]	74.1 [c]	0.57 [b]	196 [a]	5.8 [cd]	132 [cd]
Additive * 210 °C	8.51 [a]	6.44 [a]	73.4 [cd]	0.69 [a]	196 [a]	7.2 [d]	125 [d]
p	<0.01**	<0.01 **	<0.05 *	<0.01 **	≥0.05	<0.01 **	<0.05 *
Fat type * additive * heat treatment							
Lard * additive * unheated	1.72 [f]	0.82 [f]	74.3 [c]	0.25 [de]	193 [c]	0.8 [a]	162 [bc]
Lard * additive * 110 °C	1.94 [f]	1.05 [f]	74.2 [c]	0.27 [d]	193 [c]	1.5 [b]	157 [c]
Lard * additive * 130 °C	2.26 [e]	1.79 [e]	73.5 [cd]	0.34 [c]	194 [bc]	2.2 [bc]	144 [cd]
Lard * additive * 150 °C	3.15 [d]	2.36 [de]	72.1 [cd]	0.46 [bc]	195 [b]	3.4 [c]	132 [d]
Lard * additive * 170 °C	5.03 [bc]	3.09 [d]	71.9 [cd]	0.65 [b]	197 [ab]	5.1 [d]	123 [de]
Lard * additive * 190 °C	5.98 [bc]	4.26 [c]	69.7 [d]	0.77 [ab]	198 [ab]	6.7 [e]	114 [e]
Lard * additive * 210 °C	6.85 [b]	6.11 [b]	68.7 [d]	0.81 [a]	199 [a]	8.2 [ef]	103 [ef]
Goose * additive * unheated	2.13 [e]	1.12 [ef]	86.3 [a]	0.21 [de]	194 [bc]	1.4 [b]	187 [a]
Goose * additive * 110 °C	2.38 [e]	1.32 [ef]	86.0 [a]	0.24 [d]	195 [b]	1.8 [b]	182 [ab]
Goose * additive * 130 °C	3.09 [d]	1.92 [e]	85.4 [ab]	0.32 [c]	195 [b]	3.5 [c]	175 [b]
Goose * additive * 150 °C	4.31 [c]	3.85 [cd]	84.3 [ab]	0.41 [bc]	196 [b]	4.6 [cd]	166 [bc]
Goose * additive * 170 °C	6.36 [b]	4.21 [c]	82.9 [b]	0.55 [bc]	198 [ab]	7.2 [e]	152 [c]
Goose * additive * 190 °C	7.34 [ab]	6.34 [b]	81.7 [b]	0.63 [b]	199 [a]	8.6 [ef]	146 [cd]
Goose * additive * 210 °C	8.45 [a]	7.48 [a]	79.6 [bc]	0.72 [ab]	200 [a]	12.7 [f]	138 [d]
p	<0.001 ***	<0.001 ***	<0.05 *	<0.01 **	<0.05 *	<0.001 ***	<0.01 **

PV, peroxide value; TBARS, thiobarbituric acid-reactive substances; IV, iodine value; AV, acid value; SV, saponification value; TPoC total polar compounds; TPC, total phenolic content. Values are indicated as mean. Variations in the same column's letters denote statistically significant differences at $p < 0.05$ (Tukey's test). Significant differences are denoted by asterisks: * $p < 0.05$; ** $p < 0.01$; *** $p < 0.001$; $p \geq 0.05$, non-significant.

Table 2. Monitoring of quality parameters in lard (non-additivated and additivated with 0.01% burdock extract) exposed to heating at different temperatures.

Lard	Heating Temperature	PV (meq O_2 kg^{-1} Fat)	TBARS (mg MDA kg^{-1} Fat)	IV (g I_2 100 g^{-1} Fat)	AV (g Oleic Acid 100 g^{-1} Fat)	SV (mg KOH g^{-1} Fat)	TPoC (%)	TPC (mg Gallic Acid kg^{-1} Fat)
Non-additivated	Unheated	1.76 [i] ± 0.02	0.84 [k] ± 0.01	74.7 [a] ± 0.3	0.25 [h] ± 0.02	194 [e] ± 0.1	1.1 [i] ± 0.3	148 [g] ± 0.4
	110 °C	2.24 [g] ± 0.03	1.26 [f] ± 0.05	74.2 [a] ± 0.1	0.31 [h] ± 0.07	196 [d] ± 0.6	2.3 [g] ± 0.5	141 [f] ± 0.5
	130 °C	3.18 [f] ± 0.01	2.13 [e] ± 0.04	73.5 [b] ± 0.4	0.42 [f] ± 0.02	197 [d] ± 0.7	3.7 [f] ± 0.4	134 [e] ± 0.3
	150 °C	4.85 [e] ± 0.04	3.25 [d] ± 0.02	71.4 [c] ± 0.3	0.59 [e] ± 0.05	198 [c] ± 0.5	4.8 [e] ± 0.6	122 [d] ± 0.2
	170 °C	7.02 [b] ± 0.07	4.78 [c] ± 0.07	70.2 [d] ± 0.2	0.78 [c] ± 0.04	199 [c] ± 0.4	7.9 [c] ± 0.5	113 [c] ± 0.1
	190 °C	8.79 [a] ± 0.05	5.48 [b] ± 0.03	68.5 [e] ± 0.1	0.84 [b] ± 0.03	201 [b] ± 0.2	9.5 [b] ± 0.4	104 [b] ± 0.2
	210 °C	7.65 [b] ± 0.06	7.62 [a] ± 0.08	66.1 [f] ± 0.5	0.94 [a] ± 0.02	203 [a] ± 0.1	10.7 [a] ± 0.2	96 [a] ± 0.4
Additivated with 0.01% burdock extract	Unheated	1.75 [i] ± 0.02	0.83 [k] ± 0.04	74.5 [a] ± 0.4	0.24 [h] ± 0.01	194 [e] ± 0.3	0.9 [i] ± 0.3	163 [i] ± 0.2
	110 °C	2.11 [h] ± 0.01	1.02 [g] ± 0.07	74.1 [a] ± 0.5	0.28 [h] ± 0.05	194 [e] ± 0.4	1.6 [h] ± 0.2	156 [h] ± 0.3
	130 °C	2.47 [g] ± 0.03	1.83 [f] ± 0.01	73.4 [b] ± 0.3	0.36 [g] ± 0.06	195 [d] ± 0.8	2.4 [g] ± 0.5	145 [f] ± 0.5
	150 °C	3.72 [f] ± 0.04	2.41 [e] ± 0.02	72.4 [c] ± 0.2	0.48 [e] ± 0.02	196 [d] ± 0.5	3.5 [f] ± 0.6	131 [e] ± 0.4
	170 °C	5.84 [d] ± 0.07	3.26 [d] ± 0.03	71.7 [d] ± 0.1	0.64 [d] ± 0.04	198 [c] ± 0.6	5.3 [e] ± 0.4	124 [d] ± 0.1
	190 °C	6.23 [c] ± 0.05	4.52 [c] ± 0.06	69.5 [e] ± 0.6	0.78 [c] ± 0.01	200 [c] ± 0.2	6.9 [d] ± 0.1	112 [c] ± 0.2
	210 °C	7.36 [b] ± 0.02	6.38 [b] ± 0.08	68.4 [e] ± 0.1	0.82 [b] ± 0.02	201 [b] ± 0.4	8.3 [b] ± 0.8	102 [b] ± 0.4

PV, peroxide value; TBARS, thiobarbituric acid-reactive substances; IV, iodine value; AV, acid value; SV, saponification value; TPoC total polar compounds; TPC, total phenolic content. Values are expressed as mean ± standard deviation of three replicates for each parameter. Variations in the same column's letters denote statistically significant differences at $p < 0.05$ (Tukey's test).

Table 3. Monitoring of quality parameters in goose fat (non-additivated and additivated with 0.01% burdock extract) exposed to heating at different temperatures.

Goose Fat	Heating Temperature	PV (meq O_2 kg^{-1} Fat)	TBARS (mg MDA kg^{-1} Fat)	IV (g I_2 100 g^{-1} Fat)	AV (g Oleic Acid 100 g^{-1} Fat)	SV (mg KOH g^{-1} fat)	TPoC (%)	TPC (mg Gallic Acid kg^{-1} Fat)
Non-additivated	Unheated	2.28 [i] ± 0.07	1.16 [k] ± 0.03	86.8 [a] ± 0.3	0.22 [h] ± 0.03	195 [e] ± 0.4	1.4 [i] ± 0.2	164 [f] ± 0.2
	110 °C	2.76 [g] ± 0.05	1.54 [f] ± 0.02	86.1 [a] ± 0.1	0.29 [g] ± 0.02	196 [d] ± 0.8	2.8 [g] ± 0.4	160 [e] ± 0.1
	130 °C	3.96 [f] ± 0.03	2.85 [e] ± 0.03	84.3 [b] ± 0.4	0.37 [f] ± 0.07	196 [d] ± 0.6	4.3 [e] ± 0.1	152 [d] ± 0.4
	150 °C	5.94 [e] ± 0.02	3.91 [d] ± 0.06	82.7 [c] ± 0.3	0.51 [d] ± 0.02	198 [c] ± 0.2	5.8 [e] ± 0.6	143 [c] ± 0.5
	170 °C	8.58 [d] ± 0.06	5.74 [c] ± 0.04	81.3 [d] ± 0.2	0.69 [c] ± 0.05	200 [c] ± 0.3	8.6 [c] ± 0.7	135 [c] ± 0.3
	190 °C	9.71 [a] ± 0.02	7.25 [b] ± 0.01	78.6 [e] ± 0.1	0.78 [b] ± 0.01	202 [b] ± 0.1	10.3 [b] ± 0.3	128 [b] ± 0.6
	210 °C	8.32 [c] ± 0.04	8.54 [a] ± 0.07	77.4 [f] ± 0.5	0.88 [a] ± 0.02	204 [a] ± 0.5	15.1 [a] ± 0.9	115 [a] ± 0.3
Additivated with 0.01% burdock extract	Unheated	2.15 [j] ± 0.03	1.15 [k] ± 0.04	86.2 [a] ± 0.2	0.22 [h] ± 0.04	194 [f] ± 0.7	1.5 [i] ± 0.1	185 [i] ± 0.1
	110 °C	2.31 [i] ± 0.04	1.32 [g] ± 0.05	86.1 [a] ± 0.1	0.25 [g] ± 0.03	195 [e] ± 0.2	1.9 [h] ± 0.4	181 [h] ± 0.2
	130 °C	3.14 [h] ± 0.02	1.96 [f] ± 0.04	85.2 [b] ± 0.4	0.31 [f] ± 0.02	196 [d] ± 0.4	3.6 [f] ± 0.6	174 [g] ± 0.3
	150 °C	4.35 [f] ± 0.01	2.85 [e] ± 0.03	84.3 [c] ± 0.6	0.43 [e] ± 0.05	197 [e] ± 0.7	4.8 [e] ± 0.5	165 [f] ± 0.8
	170 °C	7.26 [d] ± 0.06	4.51 [d] ± 0.06	82.8 [d] ± 0.3	0.56 [d] ± 0.07	199 [c] ± 0.1	7.1 [d] ± 0.3	153 [d] ± 0.2
	190 °C	7.83 [bc] ± 0.05	6.27 [c] ± 0.02	81.4 [e] ± 0.5	0.62 [c] ± 0.04	200 [c] ± 0.4	8.7 [c] ± 0.2	145 [c] ± 0.4
	210 °C	9.05 [ab] ± 0.02	7.23 [b] ± 0.01	79.3 [f] ± 0.2	0.71 [b] ± 0.02	202 [b] ± 0.2	12.8 [b] ± 0.7	139 [b] ± 0.5

PV, peroxide value; TBARS, thiobarbituric acid-reactive substances; IV, iodine value; AV, acid value; SV, saponification value; TPoC total polar compounds; TPC, total phenolic content. Values are expressed as mean ± standard deviation of three replicates for each parameter. Variations in the same column's letters denote statistically significant differences at $p < 0.05$ (Tukey's test).

Table 4. Variations in fatty acid content (g 100 g^{-1} fat) of dietary fats exposed to heating at 210°C.

Fatty Acids	Lard			Goose Fat		
	Non-Additivated (Control)	Non-Additivated Fat Subjected to Heating at 210 °C	Burdock Extract Additivated Fat Subjected to Heating at 210 °C	Non-Additivated (Control)	Non-Additivated Fat Subjected to Heating at 210 °C	Burdock Extract Additivated Fat Subjected to Heating at 210 °C
Myristic (14:0)	0.52 ^{a} ± 0.04	0.73 ^{ab} ± 0.06	0.87 ^{b} ± 0.09	0.31 ^{a} ± 0.10	0.52 ^{b} ± 0.05	0.36 ^{a} ± 0.05
Pentadecanoic (C15:0)	0.65 ^{a} ± 0.06	0.86 ^{b} ± 0.01	0.86 ^{b} ± 0.02	0.52 ^{a} ± 0.14	0.67 ^{ab} ± 0.07	0.58 ^{a} ± 0.04
Palmitic (C16:0)	1.80 ^{a} ± 0.10	1.88 ^{a} ± 0.08	1.84 ^{a} ± 0.13	0.61 ^{a} ± 0.12	0.72 ^{a} ± 0.12	0.62 ^{a} ± 0.12
Palmitoleic (C16:1)	1.18 ^{a} ± 0.13	1.15 ^{a} ± 0.13	1.28 ^{a} ± 0.11	1.85 ^{a} ± 0.03	1.45 ^{b} ± 0.02	1.59 ^{b} ± 0.07
Hexadecadienoic (C16:2)	0.77 ^{a} ± 0.05	0.45 ^{b} ± 0.11	0.36 ^{b} ± 0.04	0.78 ^{a} ± 0.08	0.62 ^{a} ± 0.09	0.61 ^{a} ± 0.06
Hexadecatrienoic (C16:3)	0.56 ^{a} ± 0.02	0.34 ^{ab} ± 0.12	0.21 ^{b} ± 0.09	0.56 ^{a} ± 0.04	0.32 ^{b} ± 0.10	0.48 ^{a} ± 0.08
Hexadecatetranoic (C16:4)	0.19 ^{a} ± 0.06	0.17 ^{a} ± 0.01	0.13 ^{a} ± 0.01	0.59 ^{a} ± 0.11	0.37 ^{ab} ± 0.15	0.54 ^{a} ± 0.13
Heptadecanoic (C17:0)	0.86 ^{a} ± 0.03	1.03 ^{ab} ± 0.05	1.17 ^{b} ± 0.05	0.51 ^{a} ± 0.05	0.64 ^{ab} ± 0.06	0.57 ^{a} ± 0.09
Stearic (C18:0)	0.93 ^{a} ± 0.09	1.07 ^{a} ± 0.02	1.09 ^{a} ± 0.15	0.32 ^{a} ± 0.09	0.33 ^{a} ± 0.08	0.31 ^{a} ± 0.11
Oleic (C18:1)	2.24 ^{a} ± 0.14	1.86 ^{ab} ± 0.09	2.16 ^{a} ± 0.13	1.49 ^{a} ± 0.13	1.22 ^{ab} ± 0.03	1.46 ^{a} ± 0.05
Linoleic (C18:2)	0.41 ^{a} ± 0.11	0.33 ^{a} ± 0.07	0.36 ^{a} ± 0.10	0.66 ^{a} ± 0.04	0.45 ^{b} ± 0.08	0.50 ^{b} ± 0.08
Linolenic (C18:3)	0.27 ^{a} ± 0.12	0.18 ^{ab} ± 0.03	0.22 ^{a} ± 0.11	0.56 ^{a} ± 0.06	0.42 ^{b} ± 0.01	0.47 ^{ab} ± 0.01
Stearidonic acid (C18:4)	0.08 ^{a} ± 0.05	0.07 ^{a} ± 0.06	0.07 ^{a} ± 0.07	0.79 ^{a} ± 0.15	0.54 ^{b} ± 0.04	0.84 ^{a} ± 0.13
Total FA	10.58 ^{a} ± 0.88	10.07 ^{b} ± 0.67	10.75 ^{a} ± 1.09	9.58 ^{a} ± 0.95	8.29 ^{b} ± 0.90	8.41 ^{b} ± 1.03
Total SFA	4.20 ^{a} ± 0.32	4.96 ^{ab} ± 0.06	4.93 ^{ab} ± 0.44	2.32 ^{a} ± 0.32	2.58 ^{ab} ± 0.32	2.38 ^{a} ± 0.40
Total MUFA	4.07 ^{a} ± 0.27	3.65 ^{b} ± 0.22	4.03 ^{a} ± 0.23	2.37 ^{a} ± 0.16	3.08 ^{c} ± 0.05	2.63 ^{b} ± 0.12
Total PUFA	2.29 ^{a} ± 0.30	1.42 ^{c} ± 0.40	1.79 ^{b} ± 0.42	3.76 ^{a} ± 0.48	2.61 ^{c} ± 0.47	3.40 ^{b} ± 0.50

FA, fatty acids; SFA, saturated fatty acids; MUFA, monounsaturated fatty acids; PUFA, polyunsaturated fatty acids. Values are expressed as mean ± standard deviation of three replicates for each parameter. For each type of animal fat, different letters in the same row indicate statistically significant differences at $p < 0.05$ (Tukey's test).

3.1. Effect of Additivation and Heat Treatment on Peroxide Value (PV)

The number of peroxides generated during fat oxidation is shown by the peroxide value, which quantifies the primary lipid oxidation. It has been shown that lipid oxidation products such peroxides, free radicals, malonaldehyde, and other products of cholesterol oxidation promote atherosclerosis and coronary heart disease [28]. The peroxide index values were lower in unheated animal fats. The effects of additivation and heating temperature on the peroxide value were statistically significant ($p < 0.001$) (Table 1). The lowest PV was found in lard (Table 2), and the highest was observed in goose fat (Table 3).

The non-additivated fat had the highest peroxide index level, independent of the heating temperature. The highest intensity of peroxide production was observed at 180 °C, and above this temperature, decomposition was observed. This pattern could be explained by the fact that peroxides are not heat-resistant and that high temperatures reduce their concentration. Peroxides are characteristic during early heating and decompose at high temperatures. The tendency for the peroxide value (PV) in non-additivated animal fats to decrease could be caused by the melting pretreatment's hydroperoxide-generating effect, which could have been broken down by heating. The fat samples with 0.01% burdock extract presented lower values for PV compared to non-additivated samples. Therefore, a lower number of primary oxidation compounds were determined in burdock additivated fats.

Szabó et al. [29] state that peroxide breakdown takes place at a critical temperature between 190 and 200 °C. This is supported by our data, which show that peroxide generation peaked at 190 °C and that breakdown predominated above this temperature. The PVs of the lard and goose fat samples were significantly affected ($p < 0.001$) by the application of antioxidants (Table 1). According to the results, pork fat produced less peroxides when heated, and therefore, was more stable. The oxidation of both fats occurs at higher heating temperatures, leading to a greater breakdown of polyunsaturated fatty acids and the creation of oxidation products. The development of oxidation products was greatly delayed by the addition of burdock extract in a proportion of 0.01%; however, because of an increased rate of initiation reactions, antioxidant activity decreased with temperature.

A rise in PV during heating or frying of fats and oils has been reported by several researchers [30–32]. A study was conducted utilizing the Oxipres apparatus to examine temperature's effect on the antioxidant activity of α- and δ-tocopherol in pork lard [33]. The research demonstrated that the activity of α-tocopherol remained constant between 80 and 110 °C and decreased with rising temperatures. Pop and Mihalescu [24] investigated the stability of goose, duck, and chicken fats with the addition of natural antioxidants (α-tocopherol and citric acid) during refrigerated and frozen storage. The researchers showed that the results for peroxide value and the thiobarbituric acid-reactive substances test increased with temperature. Furthermore, as storage time increased, the variations in PV and TBARS also increased. The thermal oxidation of tallow was investigated by Song et al. [34], who examined the volatile chemicals produced under various oxidation circumstances, the peroxide value, acid value, and the p-anisidine value (p-AV). The researchers found that, when heated at 140 °C for two hours, the levels of beef-flavour precursors, including hexanal, (E,E)-2,4-decadienal, 1-octen-3-ol, and (E,E)-2,4-heptadienal, achieved their highest value. PV and p-AV were both at high levels while AV was relatively low at the same temperature. The impact of natural and synthetic antioxidants and synergists (green tea extract, rosemary extract, sage extract, α-tocopherol, tocopherol mixture, propyl gallate, citric acid, caffeic acid, ascorbic acid, rosmarinic acid) on the oxidative stability of goose fat was examined using the Schaal Oven test [35]. The results showed that, in goose fat, green tea extract has the strongest antioxidant activity, but rosemary extract in combination with a synergist showed a greater fat-protective factor against oxidation than that of pure rosemary extract. Also, when a blend of tocopherols was used instead of sage extract, the fat was stabilized more effectively.

3.2. Effect of Additivation and Heat Treatment on Thiobarbituric Acid-Reactive Substances Test (TBARS)

The quantity of secondary oxidation compounds (ketones, aldehydes, or other matrix compounds) present in the fat sample is measured by TBARS. As for the peroxide value, the values of thiobarbituric acid-reactive compounds increased significantly ($p < 0.001$) in all fats with increasing heating temperatures, but decreased with the addition of burdock extract in a proportion of 0.01%. The results for TBARS increased from 0.84 to 7.62 mg MDA kg^{-1} in pork lard without antioxidant and to 6.38 mg MDA kg^{-1} in burdock-additivated lard exposed to heating at 210 °C; from 1.16 to 8.52 mg MDA kg^{-1} in goose fat without antioxidant and to 7.23 mg MDA kg^{-1} in burdock-additivated goose fat subjected to heating at 210 °C. The TBARS value of non-additivated goose fat heated to 210 °C was higher than the permissible limit of 8 mg MDA kg^{-1} fat. TBARS was significantly affected ($p < 0.001$) by the additivation, type of fat, and heating temperature (Table 1). Regardless of the heating temperature, the highest level of TBARS was detected in fat without an antioxidant. There were significant positive correlations between PV and TBARS in both goose fat ($r = 0.93$; $p < 0.001$) and lard ($r = 0.96$; $p < 0.001$). As for the peroxide value, the values of thiobarbituric acid-reactive compounds increased significantly ($p < 0.001$) in all fats with increasing heating temperatures, but decreased with the addition of burdock extract in a proportion of 0.01%. This is consistent with earlier research on chicken and duck fat, which found that temperature affects the development of TBARS [6].

The potential of some phenolic compounds to prevent the oxidation of butter was evaluated by Soulti and Roussis [36]. Thiobarbituric acid-reactive substances and peroxide values were monitored during butter heating at 50 °C and at 110 °C. The results showed that butylated hydroxyanisole at 200 mg L^{-1} was equally efficient at suppressing butter oxidation at 50 °C as caffeic acid, gallic acid, and catechin at 80 mg L^{-1}. Furthermore, gallic acid was more efficient than butylated hydroxyanisole at inhibiting the oxidation of butter at 110 °C. Moslavac et al. [37] studied the effect of natural antioxidants (sage extract, rosemary extract, olive pomace extract, α-tocopherol, mixture tocopherol) on the oxidative stability of pork fat using the sustainability test at 98 °C. The results showed that rosemary extract more effectively protected pork fat against oxidation compared to the other antioxidants we tested. It was reported that heating chicken fat at 70 °C and storing it for two and four days significantly increased the amount of lipid oxidation, as shown by TBARS. When heating temperature increased, there was a significant ($p < 0.05$) decreased in both monounsaturated and polyunsaturated fatty acids [5].

The oxidative stability of duck, chicken, pork, and bovine fats was evaluated during a period of 90 day of storage at 60 °C [11]. The study showed that the TBARS value was significantly higher in duck fat at the beginning of the storage period (10 to 40 days), and that the value in chicken fat was relatively higher than in pork and bovine fats (10 to 20 days). The results also showed that the higher rate of lipid peroxidation in chicken and duck fats was due to the higher contents of PUFAs which were more susceptible to oxidation. The lower TBARS value determined in bovine fat was due to the higher SFAs content which are less vulnerable to lipid oxidation. Also, the PV levels which were higher in duck, chicken, and swine fat have caused the higher TBARS values. The unstable peroxides can be the result of polyunsaturated fatty acids decomposition; therefore; researchers suggested the addition of antioxidants to duck and chicken fats for improving their oxidative stability during storage.

3.3. Effect of Additivation and Heat Treatment on Iodine Value (IV)

The heating temperature significantly ($p < 0.01$) reduced the iodine index values for all types of fats. A decrease in iodine values from 74.7 to 66.1 g I_2 100 g^{-1} was observed in pork lard without an antioxidant and to 68.4 g I_2 100 g^{-1} in burdock-additivated lard exposed to heating at 210 °C; from 86.8 to 77.4 g I_2 100 g^{-1} in goose fat without an antioxidant and to 79.3 g I_2 100 g^{-1} in burdock-additivated goose fat exposed to heating at 210 °C. IV was significantly ($p < 0.01$) influenced by the fat type and heating temperatures, and significantly ($p < 0.05$) influenced by the additivation. Fatty acid unsaturation is reduced, as indicated by the decrease in IV levels. The decrease in iodine value during heating indicated a higher rate of oxidation, which may be ascribed to double bond oxidation and polymerization processes. Strong negative correlations between IV and TBARS were determined in lard ($r = -0.98$; $p < 0.001$) and goose fat ($r = -0.96$; $p < 0.001$). In the study conducted by Farhooshi and Moosavi [38], fish crackers were fried at 180 °C for 5 h per day for four consecutive days using an ethanolic citrus peel extract mixed to palm olein at a 0.2% concentration. Iodine value, totox value, peroxide value, and viscosity analyses showed that the orange peel extract had strong antipolymerization and antioxidant properties [38]. Gharby et al. [39] studied the effect of heat treatment on vegetable fats, and a slight decrease in iodine value was observed after 3 h of heating at 180 °C.

3.4. Effect of Additivation and Heat Treatment on Acid Value (AV)

When assessing the hydrolysis extension in fats during the heat process, free acidity is an analytical parameter that is utilized. An increase in this parameter, which is a direct result of hydrolysis and a key marker of the chemical degradation of fat, denotes a larger concentration of free fatty acids in animal fat. At the highest temperatures, the rate of growth in the AV of the fats tended to slow down. The AV increased in parallel with the intensity of heating. The non-additivated fat had the highest AV level, regardless of the heating temperature. Heating at 210 °C produced the largest amount of free fatty acids. Acid value increased from 0.25 to 0.94 g oleic acid 100 g^{-1} in pork lard without an antioxidant and to 0.82 g oleic acid 100 g^{-1} in burdock-additivated lard exposed to 210 °C during heating; from 0.22 to 0.88 g oleic acid 100 g^{-1} in goose fat without an antioxidant and to 0.71 g oleic acid 100 g^{-1} in burdock-additivated goose fat exposed to heating at 210 °C. Strong positive correlations between AV and PV were found in lard ($r = 0.98$; $p < 0.001$) and goose fat ($r = 0.96$; $p < 0.001$). Acid value was significantly ($p < 0.01$) influenced by the fat type and additivation, and significantly ($p < 0.001$) influenced by heating temperature. A rise in the acid value of goose fat and Mangalica-pig lard exposed to heating at seven temperatures (140, 150, 160, 165, 170, 175, and 180 °C) was reported by Szabó et al. [29]. Thermooxidative structural changes in fats were investigated by Gertz et al. [28] at ambient temperatures, under a frying temperature of 170 °C, and under accelerated conditions using 110 °C. The results of the study indicated that the iodine value and the Rancimat test at 110 °C had a strong negative correlation, whereas the anisidine and acid values showed a weak correlation.

3.5. Effect of Additivation and Heat Treatment on Saponification Value (SV)

The lipids' average molecular weights are indicated by the saponification value. Oxidative rancidity increases SV by oxidizing fatty acids, which forms aldehydes and ketones. Strong positive correlations between AV and SV were determined in pork lard ($r = 0.94$; $p < 0.001$), and goose fat ($r = 0.97$; $p < 0.001$). All treatments had a significant ($p < 0.01$) impact on saponification value. For both lard and goose fat, saponification value increased significantly ($p < 0.01$) with heating temperature. In a model of frying at 180 °C, Patel et al. [40] found that the stability of clarified butterfat supplemented with 0.5% commercial steam-distilled coriander extract and oleoresin has been significantly improved. The steam-distilled extract performed better than the oleoresin, according to the measurements for conjugated dienes, peroxide value, thiobarbituric acid value, saponification value, and the Rancimat at 120 °C.

3.6. Effect of Additivation and Heat Treatment on Total Polar Compounds (TPoC)

During heating, peroxides and hydroperoxides produce total polar compounds, which include alcohol, nonvolatile products, ketones, aldehydes, and short-chain fatty acids. In every sample that was subjected to heating, the TPoC values were below the limit of 25%. In comparison to pork lard, goose fat exhibited greater values of TPoC, which increased in parallel with the heating temperature. The heating temperature had a significant ($p < 0.001$) effect on the TPoC values, as shown in Table 1. The formation of TPoC was significantly ($p < 0.01$) reduced by the addition of burdock extract at a concentration of 0.01%. Qiuyu et al. [41] studied the thermal behaviour and antioxidant effect of rosemary extract on chicken fat. A synchronization was seen between the reduction in antioxidant capacity and weight loss, as the researchers discovered that rosemary extract was stable below 200 °C and that phenolics were resistant below 130 °C.

3.7. Effect of Additivation and Heat Treatment on Total Phenolic Content (TPC)

The heat treatment at 210 °C resulted in a decrease in total phenolic content from 148 to 96 mg gallic acid kg^{-1} in pork lard and from 163 to 102 mg gallic acid kg^{-1} in additivated lard under heating at 210 °C (Table 2). Total phenolic content decreased from 164 to 115 mg gallic acid kg^{-1} in goose fat exposed to heating at 210 °C and from 185 to 139 mg gallic acid kg^{-1} in additivated goose fat under heating at 210 °C (Table 3). Total phenolic content was reduced by 35% in pork lard without an antioxidant and by 23% in burdock-additivated lard under heating. For goose fat, the reduction was about 30% in non-additivated fat, and about 19% in burdock-additivated fat under heating. According to Xueqi et al. [42], most of the squalene in olive oil remained intact even after heating at 220 °C. A considerable quantity of total phenols and specific phenols, including oleocanthal, remained after heating the oil at 121 °C for 10 and 20 min. Oleocanthal had the highest temperature tolerance among the studied phenols, hydroxytyrosol the lowest, while tyrosol exhibited a smaller change with various heating techniques. The study reported a loss of total phenol content of 40.80% after 10 min of heating at 121 °C, while 78.41% of total phenol loss was observed after 10 min of heating at 220 °C.

The total phenolic content in the 70%-ethanolic extract of *A. lappa* leaves was 97.49 mg g^{-1} [43], whereas Lee et al. [44] determined that the total phenolic content of cultivated burdock was higher in the roots than in the leaves (137 and 41.4 mg 100 g^{-1} dry material, respectively). Burdock leaf fractions' effectiveness in preserving meat was assessed [45]. The findings demonstrated that *Salmonella Typhimurium* and *Escherichia coli* growth and biofilm formation were greatly suppressed by burdock leaf fraction. When compared to the pork without treatment, the shelf life of pork treated with burdock leaf fractions was increased by six days, and the pork's sensory qualities were clearly improved. Chemical composition analysis indicated that the burdock-leaf fraction consisted of caffeic acid, chlorogenic acid, p-coumaric acid, cynarin, rutin, luteolin, crocin, arctiin, and quercetin. The burdock-leaf fraction was shown to be a promising natural preservative for pork.

3.8. Effect of Heat Treatment on Fatty Acid Composition

The fatty acid contents of dietary fats (non-additivated and additivated with burdock extract) subjected to heating at 210 °C are presented in Table 4. Among the fatty acids, palmitic (C16:0), palmitoleic (C18:1), and oleic (C16:1) are the most popular in pork lard; palmitoleic (C18:1) and oleic (C16:1) are the most popular in goose fat. Saturated fatty acids (SFAs) are predominant in lard, whereas monounsaturated fatty acids (MUFAs) are predominant in goose fat. Goose fat had the highest concentration of polyunsaturated fatty acids (PUFAs), followed by the lard. For lard without an antioxidant, total SFAs increased from 4.20 (g 100 g^{-1} fat) in the control to 4.96 in lard heated to 210 °C, and for non-additivated goose fat from 2.32 (g 100 g^{-1} fat) in the control to 2.58 in goose fat heated to 210 °C. These increases were attributed to the high-temperature heating treatment. The total PUFAs decreased from 3.76 (g 100 g^{-1} fat) in the control to 2.61 in goose fat without antioxidant exposed to heating at 210 °C, and from 2.29 in the control to 1.42 in lard without antioxidant exposed to heating at 210 °C. Heat treatment at 210 °C did not significantly modify the composition of stearic and palmitic acids, while a considerable rise in the content of heptadecanoic and myristic acids was detected. Unsaturated fatty acids are altered by heat treatment of fats, they may isomerize from their cis to transform. During thermal treatment, such formation of trans fatty acids has been observed for chicken fat [46]. The heating temperature had a significant effect on total MUFAs in both lard and goose fat (to a greater extent in fat without an antioxidant). Considering that highest level of PV was found in goose fat, the oxidation of monounsaturated fatty acids may be the cause of the decrease in total MUFA. Palmitoleic acid (C18:1) significantly decreased in goose fat exposed to heating at 210 °C (with 0.39 units) and in 0.01% burdock-additivated fat exposed to heating at 210°C (with 0.21 units). From the category of polyunsaturated fatty acid, linolenic acid (C18:3), decreased significantly in goose fat exposed to heating at 210 °C (with 0.17 units) and burdock-additivated fat exposed to heating at 210 °C (with 0.12 units). The strongest correlation between heating temperature and MUFA was found in goose fat ($r = 0.94$), followed by the pork lard ($r = 0.86$). The oxidation of PUFA and MUFA was reduced by the addition of burdock extract in a 0.01% proportion.

Our results are in agreement with those reported by Niu et al. [46] who noticed an increase in total saturated fatty acid and a decrease in mono- and polyunsaturated fatty acids during thermal treatment of lard. It was reported that lard oil contained higher percentages of palmitic (35.25%) and stearic (16.99%) acids, and during thermal treatment, there were changes in the distribution of fatty acids with an increase in saturated fatty acid and a decrease in polyunsaturated fatty acids which was the most affected fraction [12]. The content of total fatty acids in chicken fat heated at different temperatures was reported [47]. Oleic (C18:1), linoleic (C18:2), and palmitic (C16:0) acids were found to be the most abundant fatty acids, and almost 60% of the total fatty acids were unsaturated fatty acids. As the heating temperature rose, the amount of saturated fatty acids generally increased while the amounts of monounsaturated and polyunsaturated fatty acids significantly ($p < 0.05$) decreased [47]. Changes in the fatty acid profile of the subcutaneous and intramuscular fat as well as a sensory evaluation of goose packaged in a modified atmosphere under different conditions and under refrigeration storage at 4 °C were investigated [8]. The results showed that a highly oxygen-modified atmosphere had a negative impact on the quality of goose meat due to a reduction in polyunsaturated fatty acids and an increase in saturated fatty acids, which indicates a significant reduction in its nutritional value. It was observed that the fatty acid composition of goose meat remains unchanged for 11 days of storage, as long as a vacuum is used to reduce the amount of oxygen exposure. Vacuum packaging proved to be a more effective technique for maintaining the fatty acid profile and flavour of the goose meat.

The fatty acid composition of duck, chicken, pork, and bovine fats was analysed by Shin et al. [11]. The major fatty acids identified in duck fat were oleic acid (C18:1), palmitic acid (C16:0), stearic acid (18:0), linoleic acid (C18:2), and arachidonic acid (C20:4). For chicken fat, oleic acid (C18:1) was the major unsaturated fatty acid, followed by linoleic acid (C18:2), palmitic acid (C16:0), and stearic acid (C18:0), but only bovine fat showed more stearic acid (C18:0) than linoleic acid (C18:2). The study showed that total SFAs and PUFAs were found to be significantly ($p < 0.05$) different across the groups. Duck fat showed the highest content of PUFAs and the lowest content of SFAs, while bovine fat presented the lowest PUFA content and the highest SFA content. The PUFA/SFA ratio of duck fat was approximately two times higher than that of bovine fat.

3.9. Effect of Heat Treatment on Microscopic View

The aim of the microscopic examination was to monitor changes that occur in different stages of the oxidation process in animal fats at the microscopic level, relating them to their physical structure.

The microscope view of the unheated animal fats showed fat cells arranged in chains, which were compact and well highlighted. Along with the increase in the exposure time, fat cells were destroyed, weakly highlighted, and irregularly distributed. At higher temperatures, fat cells were almost completely destroyed and barely visible (Figures 1 and 2). In our analysis of the literature, we have found no study that investigates the oxidation process through a microscopic examination, therefore microscopic analysis may represent a new method for the evaluation of oxidative processes and should be further studied. Animal lipids were well protected from oxidation by the burdock extract, which demonstrated its efficacy as an antioxidant; it may be used to monitor the fats oxidation and to estimate their shelf-life stability.

4. Conclusions

Heating temperature, type of fat, and antioxidant addition affected the rate of lipid oxidation in dietary animal fats. The results showed that pork lard was more resistant to heating, generating smaller amounts of peroxides. Peroxide value, thiobarbituric acid-reactive substances, acid value, total polar compounds and total phenolic content were significantly influenced by the heating temperature. The contents of primary and secondary oxidation compounds were lower in the samples with an added antioxidant. Heat treatment induced several chemical reactions such as oxidative, hydrolytic, and polymerization reactions, which modify the fatty acid composition of the heated fats. The heating temperature had a significant effect on total MUFAs in both lard and goose fat (to a greater extent in fat without antioxidant). Microscopic examination indicated differences between fresh and heated fats. It may represent a new method for the evaluation of oxidative processes, thus should be further studied. The addition of burdock extract in concentration of 0.01% significantly inhibited lipid oxidation in dietary lard and goose fat exposed to heating at varying temperatures, therefore it can be used as a potential antioxidant to increase fats' stability.

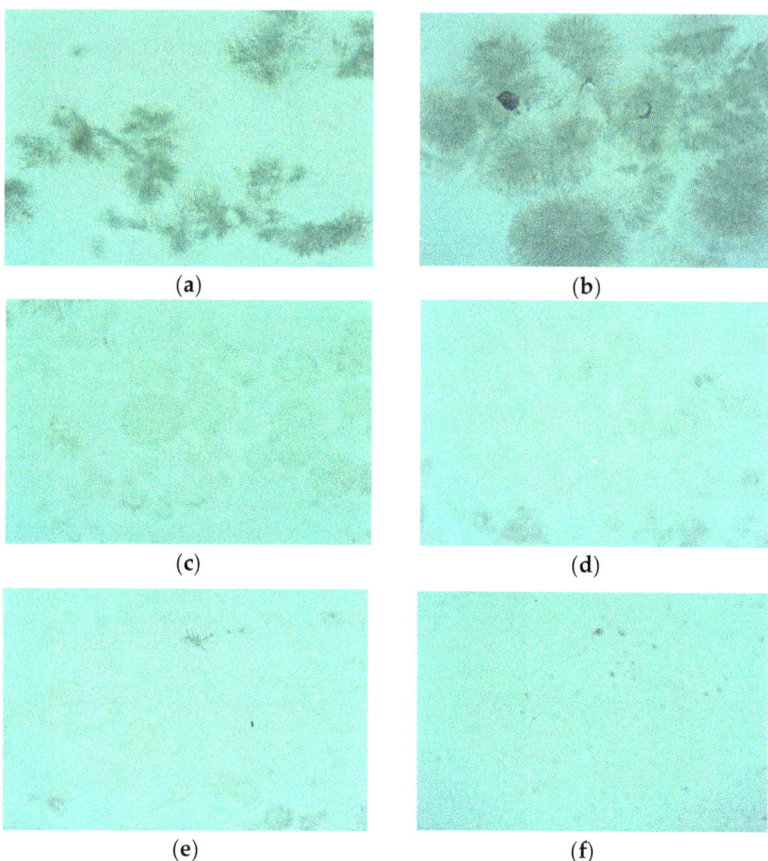

Figure 1. Microscope view of pork lard subjected to heating. (**a**)—microscope view of unheated lard; (**b**)—microscope view of lard subjected to heating at 130 °C; (**c**)—microscope view of lard subjected to heating at 150 °C; (**d**)—microscope view of lard subjected to heating at 170 °C; (**e**)—microscope view of lard subjected to heating at 190 °C; (**f**)—microscope view of lard subjected to heating at 210 °C.

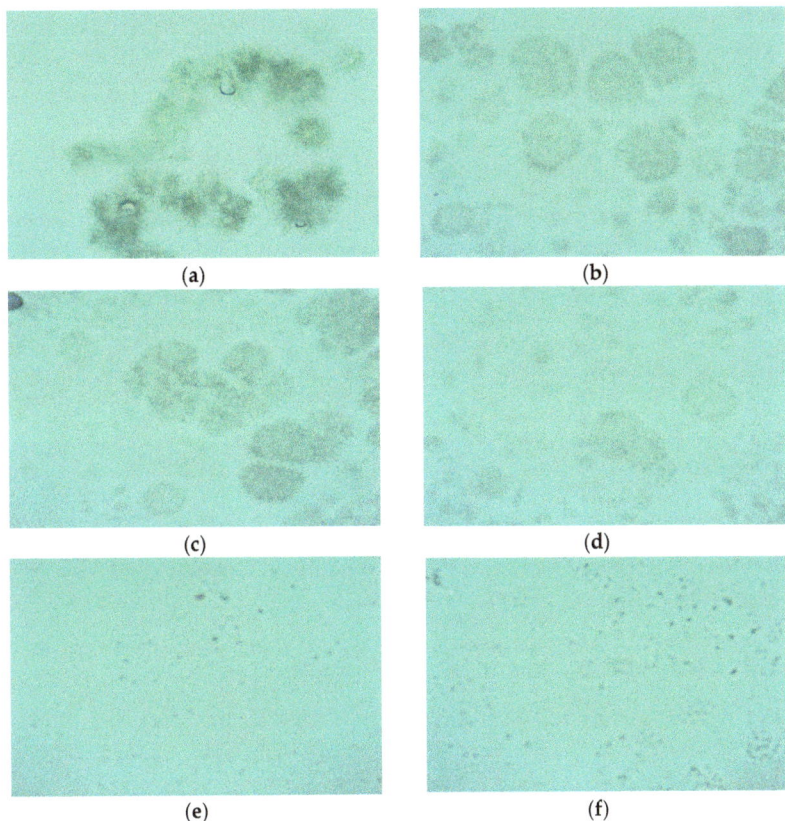

Figure 2. Microscope view of goose fat subjected to heating. (**a**)—microscope view of unheated goose fat; (**b**)—microscope view of goose fat subjected to heating at 130 °C; (**c**)—microscope view of goose fat subjected to heating at 150 °C; (**d**)—microscope view of goose fat subjected to heating at 170 °C; (**e**)—microscope view of goose fat subjected to heating at 190 °C; (**f**)—microscope view of goose fat subjected to heating at 210 °C.

Author Contributions: Conceptualization, F.P. and T.D.; methodology F.P. and T.D.; software, F.P.; validation, F.P. and T.D.; formal analysis, F.P. and T.D.; investigation, F.P.; resources, F.P. and T.D.; data curation, T.D.; writing—original draft preparation, F.P. and T.D.; writing—review and editing, F.P. and T.D.; visualization, F.P. and T.D.; supervision, F.P. All authors have read and agreed to the published version of the manuscript.

Funding: This research was funded by Technical University of Cluj-Napoca.

Institutional Review Board Statement: Not applicable.

Informed Consent Statement: Not applicable.

Data Availability Statement: Data is contained within the article.

Conflicts of Interest: The authors declare no conflicts of interest.

References

1. Aladedunye, A.F. Natural antioxidants as stabilizers of frying oils. *Eur. J. Lipid Sci. Technol.* **2014**, *116*, 688–706. [CrossRef]
2. Calvo, L.; Segura, J.; Toldrá, F.; Flores, M.; Rodríguez, A.I.; López-Bote, C.J.; Rey, A.I. Meat quality, free fatty acid concentration, and oxidative stability of pork from animals fed diets containing different sources of selenium. *Food Sci. Technol. Int.* **2017**, *23*, 716–728. [CrossRef]
3. Selani, M.M.; Herrero, A.M.; Ruiz-Capillas, C. Plant Antioxidants in Dry Fermented Meat Products with a Healthier Lipid Profile. *Foods* **2022**, *11*, 3558. [CrossRef] [PubMed]
4. Hu, F.B.; Stampfer, M.J.; Manson, J.E.; Ascherio, A.; Colditz, G.A.; Speizer, F.E. Dietary saturated fats and their food sources in relation to the risk of coronary heart disease in women. *Am. J. Clin. Nutr.* **2009**, *70*, 1001–1008. [CrossRef] [PubMed]
5. Wangxin, L.; Xianliang, L.; Ying, H.; Zhao, M.; Liu, T.; Wang, J.; Feng, F. Influence of cooking techniques on food quality, digestibility, and health risks regarding lipid oxidation. *Food Res. Int.* **2023**, *167*, 112685. [CrossRef]
6. Ianni, A.; Bennato, F.; Martino, C.; Odoardi, M.; Sacchetti, A.; Martino, G. Qualitative Attributes of Commercial Pig Meat from an Italian Native Breed: The Nero d'Abruzzo. *Foods* **2022**, *11*, 1297. [CrossRef]
7. Wołoszyn, J.; Haraf, G.; Okruszek, A.; Werenska, M.; Goluch, Z.; Teleszko, M. The protein and fat quality of thigh muscles from Polish goose varieties. *Poultry Sci.* **2020**, *99*, 1216–1229. [CrossRef]
8. Orkusz, A.; Michalczuk, M. Effect of packaging atmosphere on the fatty acid profile of intramuscular, subcutaneous fat, and odor of goose meat. *Poultry Sci.* **2020**, *99*, 647–652. [CrossRef]
9. Daniali, G.; Jinap, S.; Hajeb, P.; Sanny, M. Acrylamide formation in vegetable oils and animal fats during heat treatment. *Food Chem.* **2016**, *212*, 244–249.
10. El Yamani, M.; Sakar, E.H.; Mansouri, F.; Serghini-Caid, H.; Elamrani, A.; Rharrabti, Y. Effect of pigments and total phenols on oxidative stability of monovarietal virgin olive oil produced in Morocco. *Riv. Ital. Sostanze Grasse* **2019**, *96*, 17–24.
11. Shin, D.; Kim, D.Y.; Yune, J.; Kwon, H.C.; Kim, H.J.; Seo, H.G.; Han, S. Oxidative Stability and Quality Characteristics of Duck, Chicken, Swine and Bovine Skin Fats Extracted by Pressurized Hot Water Extraction. *Food Sci. Anim. Resour.* **2019**, *39*, 446–458. [CrossRef] [PubMed]
12. Tejerina, D.; García-Torres, S.; De Vaca, M.C.; Vázquez, F.M.; Cava, R. Effect of production system on physical–chemical, antioxidant and fatty acids composition of Longissimus dorsi and Serratus ventralis muscles from Iberian pig. *Food Chem.* **2021**, *133*, 293–299. [CrossRef] [PubMed]
13. Gruffat, D.; Bauchart, D.; Thomas, A.; Parafita, E.; Durand, D. Fatty acid composition and oxidation in beef muscles as affected by ageing times and cooking methods. *Food Chem.* **2021**, *343*, 128476. [CrossRef]
14. Márquez-Ruiz, G.; Ruiz-Méndez, V.; Velasco, J. Antioxidants in frying: Analysis and evaluation of efficacy. *Eur. J. Lipid Sci. Technol.* **2014**, *116*, 1441–1450. [CrossRef]
15. Petkova, N.; Hambarlyiska, I.; Tumbarski, Y.; Vrancheva, R.; Raeva, M.; Ivanov, I. Phytochemical Composition and Antimicrobial Properties of Burdock (*Arctium lappa* L.) Roots Extracts. *Biointerface Res. App Chem.* **2022**, *12*, 2826–2842.
16. De Leonardis, A.; Macciola, V.; Lembo, G.; Aretini, F.; Nag, A. Studies on oxidative stabilisation of lard by natural antioxidants recovered from olive-oil mill wastewater. *Food Chem.* **2007**, *100*, 998–1004. [CrossRef]
17. Réblová, Z. Effect of temperature on the antioxidant activity of phenolic acids. *Czech J. Food Sci.* **2012**, *30*, 171–175. [CrossRef]
18. Petcu, C.D.; Mihai, O.D.; Tăpăloagă, D.; Gheorghe-Irimia, R.A.; Pogurschi, E.N.; Militaru, M.; Borda, C.; Ghimpețeanu, O.M. Effects of Plant-Based Antioxidants in Animal Diets and Meat Products: A Review. *Foods* **2023**, *12*, 1334. [CrossRef]
19. Lehukov, K.A.; Tsikin, S. A study on an effect of the green tea extract on quality and shelf life of animal fats during storage. *Meat Process.* **2020**, *1*, 32–42.
20. Kalinowska, I.W.; Górska-Horczyczak, E.; Stelmasiak, A.; Marcinkowska-Lesiak, M.; Onopiuk, A.; Wierzbicka, A.; Półtorak, A. Effect of Temperature and Oxygen Dose During Rendering of Goose Fat to Promote Fatty Acid Profiles. *Eur. J. Lipid Sci. Technol.* **2021**, *123*, 2100085. [CrossRef]
21. Kulisic, T.; Radonic, A.; Milos, M. Inhibition of lard oxidation by fractions of different essential oils. *Grasas y Aceites* **2005**, *56*, 284–291.
22. Sekretár, S.; Schmidt, S.; Vajdák, M.; Zahradníková, L.; Annus, J. Antioxidative and Antimicrobial Effects of Some Natural Extracts in Lard. *Czech J. Food Sci.* **2004**, *22*, 215–218.
23. Ameer, K.; Shahbaz, H.M.; Kwon, J.H. Green extraction methods for polyphenols from plant matrices and their byproducts: A review. *Compr. Rev. Food Sci. Food Saf.* **2017**, *16*, 295–315. [CrossRef] [PubMed]
24. Pop, F.; Mihalescu, L. Effects of α-tocopherol and citric acid on the oxidative stability of alimentary poultry fats during storage at low temperatures. *Int. J. Food Prop.* **2017**, *20*, 1085–1096.
25. Pop, F. Effect of microwave heating on quality and fatty acids composition of vegetable oils. *Studia UBB Chem.* **2018**, *63*, 43–52. [CrossRef]
26. Mlcek, J.; Druzbikova, H.; Valasek, L.; Sochor, J.; Jurikova, T.; Borkovcova, M.; Baron, M.; Balla, S. Assessment of total polar materials in frying fats from Czech restaurants. *Ital. J. Food Sci.* **2015**, *27*, 32–37.
27. Singleton, V.L.; Orthofer, R. Analysis of total phenols and other oxidation substrates and antioxidants by means of Folin-Ciocalteu reagent. In *Methods in Enzymology*; Abelson, J.N., Simon, M.I., Sies, H., Eds.; Academic Press: Burlington, MA, USA, 1999; Volume 299, pp. 152–178. [CrossRef]

28. Gertz, C.; Klostermann, S.; Kochhar, S.P. Testing and comparing oxidative stability of vegetable oils and fats at frying temperature. *Eur. J. Lipid Sci. Technol.* **2002**, *102*, 543–551. [CrossRef]
29. Szabó, A.; Bázár, G.; Locsmándi, L.; Romvári, R. Quality alterations of four frying fats during long-term heating (conventional analysis and NIRS calibration). *J. Food Quality* **2010**, *33*, 42–58. [CrossRef]
30. Caponio, F.; Pasqualone, A.; Gomes, T. Effects of conventional and microwave heating on the degradation of olive oil. *Eur. Food Res. Technol.* **2002**, *215*, 114–117.
31. Wroniak, M.; Krygier, K.; Kaczmarczyk, M. Comparison of the quality of cold pressed and virgin rapeseed oils with industrially obtained oils. *Polish J. Food Nutr. Sci.* **2008**, *58*, 85–89.
32. Azadmard-Damirchi, S.; Habibi-Nodeh, F.; Hesari, J.; Nemati, M.; Achachloei, B.F. Effect of pretreatment with microwaves on oxidative stability and nutraceuticals content of oil from rapeseed. *Food Chem.* **2010**, *121*, 1211–1215. [CrossRef]
33. Réblová, Z. The effect of temperature on the antioxidant activity of tocopherols. *Eur. J. Lipid Sci. Technol.* **2006**, *108*, 858–863. [CrossRef]
34. Song, S.; Zhang, X.; Hayat, K.; Liu, P.; Chengsheng, J.; Shuqin, X.; Xiao, Z.; Tian, H.; Niu, Y. Formation of the beef flavour precursors and their correlation with chemical parameters during the controlled thermal oxidation of tallow. *Food Chem.* **2010**, *124*, 203–209.
35. Moslavac, T.; Jokić, S.; Flanjak, I. Stabilization of goose fat with antioxidants and synergists. *MESO First Croat. Meat J.* **2022**, *24*, 436–446. [CrossRef]
36. Soulti, K.; Roussis, I.G. Inhibition of butter oxidation by some phenolics. *Eur. J. Lipid Sci. Technol.* **2007**, *109*, 706–709.
37. Moslavac, T.; Jokić, S.; Šubarić, D.; Aladić, K. The influence of antioxidants on the oxidative stability of pork fat. *MESO: First Croat. Meat J.* **2018**, *20*, 317–324. [CrossRef]
38. Farhooshi, R.; Moosavi, S.M.R. Evaluating the performance of peroxide and conjugated diene values in monitoring quality of used frying oils. *J. Agric. Food Chem.* **2009**, *11*, 173–179.
39. Gharby, S.; Harhar, H.; Boulbaroud, S.; Bouzouba, Z.; Madani, N.; Chafchaouni, I.; Charrouf, Z. The stability of vegetable oils (sunflower, rapeseed and palm) sold on the Moroccan market at high temperature. *Int. J. Chem. Biochem. Sci.* **2014**, *5*, 47–54.
40. Patel, S.; Shende, S.; Arora, S.; Singh, A.K. An assessment of the antioxidant potential of coriander extracts in ghee when stored at high temperature and during deep fat frying. *Int. J. Dairy. Technol.* **2013**, *66*, 207–213. [CrossRef]
41. Qiuyu, L.; Yehui, Z.; Wenjuan, J.; Liyan, Z. Study of the thermal behavior of rosemary extract and its temperature-related antioxidant effect on chicken fat. *J. Food Proces. Preserv.* **2022**, *46*, 16793. [CrossRef]
42. Xueqi, L.; Bremer, C.; Kristen, N.C.; Courtney, N.; Quyen, A.T.; Shengling, P.; Wang, M.; Ravetti, L.; Guillaume, C.; Yichuan, W.; et al. Changes in chemical compositions of olive oil under different heating temperatures similar to home cooking. *J. Food Chem. Nutr.* **2016**, *4*, 7–15.
43. Skowrońska, W.; Granica, S.; Dziedzic, M.; Kurkowiak, J.; Ziaja, M.; Bazylko, A. *Arctium lappa* and *Arctium tomentosum*, Sources of *Arctii radix*: Comparison of Anti-Lipoxygenase and Antioxidant Activity as well as the Chemical Composition of Extracts from Aerial Parts and from Roots. *Plants* **2021**, *10*, 78. [CrossRef] [PubMed]
44. Lee, C.J.; Park, S.K.; Kang, J.Y.; Kim, J.M.; Yoo, S.K.; Han, H.J.; Kim, D.O.; Heo, H.J. Melanogenesis regulatory activity of the ethyl acetate fraction from *Arctium lappa* L. leaf on α-MSH–induced B16/F10 melanoma cells. *Ind. Crops Prod.* **2019**, *138*, 111581. [CrossRef]
45. Zaixiang, L.; Li, C.; Kou, X.; Yu, F.; Wang, H.; Smith, G.M.; Zhu, S. Antibacterial, Antibiofilm Effect of Burdock (*Arctium lappa* L.) Leaf Fraction and Its Efficiency in Meat Preservation. *J. Food Prot.* **2016**, *79*, 1404–1409. [CrossRef]
46. Niu, Y.; Wu, M.; Xiao, Z.; Chen, F.; Zhu, J.; Zhu, G. Effect of fatty acids profile with thermal oxidation of chicken fat on characteristic aroma of chicken flavors assessed by gas chromatography-mass spectrometry and descriptive sensory analysis. *Food Sci. Technol. Res.* **2016**, *22*, 245–254.
47. Charuwat, P.; Boardman, G.; Bott, C.; Novak, J.T. Thermal degradation of long chain fatty acids. *Water Environ. Res.* **2018**, *90*, 278–287. [CrossRef]

Disclaimer/Publisher's Note: The statements, opinions and data contained in all publications are solely those of the individual author(s) and contributor(s) and not of MDPI and/or the editor(s). MDPI and/or the editor(s) disclaim responsibility for any injury to people or property resulting from any ideas, methods, instructions or products referred to in the content.

Article

Study on the Effect of Different Concentrations of SO$_2$ on the Volatile Aroma Components of 'Beibinghong' Ice Wine

Baoxiang Zhang [1], Weiyu Cao [1], Changyu Li [1], Yingxue Liu [1], Zihao Zhao [2], Hongyan Qin [1], Shutian Fan [1], Peilei Xu [1], Yiming Yang [1] and Wenpeng Lu [1,*]

[1] Institute of Special Animal and Plant Sciences of Chinese Academy of Agricultural Sciences, Changchun 130112, China; zbx0319@126.com (B.Z.); 82101202231@caas.cn (W.C.); lichangyu@caas.cn (C.L.); liuyingxue@caas.cn (Y.L.); qinhongyan@caas.cn (H.Q.); fanshutian@caas.cn (S.F.); xupeilei@caas.cn (P.X.); yangyiming@caas.cn (Y.Y.)

[2] School of Foreign Languages, Jilin Science and Technology Vocational College, Changchun 130123, China; 13504486086@163.com

* Correspondence: luwenpeng@caas.cn

Abstract: SO$_2$ plays an important role in wine fermentation, and its effects on wine aroma are complex and diverse. In order to investigate the effects of different SO$_2$ additions on the fermentation process, quality, and flavor of 'Beibinghong' ice wine, we fermented 'Beibinghong' picked in 2019. We examined the fermentation rate, basic physicochemical properties, and volatile aroma compound concentrations of 'Beibinghong' ice wine under different SO$_2$ additions and constructed a fingerprint of volatile compounds in ice wine. The results showed that 44 typical volatile compounds in 'Beibinghong' ice wine were identified and quantified. The OAV and VIP values were calculated using the threshold values of each volatile compound, and t the effect of SO$_2$ on the volatile compounds of 'Beibinghong' ice wine might be related to five aroma compounds: ethyl butyrate, ethyl propionate, ethyl 3-methyl butyrate-M, ethyl 3-methyl butyrate-D, and 3-methyl butyraldehyde. Tasting of 'Beibinghong' ice wine at different SO$_2$ additions revealed that the overall flavor of 'Beibinghong' ice wine was the highest at an SO$_2$ addition level of 30 mg/L. An SO$_2$ addition level of 30 mg/L was the optimal addition level. The results of this study are of great significance for understanding the effect of SO$_2$ on the fermentation of 'Beibinghong' ice wine.

Keywords: HS-GC-IMS; volatile components; OAV value; VIP value; sensory evaluation

1. Introduction

Ice wine is a sweet wine that is made by fermenting the naturally frozen grape juice from the vines when the temperature drops to −7~−8 °C [1]. Canada and Germany are the leading producers of ice wine, while China, Austria, and the United States also produce ice wine in large quantities. In recent years, the annual production of ice wine in China has reached 300 million liters, especially in Huanren County, Liaoning Province, and the Yalu River Basin, Jilin Province, where the ice wine industry is developing rapidly [2]. The 'Beibinghong' grape is the world's first wild grape variety (*Vitis amurensis* Rupr) that can make ice wine and is very popular in northeast China. It has the advantages of high cold tolerance and stable yield. Compared to unfrozen grapes, frozen grapes contain high concentrations of sugar, aroma, and flavor compounds, giving the resulting ice wine a rich, fruity flavor. After alcoholic fermentation, ice wine still contains a rich concentration of residual sugar, which gives it a solid, sweet flavor. The 'Beibinghong' ice wine has a more rounded taste and mellow aroma than the 'Beibinghong' dry wine, and ice wine has received a lot of attention because of its unique flavor.

'Beibinghong' is one of the famous high-quality grape varieties; *Vitis amurensis Rupr* belong to the East Asian grape family. 'Beibinghong' is an interspecific cross between 'Zuoyouhong' as the mother and '84-26-53'—a mountain-European F2 grape variety with low acid and high sugar content, thick skin and large bunches—as the father. This interspecific hybridization from the F5 generation to select and breed a new variety of ice wine brewing was first made in 1995. [3]. The 'Beibinghong' is the first *Vitis amurensis Rupr* variety cultivated at home and abroad to produce ice wine, and its preciousness and rarity exceed that of existing varieties on the market. It is cultivated in Inner Mongolia, Shaanxi, Gansu, and Northeastern provinces, with a planting area of about 8600 hm^2 [4], and it is the first *Vitis amurensis Rupr* variety cultivated at home and abroad to produce ice wine. Its main planting area is in Jilin Province, and it occupies an essential position in cultivating unique plants in Jilin Province [5]. In recent years, 'Beibinghong' has been widely cultivated in the "Changbai Mountain" region of Jilin Province and the "Huanren" region of Liaoning Province and is the most popular variety of ice wine. Local breweries develop ice wine products and have distinctive aromas of "sweet", "honey", "roasted", and "caramel".

Volatile compounds are essential for wine quality, determining the characteristics of specific varieties and reflecting the effects of environmental conditions and viticultural management [6]. SO_2 is an indispensable additive in the wine-making process. It has the following leading roles in the production and preservation of wine: inhibiting the growth of harmful bacteria and yeasts, eliminating dissolved oxygen, inhibiting polyphenol oxidizing enzymes and the infestation of stray bacteria, protecting the hygienic quality and stability of wines, and the addition of SO_2 in appropriate quantities can attenuate the undesirable flavor of wines [7–9]. It prevents oxidative deterioration of wines and maintains their color, aroma, and flavor. Increases the acidity of wines, improves the balance and freshness of wines, promotes clarification and stabilization of wines, and reduces cloudiness and sedimentation of wines. SO_2 should also be used in moderation, as excessive SO_2 can adversely affect the quality of the wine and human health [10], such as reducing the aromatic intensity of the wine and masking its character and terroir, producing irritating odors that affect the taste and enjoyment of the wine. This causes allergic reactions such as headaches, breathing difficulties, and skin rashes, which are inappropriate for some people [11].

The SO_2 content of wines must be strictly controlled, and each country and region must have its legal regulations and standards. According to EU regulations, the SO_2 content in red wine should not exceed 150 mg/L, and white and pink wine should not exceed 200 mg/L. In China, according to The national standard GB 2760-2014 [12], $SO_2 \leq 250$ mg/L, and according to The national standard GB 7718-2011 [13], as long as SO_2 is used in food, it has to be marked on the food label [14]. The overall goal of SO_2 addition prior to wine consumption is to achieve the desired level of free SO_2 at the lowest possible total SO_2 [15]. The amount and timing of the addition of SO_2 can affect the aroma of the wine in different ways. The right amount of SO_2 protects the fruity and floral aromas of the wine from oxidative deterioration and increases the complexity and aging potential of the wine. The timing of the addition of SO_2 is also essential, and in general, the earlier it is added, the more significant the impact on the aroma of the wine. For example, adding SO_2 before wine fermentation can inhibit the growth of non-winemaking yeasts and maintain the cleanliness and purity of the wine. However, it can also reduce the aromatic diversity and complexity of the wine, and the addition of SO_2 after wine fermentation can inhibit the growth of lactic acid bacteria and prevent the contamination of acetic acid bacteria. However, it also affects the taste and style of the wine. The study not only analyzed the chemical effects of SO_2 on flavor substances only at the level of the sensory evaluation but also comprehensively from the point of view of the sensory evaluation, although the use of SO_2 in winemaking is well known [16–20]. However, few articles have explored its effect on wine flavor in-depth, and even fewer studies have investigated the effect of different SO_2 additions on the aroma of 'Beibinghong' ice wine, a special ice wine variety in the Jilin region. The study of 'Beibinghong' ice wine, which is a unique ice wine variety from the Jilin region, can improve the flavor quality of the wine by adjusting the strategy of using SO_2, assessing the effect of SO_2 concentration on the flavor of the wine in a more precise way, exploring how SO_2 affects the fermentation process and aroma characteristics of the wine, and identifying the volatile compounds that play a vital role in the different amounts of SO_2 added to the wine.

This study measured the fermentation start and end times, basic physicochemical properties, and volatile aroma compounds of 'Beibinghong' ice wine harvested in 2019. The fingerprints of volatile compounds of ice wine brewed with different concentrations of SO_2 were established, and its brewed ice wine was tasted. In the experiment, SO_2 was added by adding solid potassium metabisulfite (PMS), and each gram of PMS produced 0.56 g of SO_2, i.e., 10 mg of SO_2 required 17.86 µg of PMS, which was then added in eight treatments, i.e., 10 mg/L, 20 mg/L, 30 mg/L, 40 mg/L, 50 mg/L, 60 mg/L, 80 mg/L, and 100 mg/L, in the following order. Eight treatments were performed to determine the optimal amount of SO_2 addition at the same temperature and under the same yeast strain and enzyme treatment conditions.

1.1. Materials and Reagents

1.1.1. Experimental Materials

'Beibinghong' ice grapes harvested from vineyards of Yujiang Valley Winery Co., Ltd. (Ji'an, China) in Ji'an City for 'Beibinghong' ice wine production; yeasts BV818 and CEC01 (Angie's Yeast Co., Ltd., Yichang, Hubei, China); pectinases RF and RCO (AB Enzymes, Darmstadt, Germany); and potassium metabisulphite (SAS SOFRELAB OENOFRANCE).

1.1.2. Reagents

Analytical purity: sulfuric acid, sodium chloride, potassium chloride, sodium bicarbonate (Beijing Chemical Plant, Beijing, China); tannic acid (Tianjin Guangfu Fine Chemical Research Institute, Tianjin Fine Chemical Research Institute, Tianjin, China); Folin–Denis reagent (US Sigma, St. Louis, MO, USA); anhydrous sodium carbonate (Tianjin Hengxing Chemical Reagent Manufacturing Co., Ltd., Tianjin, China); glacial acetic acid, hydrochloric acid, anhydrous ethanol, sodium hydroxide, phosphoric acid (Beijing Chemical Plant); potassium hydrogen phthalate, anthracene ketone (Sinopharm Chemical Reagent Co., Ltd., Shanghai, China); anhydrous sodium acetate (Shanghai Hubtest Chemical Co., Ltd., Shanghai, China); dextrose (Tianjin Hengxing Chemical Reagent Manufacturing Co., Ltd., Tianjin, China).

Chromatographic purity: methanol (TEDIA Reagents, Fairfield, OH, USA), 4-methyl-2-pentanol (Shanghai Lianshuo Biotechnology Co., Ltd., Shanghai, China).

Fermentation auxiliaries: CEC01 active dry yeast (Angel Yeast Co., Ltd., Yichang, China); potassium metabisulphite (Yantai Dibs Homebrewer Co., Ltd., Yantai, China).

1.2. Instruments and Equipment

Agilent High Performance Liquid Chromatograph (Agilent Technologies Ltd., Santa Clara, CA, USA); FlavourSpec® Flavor Analyzer (Shandong Haineng Scientific Instrument Co., Ltd., Zibo, Shandong, China); BSA224S-CW Sartorius Electronic Balance (Sartorius Scientific Instruments Co., Ltd., Göttingen, Germany); PAL-1 Digital Hand-held Refractometer (ATAGO, Tokyo, Japan), CJJ-931 Dual-link Magnetic Heating Stirrer (Jiangsu Jintan Jincheng Guosheng Experimental Instrument Factory, Changzhou, Jiangsu, China); HWS-12 type electric thermostatic water bath (Shanghai Yiheng Scientific Instrument Co., Ltd., Shanghai, China); KQ-300E type ultrasonic cleaner, snowflake ice machine (Beijing Changliu Scientific Instrument Co., Ltd., Beijing, China), FA1004B electronic balance (Shanghai Yue Ping Scientific Instrument Co., Ltd., Shanghai, China), DHG-9240 constant temperature drying oven (Shanghai Yihang Scientific Instrument Co., Ltd., Shanghai, China), WAX chromatography columns (U.S. RESTEK, Bellefonte, PA, USA); Milli-Q Advantage A1 ultrapure water apparatus (Millipore Corporation, USA); Cary60UV-Vis UV spectrophotometer (Agilent Technologies Ltd., Santa Clara, CA, USA).

1.3. Methods

1.3.1. Process Flow of Brewing 'Beibinghong' Ice Wine

The process of ice wine production involves harvesting the fruit, screening, and de-stemming. The harvested fruit is then granulated, pressed, and preserved by adding different concentrations of SO_2. Three sets of replicate brewing experiments were conducted using three fermenters per treatment to ensure the reproducibility of the experiments. Gum reduction is carried out, followed by a low-temperature maceration at 2–4 °C. The wine is then post-tempered to 15 °C. Subsequently, the post-temperature was adjusted to 15 °C, and controlled fermentation was carried out with the addition of CECO1 yeast at a dosage of 250 mg/Kg. After completion of fermentation, fermentation was stopped, and crude filtration was carried out to obtain the original wine. After controlled aging, the wine is then fine-filtered and sterilized. After passing quality tests, the final product is bottled, sealed, and labeled as ice wine.

1.3.2. Sample Labeling

The amount of SO_2 used according to the quality of treated iced grape juice was as follows: sample No. 1 (10 mg/L), sample No. 2 (20 mg/L), sample No. 3 (30 mg/L), sample No. 4 (40 mg/L), sample No. 5 (50 mg/L), sample No. 6 (60 mg/L), sample No. 7 (80 mg/L), sample No. 8 (100 mg/L).

1.3.3. Detection Methods of Basic Physical and Chemical Indexes

Soluble solids were determined by handheld refractometer, and titrable acid content of wine was determined by indicator method according to The GB/T 15038-2006 [21]. The alcohol content is measured by the alcohol meter method. The total sugar content in wine was determined by anthrone and sulfuric acid colorimetric method, and standard koji was prepared by standard glucose solution. The total anthocyanin content in grape juice was determined by pH difference method, i.e., anthocyanin reacted with potassium chloride buffer (0.025 M, pH = 1) and acetic acid buffer (0.4 M, pH = 4.5), and then the difference of 520 nm and 700 nm was calculated. Total phenol content—Folin–Ciocalteu colorimetric method. Dry extract content: refer to the dry extract test method in the national standard (GB/T 15038-2006).

1.3.4. Quantification of Volatile Compounds in 'Beibinghong' Ice Wine by HS-GC-IMS

Headspace-gas chromatography-ion migration spectrometry (HS-GC-IMS) determined volatile substances in wine. The FlavourSpec® flavor analyzer was used to take a 1 mL sample, place it into a 20 mL headspace bottle, add 20 ppm 4-methyl-2-amyl alcohol 10 µL, incubate at 60 °C for 15 min, and then inject it into the sample.

Chromatographic conditions: the column was WAX column (15 m × 0.53 mm,1 µm), column temperature was 60 °C, carrier gas was N2, IMS temperature was 45 °C, and chromatographic conditions were shown in Table 1.

Table 1. Gas chromatography conditions.

Time (Min:Sec)	E1	E2	Recording
00:00,000	150 mL/min	2 mL/min	rec
02:00,000	150 mL/min	2 mL/min	-
10:00,000	150 mL/min	10 mL/min	-
20:00,000	150 mL/min	100 mL/min	-
30:00,000	150 mL/min	100 mL/min	stop

The conditions of automatic headspace injection were as follows: injection volume of 100 µL, incubation time of 10 min, incubation temperature of 60 °C, injection needle temperature of 65 °C, and incubation speed of 500 rpm; 4-methyl-2-pentanol was used as the internal standard for the analysis, and the concentration of 198 ppb, the signal peak volume of 493.34, and the intensity of each signal peak was about 0.401 ppb.

Quantitative calculation formula:

$$Ci = \frac{Cis * Ai}{Ais}$$

Ci is the calculated mass concentration of any component in μg/L, Cis is the mass concentration of the internal standard used in μg/L, and Ai/AIS is the volume ratio of any signal peak to the signal peak of the internal standard. The NIST database and IMS database are built into the software for the qualitative analysis of the substances.

1.3.5. Odor Activity Value (OAV) Calculation

OAV was used to assess the contribution of volatile compounds to the overall aroma of the wine. The concentration of volatile compounds was divided by the odor threshold (OT) to calculate the OAV value. Volatile compounds with OAV > 1 were considered to be types of aroma-active compounds, and the larger the OAV value, the more significant the contribution of components to the flavor; the OAV value can help to determine the critical aroma substances in food or plants [22–24], analyze the causes of flavor differences, and optimize the flavor quality of wine aroma characteristics formation plays an important role.

1.4. Sensory Evaluation

Sensory evaluation methodology: the wines were subjected to quantitative descriptive analysis (QDA) by a trained sensory panel of 17 tasters (10 women and 7 men, aged 24 to 52 years, with an average of 33 years). These experts were recruited based on their motivation and availability, having been trained according to the national standards ISO 6658 [25] and ISO 8586 [26] prior to the sensory evaluation. According to the definitions in the published literature, according to the definitions in The national standard GB 15038-2006 [27,28], and based on the discussion results, specify the development of a sensory evaluation form (Table 2). The samples were marked with three numbers and submitted to the tasters randomly.

Table 2. Sensory score.

Item	Percentage	Features	Full Marks
Color	10%	Chroma and hue	10
Clarification	10%	Degree of clarification	10
Aroma	30%	Finesse	5
		Richness	5
		Coherence	5
		Variety characteristics	5
		Duration	5
		Variation and complexity (multiple levels of aroma)	5
Taste	40%	Balance and coordination	10
		Body and fullness (weightiness in the mouth)	10
		Texture and structure	5
		Continuity and layers	5
		Flavor quality and persistence	5
		Lingering flavor	5
Typicality	10%	Synthesize and evaluate	10
Totals			100

1.5. Statistical Analysis of Data

Excel 2010 was used to organize the experimental data statistically, and an analysis of variance (ANOVA) was performed using SPSS (version 22.0, IBM, Armonk, NY, USA). Statistical). Statistical analyses were performed on the experimental data to check for significant differences in the individual results, and all the data were expressed as mean ± standard deviation. Differences between the two groups were considered significant at $p < 0.05$. Simca 14.1 software was used for OPLS-DA and VIP value analysis; GC-IMS assay was done with Savitzky Golay for smoothing and denoising, and migration time normalization was done by setting the RIP position as 1, i.e., dividing the actual migration time by the peak out time of the RIP to obtain the approximate migration time. The Reporter plug-in was used to directly compare the spectral differences between the samples, and the Gallery Plot plug-in was used to compare the fingerprints visually and quantitatively to compare the differences of VOCs between different samples. Heat map and correlation analysis were performed using the OmicShare tools, a free online platform for data analysis (https://www.omicshare.com/tools (accessed on 27 July 2020)).

2. Results and Analysis

2.1. Fermentation Process and Basic Indexes of 'Beibinghong' Ice Wine with Different SO_2 Additions

The yeast treatment time in this experiment was 11 December 2019, and the initial solid content of 'Beibinghong' grape juice was 40.3%. It can be seen from Table 3 that the fermentation time was 24 h at 10 mg/L and 48 h at 30, 40, and 50 mg/L. The fermentation time of 10 mg/L, 40 mg/L, and 50 mg/L was 23, 23, and 21 days, but the red wine was suitable for slow fermentation at low temperatures. This is in line with the findings of Sun Hening and others [29] that the addition of SO_2 prior to fermentation inhibits yeast activity in the short term and increases the delay in yeast multiplication, leading to a delay in fermentation; however, during this period, it allows the must to remain static and encourages the precipitation of impurities, colloidal substances, and decomposed tartaric acid, which is highly susceptible to the formation of tartar in wines. Wang et al. [30] studied the effect of SO_2 on the fermentation process and quality of pineapple wine and found that SO_2 had a significant effect on the time of starting fermentation when the concentration of SO_2 was 150 mg/L or less, the time of starting fermentation was around ten h. With the increasing concentration of SO_2, the time of starting fermentation was delayed significantly, and when the concentration of SO_2 was 250 mg/L, the time of starting the fermentation time was more than 3 d at a concentration of 250 mg/L, which was the same as our results, indicating that the higher concentration of SO_2 affects the fermentation of fruit wines. The fermentation time of fruit wines should not be too long, so it should be considered comprehensively. The total sugar of ice wine fermented with different amounts of SO_2 was found to be above 160.0 g/L, which indicates that different amounts of SO_2 do not have much effect on the total sugar, which is the same as the results of the study by Mou Jingxia et al. [10]. The total acid content of SO_2 was relatively low at 10 mg/L, 20 mg/L, 30 mg/L, 40 mg/L. The level of dry leachate index is closely related to the raw materials, production process, and storage method of wine, and it is one of the essential symbols of the quality of wine [31]. Dry leachate was higher at SO_2 additions of 30 mg/L, 40 mg/L; From the above table, adding 30 mg/L and 40 mg/L SO_2 is more appropriate.

Table 3. Fermentation process and basic physicochemical indexes of 'Beibinghong' ice wine with different SO_2 additions.

SO_2 Concentration (mg/L)	Start of Fermentation (h)	End of Fermentation (days)	Soluble Solids (%)	Total Sugar (g/L)	Total Acid (g/L)	Dry Extract (g/L)	Alcohol Content (v/v)
10	24 ± 0 f	23 ± 1 e	26.3 ± 0.5 b	162.66 ± 13.2 d	11.36 ± 0.41 bc	151.24 ± 15.61 d	11.5 ± 0.50 bc
20	72 ± 0 d	34 ± 3 d	26.5 ± 2.0 b	163.57 ± 8.06 c	11.28 ± 1.10 c	150.33 ± 7.58 e	11.5 ± 0.36 bc
30	48 ± 0 e	34 ± 3 d	25.4 ± 1.7 c	162.19 ± 5.13 d	11.26 ± 1.65 c	153.71 ± 6.42 b	12 ± 1.0 a
40	48 ± 0 e	23 ± 1 e	26.5 ± 1.0 b	160.96 ± 10.7 e	11.32 ± 0.26 bc	156.94 ± 13.33 a	11.5 ± 0.72 bc
50	48 ± 0 e	21 ± 2 f	26.7 ± 2.0 b	161.08 ± 4.5 e	11.48 ± 0.15 abc	152.82 ± 17.21 c	11.5 ± 0.21 bc
60	96 ± 0 c	35 ± 4 c	27.8 ± 1.0 a	164.69 ± 5.57 b	11.65 ± 1.03 abc	149.21 ± 4.31 f	11 ± 0.06 cd
80	168 ± 0 b	64 ± 9 a	28.3 ± 0.8 a	168.73 ± 11.02 a	12.06 ± 1.02 ab	145.17 ± 15.30 g	10.5 ± 0.42 d
100	192 ± 0 a	38 ± 5 b	28.5 ± 1.7 a	168.92 ± 19.20 a	12.15 ± 0.78 a	144.98 ± 7.06 g	10.5 ± 0.11 d

Means with different letters in the same column express significant differences (Duncan's test $p < 0.05$).

2.2. Changes in Anthocyanin Content of 'Beibinghong' Ice Wine Brewed with Different SO_2 Additions

Color is one of the most critical indicators affecting the sensory quality of red wine, and anthocyanin, a class of flavonoid compounds with a benzopyran structure, is an essential water-soluble natural pigment in red wine, as well as a crucial color-presenting substance, with a variety of critical physiological functions and biological activities [32,33]. During grape growth and development, anthocyanosides are biosynthesised via the phenylpropane-flavonoid pathway. In wine, there is an equilibrium between the various states of anthocyanins, and their color expression is closely related to the structure and morphology of the anthocyanin molecule [34]. After human consumption of wine, wine anthocyanosides are mainly metabolized and absorbed by intestinal flora in the colon. The type, state, and content of anthocyanosides are essential in red wines' color characteristics and aging potential. Changes in the content of anthocyanosides in 'Beibinghong' ice wine brewed with different SO_2 additions are shown in Table 3.

As shown in Table 4, the content of anthocyanin was higher when SO_2 was added at 30 mg/L, 60 mg/L, and 80 mg/L, especially at 80 mg/L, which reached the highest value. However, the total acid content was slightly higher because 'Beibinghong' belongs to the *Vitis amurensis Rupr* variety. From the taste and quality perspective, the SO_2 addition was not too high, so 30 and 60 mg/L were more appropriate. A moderate addition of SO_2 has a protective effect on the anthocyanins in wine [35]. SO_2 inhibits the action of oxidative enzymes and prevents oxidation of the raw material, which helps to maintain the stability and color vividness of the anthocyanosides. The addition of SO_2 also helps to select the fermentation microorganisms, clarify the fermentation matrix, and regulate the acidity of the fermentation matrix, which indirectly affects the solubilization and stability of the anthocyanosides [36–39]. J Bakker et al. [40] found that as the level of SO_2 increased during winemaking, more anthocyanins were extracted. During maturation, all wines lost color and increased brownness. Wines without added SO2 browned more severely than wines with added SO_2. This also shows that SO2 is essential to maintain color stability in wine production.

Table 4. Content of anthocyanin in 'Beibinghong' ice wine with different SO_2 additions.

SO_2 (mg/L)	10	20	30	40	50	60	80	100
Total anthocyanin (μg/L)	126.63 ± 11.26 h	133.87 ± 7.48 f	155.86 ± 2.51 c	135.26 ± 9.06 e	140.27 ± 3.29 d	172.56 ± 11.38 b	194.26 ± 26.93 a	132.76 ± 18.71 g

Means with different letters in the same column express significant differences (Duncan's test $p < 0.05$).

However, when the added SO_2 level is too high, it can adversely affect anthocyanin [41]. This is in line with our findings that an addition of sulphur dioxide that is too high affects the content of anthocyanosides. High concentrations of SO_2 may generate sulfites in acidic environments, which are capable of reacting with anthocyanoside molecules, leading to the formation of anthocyanoside sulfites, which are colorless, and therefore reduce the anthocyanoside content of wines, thus affecting the color of the wines. SO_2 has both protective and potentially damaging effects on the anthocyanosides in wine, depending on the amount added and the conditions of use. The amount of SO_2 added during the winemaking process needs to be precisely controlled to maximize the positive effects and minimize the negative effects on the anthocyanins, thus ensuring optimal wine quality and taste.

2.3. HS-GC-IMS Analysis of 'Beibinghong' Ice Wine Brewed with Different SO_2 Additions

Fingerprints of Volatile Components of 'Beibinghong' Ice Wine Brewed with Different SO_2 Additives

In order to analyze the variability of volatile substances in ice wine brewed with different concentrations of SO_2, we constructed a fingerprint of volatile flavor compounds based on all the signal peaks in the two-dimensional HS-GC-IMS spectra (Figure 1). Each sample was measured three times in parallel, and the darker color indicated a greater peak intensity and higher content. The composition and differences of volatile flavor compounds in ice wine brewed with different concentrations of SO_2 were revealed from the fingerprints.

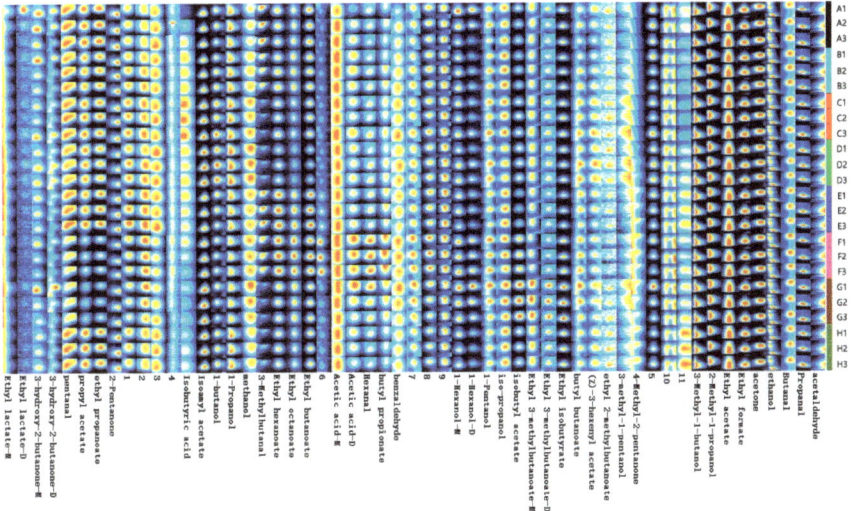

Figure 1. Fingerprints of volatile compounds of 'Beibinghong' ice wine under different treatments (Note: A–H are the amounts of SO_2 added, in the order of 10 mg/L, 20 mg/L, 30 mg/L, 40 mg/L, 50 mg/L, 60 mg/L, 80 mg/L, and 100 mg/L. Same below).

As shown in Figure 2, the volatile compounds in the eight wine samples were well separated, and the aroma fingerprints of the wine samples differed significantly, mainly in the content of volatile compounds.

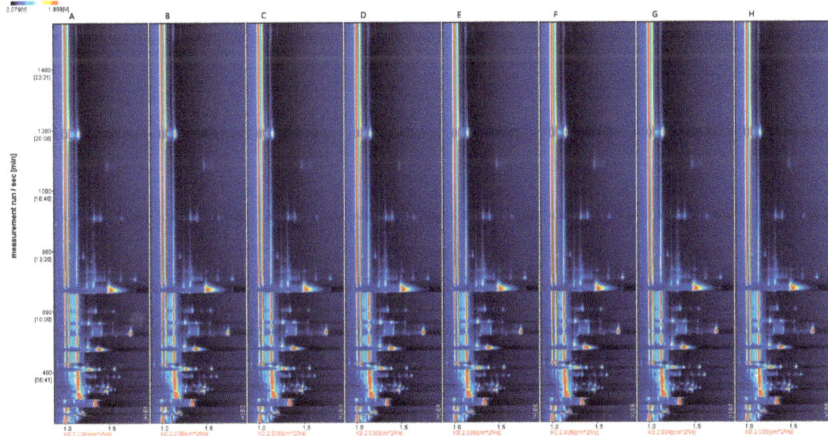

Figure 2. HS-GC-IMS 2D spectrum (top view). Note: The background of the whole graph is blue, and the red vertical line at horizontal coordinate 1.0 is the RIP peak (reactive ion peak, normalized). The vertical coordinate represents the retention time (s) of the gas chromatogram, and the horizontal coordinate represents the ion migration time (normalized). Each point on both sides of the RIP peak represents a volatile organic compound.

The 'Beibinghong' ice wine from the first treatment group was used as a reference, and the rest of the spectra were subtracted from the signal peaks in the first treatment to obtain the difference spectra (Figure 3).

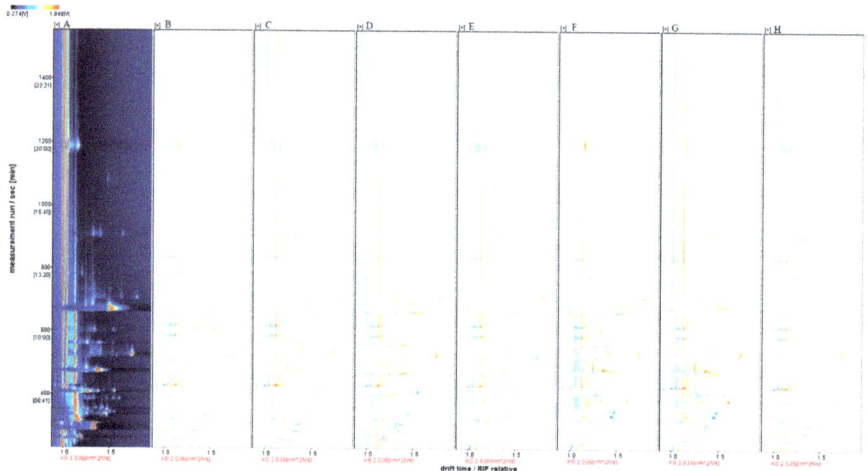

Figure 3. HS-GC-IMS difference spectrum of the sample. Note: Using A as a reference, the rest of the spectrum subtracts the signal peaks in A to obtain the difference spectrum between the two. Blue areas indicate less substance in this sample than in A. Red areas indicate more substance in this sample than in A. The darker the color, the more significant the difference. The difference between the samples can be seen from the above graph.

2.4. Analysis of Volatile Components

Aroma is an essential index for evaluating the quality of wine, and the study of wine aroma mainly focuses on the aromatic substances that positively affect it. The critical role of SO_2 in winemaking also includes its role in the aroma. SO_2 also significantly affects the aroma of wine, with the addition of SO_2 affecting the concentration of volatile compounds in the wine by between 33% and 43% [42]. Aromatic substances in wine can be categorized into terpenoids, aliphatic compounds, and aromatic compounds according to their chemical structure, of which aliphatic compounds include alcohols, acids, ketones, and esters. The analytical spectra and data were viewed with VOCal for qualitative and quantitative analysis, and the volatile components in the wine samples were characterized by the built-in NIST and IMS databases of HS-GC-IMS. A total of 44 typical volatile compounds were detected (Table 5), and the most significant number of species was 18 esters, 11 alcohols, 3 acids were detected, 5 ketones, and 11 uncharacterized volatile compounds. The volatile aroma compounds detected in the ice wine samples from the eight treatments were the same type, but the contents were significantly different. Among the eight treatments, the highest total volatile compound content was found in the seventh treatment group with 80 mg/L of added SO_2, with a total volatile compound content of 81,930.42256 µg/L. The following treatments with the highest total volatile compound content in descending order were the treatment group with 50 mg/L of added SO_2, with a volatile compound concentration of 81,394.73104 µg/L, treatment group with 70 mg/L SO_2 addition 81,328.35368 µg/L, treatment group with 40 mg/L SO_2 addition 81,182.37288 µg/L, treatment group with 60 mg/L SO_2 addition 79,569.26936 µg/L, and treatment group with 90 mg/L SO_2 addition 79,215.75989 µg/L. 79,569.26936 µg/L for 60 mg/L of SO_2, 79,215.75984 µg/L for 90 mg/L of SO_2, 78,921.88528 µg/L for 30 mg/L of SO_2, and 77,818.71496 µg/L for 20 mg/L of SO_2. Table 5 shows that alcohols accounted for the most significant proportion of 62.9–64.73%, followed by esters at 24–25.82%, and alcohols and esters were the main aroma compounds in the wine samples.

Table 5. Composition of volatile compounds in 'Beibinghong' ice wine treated with different concentrations of SO_2.

Serial No.	Retention Time (sec)	Substances	Aroma Description	Substance Content (μg/L)							
				Sample No. 1	Sample No. 2	Sample No. 3	Sample No. 4	Sample No. 5	Sample No. 6	Sample No. 7	Sample No. 8
1	1080.89	Ethyl octanoate	Apricot, Brandy, Fat, Floral, Pineapple	322.722 ± 19.384	347.309 ± 13.027	296.373 ± 17.554	248.987 ± 13.003	376.587 ± 32.87	419.786 ± 8.103	360.532 ± 17.447	310.012 ± 6.096
2	886.48	Ethyl lactate-M	Cheese, Floral, Fruit, Pungent, Rubber	357.994 ± 50.581	330.748 ± 9.635	345.173 ± 6.24	393.777 ± 9.815	378.219 ± 11.958	302.744 ± 2.979	333.986 ± 37.971	318.373 ± 5.942
3	885.16	Ethyl lactate-D	Cheese, Floral, Fruit, Pungent, Rubber	41.248 ± 10.967	27.997 ± 2.465	30.767 ± 4.564	33.067 ± 6.445	28.271 ± 3.16	29.047 ± 5.952	31.753 ± 3.379	29.832 ± 3.881
4	707.83	Ethyl hexanoate	Cheese, Floral, Fruit, Pungent, Rubber	408.356 ± 25.405	414.364 ± 31.845	360.863 ± 18.269	322.325 ± 11.466	548.61 ± 17.016	555.558 ± 18.26	394.243 ± 29.17	382.963 ± 9.191
5	526.01	Isoamyl acetate	Apple, Banana, Pear	2867.294 ± 58.401	2775.802 ± 94.577	3217.441 ± 64.118	3539.672 ± 96.581	2949.724 ± 2.712	3181.046 ± 71.388	3621.83 ± 67.244	3319.403 ± 54.505
6	409.12	Ethyl butanoate	pineapple flavor	712.02 ± 17.695	653.213 ± 6.981	577.538 ± 7.574	417.204 ± 11.829	765.006 ± 29.341	725.577 ± 20.563	580.78 ± 6.748	617.956 ± 16.818
7	383.33	isobutyl acetate	Apple, Banana, Floral, Herb	266.451 ± 3.555	255.106 ± 3.543	297.941 ± 9.633	308.365 ± 5.877	224.468 ± 3.165	377.387 ± 14.584	512.87 ± 10.248	302.304 ± 2.072
8	350.14	propyl acetate	Celery, Floral, Pear, Red Fruit	589.059 ± 20.017	535.422 ± 7.295	571.309 ± 0.436	561.966 ± 4.026	554.909 ± 11.265	327.439 ± 5.638	295.958 ± 12.592	600.24 ± 9.152
9	336.31	ethyl propanoate	Apple, Pineapple, Rum, Strawberry	904.927 ± 50.425	783.426 ± 25.267	827.193 ± 14.304	760.554 ± 30.164	802.542 ± 14.539	479.4 ± 17.647	460.67 ± 32.085	845.108 ± 17.746
10	340.92	Ethyl isobutyrate	fruit	189.551 ± 5.722	173.682 ± 1.248	184.696 ± 6.095	174.112 ± 6.533	163.053 ± 6.197	187.054 ± 6.442	233.276 ± 8.061	192.519 ± 1.879
11	292.51	Ethyl acetate	Aromatic, Brandy, Grape	10,270.327 ± 79.702	10,043.96 ± 35.944	10,070.65 ± 23.923	9856.912 ± 39.493	9926.147 ± 32.15	9879.314 ± 112.062	9994.921 ± 63.219	10,158.871 ± 27.223
12	263.7	Ethyl formate	Pungent	2485.568 ± 36.304	2385.056 ± 11.981	2380.013 ± 13.785	2164.664 ± 5.431	2485.498 ± 36.882	2487.519 ± 7.555	2639.144 ± 16.111	2418.238 ± 6.044
13	732.02	Butyl butanoate	Apple, pineapple flavor	314.625 ± 20.591	339.229 ± 24.721	346.936 ± 5.068	377.007 ± 20.819	372.044 ± 14.43	362.305 ± 5.044	343.359 ± 12.26	331.742 ± 9.67
14	767.45	(Z)-3-Hexenyl acetate	Banana, floral	51.247 ± 2.123	50.296 ± 1.656	57.74 ± 4.47	60.325 ± 4.389	50.99 ± 3.059	49.451 ± 1.981	52.637 ± 7.373	52.258 ± 2.257
15	516.1	Butyl propionate	Fruit	88.274 ± 3.05	99.196 ± 7.891	98.357 ± 5.902	88.606 ± 7.85	100.681 ± 5.435	140.951 ± 4.581	114.616 ± 7.301	90.438 ± 1.154
16	424.71	Ethyl3-methylbutanoate-M	Apple, Mulberry Aroma	119.992 ± 2.944	113.215 ± 4.203	123.41 ± 5.895	121.37 ± 5.868	114.138 ± 3.706	98.191 ± 6.187	128.856 ± 1.569	119.214 ± 1.785
17	424.09	Ethyl3-methylbutanoate-D	Apple, Mulberry Aroma	63.356 ± 8.178	66.816 ± 14.996	61.928 ± 9.264	58.284 ± 8.347	70.701 ± 3.646	73.035 ± 5.266	71.903 ± 6.921	67.135 ± 8.382
18	444.3	Ethyl 2-methylbutanoate	Apple, Ester, Green Apple, Kiwi, Strawberry	36.115 ± 0.756	43.242 ± 5.611	49.203 ± 7.374	45.902 ± 3.926	54.997 ± 4.229	58.865 ± 4.477	46.289 ± 4.716	44.875 ± 2.32
No. of ester species 18		Total %		20,089.12584 25.82	19,438.08048 24.63	19,897.53192 24.51	19,533.09736 24	19,966.58664 25.09	19,734.66768 24.27	20,217.62344 24.68	20,201.48088 25.5
1	908.81	1-Hexanol-M	Banana, Flower, Grass, Herb	761.642 ± 160.852	664.524 ± 32.62	709.027 ± 50.629	695.845 ± 76.075	637.454 ± 59.986	802.564 ± 14.503	823.121 ± 75.245	650.641 ± 30.475
2	907.5	1-Hexanol-D	Banana, Flower, Grass, Herb	230.981 ± 77.158	172.311 ± 15.172	194.661 ± 16.176	197.904 ± 33.232	181.147 ± 34.356	246.892 ± 10.754	286.767 ± 50.847	173.555 ± 14.727
3	672.36	3-Methyl-1-butanol	brandy	11,466.999 ± 322.872	11,589.634 ± 344.147	12,161.029 ± 151.633	12,360.758 ± 264.311	11,737.75 ± 270.254	11,990.457 ± 38.299	12,405.587 ± 280.962	11,910.102 ± 94.116
4	560.94	1-Butanol	Fruit	566.218 ± 48.597	603.987 ± 42.987	747.541 ± 51.003	813.903 ± 68.365	606.207 ± 54.248	552.356 ± 4.052	549.573 ± 59.427	656.573 ± 8.621
5	472.65	2-Methyl-1-propanol	pungent odor	5164.863 ± 71.084	5345.103 ± 89.543	5632.972 ± 81.471	5768.918 ± 91.136	5318.293 ± 89.947	5916.619 ± 59.966	6204.353 ± 51.918	5508.228 ± 54.195

45

Table 5. Cont.

Serial No.		Retention Time (sec)	Substances	Aroma Description	Substance Content (μg/L)							
					Sample No. 1	Sample No. 2	Sample No. 3	Sample No. 4	Sample No. 5	Sample No. 6	Sample No. 7	Sample No. 8
6		408.34	1-Propanol	Alcohol, Candy, Pungent	2975.777 ± 47.018	3039.996 ± 39.433	3196.466 ± 54.537	3380.095 ± 61.027	3033.742 ± 44.128	2616.758 ± 31.35	2550.967 ± 71.881	3153.482 ± 10.22
7		355.13	Iso-propanol	pungent odor	205.871 ± 14.473	209.079 ± 14.817	222.704 ± 19.452	202.013 ± 14.932	183.215 ± 10.323	187.335 ± 19.493	207.814 ± 30.254	204.116 ± 8.551
8		316.72	Ethanol	alcoholic flavor	27,126.098 ± 103.789	27,986.877 ± 144.839	28,515.642 ± 295.583	28,715.711 ± 181.66	28,281.016 ± 139.473	29,096.813 ± 127.829	28,853.397 ± 710.693	27,325.985 ± 86.132
9		304.42	Methanol	alcoholic flavor	226.817 ± 10.88	260.779 ± 9.167	270.235 ± 1.287	295.837 ± 5.807	277.732 ± 4.16	285.157 ± 7.154	288.702 ± 19.969	255.843 ± 5.281
10		856.39	3-methyl-1-pentanol	Fruit	19.148 ± 1.42	20.134 ± 0.737	24.437 ± 1.327	26.431 ± 2.355	20.43 ± 1.558	20.347 ± 1.472	22.287 ± 1.954	21.038 ± 0.308
11		734.1	1-Pentanol	alcoholic flavor	207.371 ± 15.316	214.742 ± 8.932	226.28 ± 9.042	230.289 ± 15.029	209.406 ± 6.591	221.346 ± 10.837	226.752 ± 13.725	207.941 ± 5.393
No. of alcohol species	11			Total	48,951.784	50,107.1648	51,900.99208	52,687.70472	50,486.39344	51,936.64448	52,419.32192	50,067.50616
				%	62.9	63.49	63.93	64.73	63.45	63.86	63.98	63.2
1		1184.66	Acetic acid-M	Acid, Fruit, Pungent, Sour, Vinegar	2054.497 ± 79.013	2128.91 ± 70.701	2107.461 ± 50.139	2023.938 ± 65.237	2100.214 ± 30.593	2214.206 ± 13.902	2077.372 ± 158.519	2037.589 ± 46.317
2		1187.29	Acetic acid-D	Acid, Fruit, Pungent, Sour, Vinegar	1451.248 ± 398.813	1903.677 ± 96.528	1755.035 ± 222.202	1736.671 ± 70.548	1608.712 ± 60.058	2177.888 ± 122.534	1989.475 ± 388.328	1604.389 ± 116.831
3		1528.26	Isobutyric acid	Burnt, Butter, Cheese, Sweat	175.997 ± 74.937	277.775 ± 5.854	311.54 ± 1.733	261.381 ± 20.018	246.378 ± 18.058	212.109 ± 35.442	179.723 ± 16.096	206.798 ± 15.601
No. of acid species	3			Total	3681.7424	4310.36144	4174.0356	4021.99056	3955.30352	4604.20464	4246.57128	3848.77752
				%	4.73	5.46	5.14	4.94	4.97	5.66	5.18	4.86
1		518.25	Hexanal	Apple, Fat, Fresh, Green, Oil	81.55 ± 4.566	87.068 ± 0.793	87.762 ± 6.23	90.012 ± 7.245	84.554 ± 2.101	108.163 ± 4.545	103.644 ± 10.978	82.262 ± 0.496
2		355.52	Pentanal	pungent odor	234.398 ± 19.219	230.222 ± 8.562	236.699 ± 6.672	225.716 ± 6.372	234.86 ± 3.745	193.785 ± 7.765	208.26 ± 7.021	227.682 ± 3.167
3		230.52	Acetaldehyde	Floral, Green Apple	591.445 ± 10.713	610.715 ± 20.006	638.589 ± 7.464	562.302 ± 17.218	614.15 ± 53.505	638.151 ± 24.136	609.98 ± 55.699	552.894 ± 24.171
4		265.32	Propanal	pungent odor	492 ± 27.458	491.237 ± 23.69	517.149 ± 20.744	526.488 ± 27.194	481.319 ± 11.761	489.426 ± 23.396	506.841 ± 16.626	504.963 ± 11.44
5		299.69	Butanal	lemon scent	89.624 ± 4.037	87.713 ± 3.454	89.178 ± 1.06	88.058 ± 6.826	83.719 ± 1.95	87.988 ± 3.23	89.737 ± 2.049	86.741 ± 1.879
6		1306.57	Benzaldehyde	Bitter Almond, Burnt Sugar, Cherry, Malt, Roasted Pepper	115.298 ± 12.179	136.866 ± 11.171	155.016 ± 16.2	136.091 ± 9.857	134.335 ± 5.562	139.781 ± 10.206	134.194 ± 7.058	129.152 ± 3.039
7		309.35	3-Methylbutanal	apple flavor	72.15 ± 0.211	57.976 ± 0.993	34.274 ± 3.734	38.078 ± 3.222	73.824 ± 4.135	73.536 ± 3.911	39.255 ± 3.173	44.275 ± 1.451
No. of aldehyde species	7			Total	1676.46248	1701.79744	1758.6688	1666.7448	1706.76016	1730.82896	1691.9112	1627.96816
				%	2.15	2.16	2.17	2.05	2.14	2.13	2.07	2.06
1		782.71	3-Hydroxy-2-butanone-M	Buttery	166.582 ± 8.812	151.271 ± 30.591	143.925 ± 20.161	144.25 ± 14.122	130.308 ± 3.805	128.757 ± 18.812	131.074 ± 31.528	125.517 ± 5.321
2		781.39	3-Hydroxy-2-butanone-D	Buttery	106.295 ± 15.869	97.755 ± 7.259	117.172 ± 15.478	133.046 ± 15.239	93.92 ± 9.742	93.471 ± 4.586	123.232 ± 21.554	102.916 ± 3.284
3		258.7	Acetone	Butter, Creamy, Green Pepper	1019.321 ± 12.794	988.156 ± 5.67	1014.408 ± 10.468	1002.652 ± 12.949	1002.012 ± 6.228	1025.531 ± 11.472	1002.946 ± 8.749	1035.26 ± 10.713
4		381.32	4-Methyl-2-pentanone	ketone odor	73.862 ± 2.163	79.874 ± 1.468	80.804 ± 1.759	88.907 ± 4.275	101.28 ± 4.113	103.437 ± 1.217	102.252 ± 2.736	104.469 ± 1.763
5		356.18	2-Pentanone	Fruit, Pungent	209.642 ± 4.206	202.415 ± 1.836	193.008 ± 4.084	160.568 ± 5.345	184.14 ± 2.373	170.299 ± 4.524	190.478 ± 3.963	204.548 ± 3.183
No. of ketone species	5			Total	1575.7028	1519.4704	1549.31728	1529.42328	1511.65784	1521.49704	1549.97976	1572.70792
				%	2.02	1.93	1.88	1.88	1.9	1.87	1.89	1.99
1		350.14	1		216.928 ± 6.115	233.045 ± 1.731	247.923 ± 7.816	247.363 ± 4.28	248.273 ± 5.335	202.989 ± 9.454	193.582 ± 8.193	233.392 ± 2.338
2		335.54	2		353.637 ± 9.007	379.626 ± 7.302	399.756 ± 16.515	395.541 ± 21.29	407.613 ± 1.878	375.575 ± 10.84	362.538 ± 14.597	379.164 ± 5.413
3		335.92	3		378.483 ± 11.402	377.544 ± 1.911	379.778 ± 14.451	386.267 ± 18.497	377.753 ± 17.064	302.47 ± 5.513	314.602 ± 16.398	385.479 ± 8.25
4		846.77	4		45.751 ± 10.74	39.73 ± 3.065	36.079 ± 1.483	42.046 ± 3.424	42.802 ± 1.533	41.685 ± 3.221	36.582 ± 8.861	42.381 ± 3.375
5		780.1	5		316.261 ± 31.987	277.719 ± 49.572	291.323 ± 53.825	315.245 ± 28.833	278.636 ± 8.194	252.474 ± 40.115	301.364 ± 77.458	253.721 ± 10.992

Table 5. *Cont.*

Serial No.	Retention Time (sec)	Substances	Aroma Description	Substance Content (μg/L)							
				Sample No. 1	Sample No. 2	Sample No. 3	Sample No. 4	Sample No. 5	Sample No. 6	Sample No. 7	Sample No. 8
6	438.27	6		45.653 ± 2.394	46.319 ± 1.665	50.52 ± 3.754	49.329 ± 2.226	59.937 ± 0.194	78.5 ± 5.716	50.037 ± 0.974	50.825 ± 0.44
7	391.08	7		125.533 ± 8.327	133.177 ± 4.903	143.463 ± 11.165	141.022 ± 3.036	124.45 ± 2.218	133.933 ± 15.61	136.195 ± 8.149	130.383 ± 2.378
8	589.68	8		73.228 ± 2.132	77.55 ± 2.724	75.599 ± 3.334	74.943 ± 1.968	76.667 ± 0.702	84.907 ± 3.385	73.308 ± 3.223	72.36 ± 3.653
9	668.96	9		115.615 ± 1.826	120.402 ± 6.387	122.84 ± 4.661	123.177 ± 4.685	126.067 ± 4.301	128.344 ± 4.744	120.47 ± 5.171	119.037 ± 0.98
10	254.41	10		71.712 ± 6.866	66.46 ± 3.262	69.208 ± 3.714	67.92 ± 3.427	66.642 ± 1.678	69.151 ± 1.258	70.994 ± 2.404	71.019 ± 1.878
11	383.56	11		101.099 ± 13.283	93.438 ± 2.733	85.338 ± 5.32	112.917 ± 11.088	133.728 ± 7.189	130.483 ± 9.075	145.343 ± 8.451	159.559 ± 1.114
No. of other categories 11		Total		1843.89744	1845.01072	1901.8272	1955.77032	1942.56776	1800.51088	1805.01496	1897.3192
		%		2.37	2.34	2.34	2.4	2.44	2.26	2.2	2.4
		Total		77,818.71496	78,921.88528	81,182.37288	81,394.73104	79,569.26936	81,328.35368	81,930.42256	79,215.75984

Note: Compound flavor description from the Flavornet database (https://www.femaflavor.org); http://www.flavornet.org; accessed on 6 June 2020.

2.4.1. Esters

The esters in wine are mainly produced by acyl-coenzyme A and fatty acids and alcohols in yeast cells under the catalytic action of relevant enzymes during alcoholic fermentation, and they have the aroma of fruits or flowers, which play a vital role in the aroma of wine [43,44]. Esters give wines a unique and complex fruity flavor, which is a critical component of their aroma composition, and the description of the aroma of the detected ester compounds also shows that the esters are mainly dominated by fruity aroma. The various ester compounds present in wines have coordinated compositional ratios and have synergistic effects on the formation of aroma [43,45]. The esters in grapes are mainly found in grape skins, which are fully macerated during fermentation, releasing the variety's unique fruity and floral aromas. As can be seen in Table 4, esters were the compounds with the highest percentage of content in the assay, and from the esters detected, the esters that provided a higher concentration of aromas were isoamyl acetate, ethyl acetate, methyl formate, ethyl butyrate, and Ethyl propionate, the highest content of esters detected among the eight treatments was in treatment group 7 (SO_2 addition of 80 mg/L) at 20,217.62344 µg/L, followed by ester volatile compounds in descending order by treatment group 8 (SO_2 addition of 90 mg/L) at 20,201.48088 µg/L; treatment group 1 (SO_2 addition of 20 mg/L) was 20,089.12584 µg/L; treatment group 5 (SO_2 addition of 60 mg/L) was 19,966.58664 µg/L; treatment group 3 (SO_2 addition of 40 mg/L) was 19,897.53192 µg/L; and treatment group 6 (SO_2 addition of 70 mg/L) was 19,734.66768 µg/L; the fourth treatment group (SO_2 addition of 50 mg/L) was 19,533.09736 µg/L; and the second treatment group (SO_2 addition of 30 mg/L) was 19,438.08048 µg/L. A study by Teresa Garde-Cerdán et al. found [46] that when volatile aroma compounds were examined in wines with or without the addition of SO_2, it was found that the concentration of SO2 did not have a significant effect on the total esters in the wines, which is the same as our findings, and this is probably because the wines are rich in unsaturated fatty acids. Therefore, the concentration of oxygen in the medium did not affect the formation of these compounds. Liu et al. [47] found that the addition of an appropriate concentration of SO_2 can also increase the content of isoamyl acetate, ethyl acetate, and ethyl octanoate in wine to a certain extent, which can bring pleasant floral and fruity aroma to wine and increase the complexity of wine aroma. The high concentration of SO_2 can lead to the increase of ethyl acetate, which can hurt the aroma quality of the wine. Therefore, when choosing the concentration of SO_2, it is essential to consider not only the concentration of the aroma but also the negative effect of the aroma content on the wine, and the amount added should be manageable [42,48,49]. Moderate amounts of SO_2 can help stabilize the aroma components in wine and contribute to forming certain aromas. However, excessive amounts of SO_2 may react with the aroma components of the wine, resulting in a change or loss of aroma. Such changes may be manifested as a diminution or loss of certain aromas in the wine or the production of some unpleasant off-flavors.

2.4.2. Alcohols

Alcohols accounted for the most significant percentage of 62.9–64.73% of the wines, and their aroma characteristics were mainly pungent or grassy. The content of alcohols between different treatment groups in descending order was 52,687.70472 µg/L in the fourth treatment group (SO_2 addition of 50 mg/L); 52,419.32192 µg/L in the seventh treatment group (SO_2 addition of 80 mg/L); 51,936.64448 µg/L in the sixth treatment group (SO_2 addition of 70 mg/L); third treatment group (SO_2 addition of 40 mg/L) at 51,900.99208 µg/L; fifth treatment group (SO_2 addition of 60 mg/L) at 50,486.39344 µg/L; second treatment group (SO_2 addition of 30 mg/L) at 50,107.1648 mg/L; the eighth treatment group (SO_2 addition of 90 mg/L) with 50,067.50616 µg/L; and the first treatment group (SO_2 addition of 20 mg/L) with 48,951.784 µg/L. The highest ethanol content was found in the treatment group with the highest SO_2 addition of 70 mg/L, and the lowest was found in the treatment group with the lowest SO_2 addition of 20 mg/L. The highest ethanol content was found in the treatment group with the highest SO_2 addition of 70 mg/L and the lowest in the treatment group with the lowest SO_2 addition of 20 mg/L. The highest ethanol content was in the treatment group with 70 mg/L of SO_2, and the lowest was in the treatment group with 20 mg/L of SO_2; the highest methanol content was in the treatment group with 50 mg/L of SO_2, and the lowest methanol content was in the treatment group with 20 mg/L of SO_2;

2.4.3. Others

Aldehydes, acids, and ketones accounted for a relatively small percentage of the wine. The percentage of aldehydes in the volatile compounds of ice wine was 2.06–2.17%. Among the treatment groups, the most considerable aldehydes content was in the third treatment group (SO_2 addition of 40 mg/L) with 1758.6688. The lowest content was in the eighth treatment group (SO_2 addition of 90 mg/L) with 1627.96816 µg/L. The proportion of ketones in the volatile compounds of ice wine ranged from 1.87% to 2.02%, with the most considerable ketone content in the first treatment group (20 mg/L SO_2) at 1575.7028 µg/L and the lowest in the fifth treatment group (60 mg/L SO_2) at 1511.65784 µg/L. The proportion of acid compounds in the volatile compounds of ice wine ranged from 1.87% to 2.02%. The percentage of acid compounds in the volatile compounds of ice wine ranged from 4.73% to 5.66%, with the most significant amount of acid compounds in treatment group 6 (70 mg/L of SO_2 addition) at 4604.20464 µg/L and the lowest in treatment group 1 (20 mg/L of SO_2 addition) at 3%.

Among the treatments, regarding the total content of volatile compounds, the SO_2 addition of 30 mg/L had a higher content of total aroma substances. Methanol and other substances harmful to the aroma components of 'Beibinghong' ice wine were fewer. The content of the components that impacted the quality was lower, so the quality of the wine was better than that of the other treatment groups from a comprehensive point of view.

2.5. Principal Component Analysis (PCA) of Wine Samples

In order to better present and distinguish the differences between the different treatment ice wine samples, the known volatile compounds identified by GC-IMS were analyzed by PCA (Figure 4). Eight treatment groups of samples were well differentiated according to their aroma characteristics. PCA was performed on the samples to discriminate the magnitude of variability between the samples of the groups of different wines, between subgroups, and between samples within groups. The contribution rate of PC1 was 28.4%, and that of PC2 was 21.1%, and the eight groups of samples showed apparent separation trends on the two-dimensional graph, with no outlier samples. The samples of the same kind of wines were clustered well, with high experimental reproducibility. The PCA results show that the differences in aroma substances among the eight groups of samples are significant and clearly distinguishable from other samples. As shown in the figure, it can be seen that the four treatments of A, B, H, and E clustered together, the treatment groups of C and D clustered together, and the groups of G and F clustered together, which indicated that the volatile compounds were similar between these treatment groups. The concentration of the aroma compounds was also somewhat different between the different treatments.

2.5.1. OAV Analysis of Major Aroma Compounds of Different Wine Samples

Generally, the components with OAV greater than one directly impacts the overall flavor and are the main components providing the flavor [50–52]. Based on the qualitative and quantitative results of GC–IMS, the threshold values of corresponding aroma compounds in water were found in the literature, and their OAV values were calculated. As shown in Table 6, a total of 21 aroma compounds with OAV values greater than one were detected in different ice wine samples, which were ethyl caprylate, 1-hexanol-M, ethyl caproate, 3-methyl-1-butanol, isoamyl acetate, hexanal, 2-methyl-1-propanol, ethyl butyrate, isobutyl acetate, pentanal, ethyl acrylate, ethyl isobutyrate, ethyl acetate, acetone, acetaldehyde, propionaldehyde, (Z)-3-hexenyl acetate, ethyl 3-methyl butanoate-M, ethyl 3-methyl butanoate-D, butyraldehyde, and 3-methyl butanal; studies have shown that the OAV values are directly proportional to the contribution of aroma [53,54]. Among the compounds with OAV values greater than 1, esters accounted for the most significant proportion of ten, and the OAV values of esters were significantly higher than those of other types of compounds, indicating that esters contribute to the prominent aroma of 'Beibinghong' ice wine. This is also in line with the GC–IMS results, where esters contributed the primary aromas in 'Beibinghong' ice wine, dominated by fruity and floral notes [55], followed by aldehydes, with six aldehydes having OAVs greater than 1, indicating that the grassy aroma of aldehydes is also a significant contributor to the aroma of 'Beibinghong' ice wine; there was also a ketone compound of acetone among the compounds with an OAV value of greater than 1; and there were three alcohols, which is also a significant contributor to the iconic wine flavor of the 'Beibinghong' ice wine aroma.

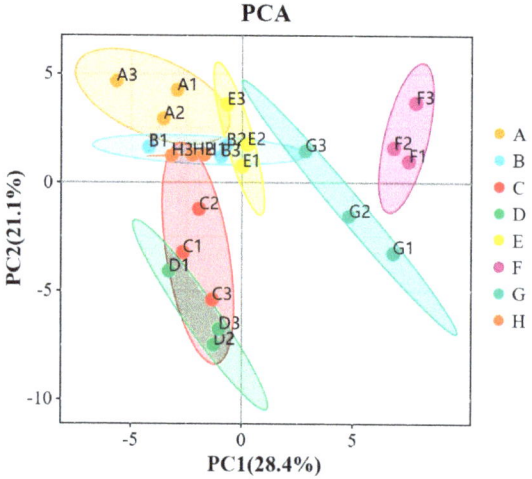

Figure 4. PCA analysis of the sample.

Principal component analysis (PCA) is a multivariate statistical analysis technique. Many complex and hard-to-find variables in the original sample are represented by identifying several principal component factors. The regularity and variability between samples are assessed based on the contribution of the principal component factors in different samples [56]. The PCA results clearly showed that two principal components were extracted from the PCA analysis of the concentration of volatile aroma compounds with OVA values greater than 1 in different treatments of 'Beibinghong' ice wine samples in a relatively independent space (Figure 5a), with a contribution of 32.9% for PC1 and 29.3% for PC2. Among the different varietal treatments, D, C, and H were located at the junction of one and four quadrants and were positive on PC2. G and D were located in the first quadrant of the score and were positive on PC1 and PC2. F was located in the second quadrant and was positive on PC1 and negative on PC2. E was located in the third quadrant and was negative on PC1 and PC2, and A and B were located in the junction of the third and fourth quadrants and were negative on PC1. This indicates significant differences in the volatile aroma compounds with OVA values greater than 1 in the 'Beibinghong' ice wine samples from different treatments. However, specific treatment groups also showed similarities in the aroma compounds with OAV values greater than 1.

Table 6. OAV analysis of major aroma compounds of 'Beibinghong' ice wine treated with different SO_2 additions.

Serial No.	Substance	A	B	C	D	E	F	G	H
1	Ethyl octanoate	3.508 ± 0.211	3.775 ± 0.142	3.221 ± 0.191	2.706 ± 0.141	4.093 ± 0.357	4.563 ± 0.088	3.919 ± 0.19	3.37 ± 0.066
2	1-Hexanol-M	1.523 ± 0.322	1.329 ± 0.065	1.418 ± 0.101	1.392 ± 0.152	1.275 ± 0.12	1.605 ± 0.029	1.646 ± 0.15	1.301 ± 0.061
3	Ethyl hexanoate	81.671 ± 5.081	82.873 ± 6.369	72.173 ± 3.654	64.465 ± 2.293	109.722 ± 3.403	111.112 ± 3.652	78.849 ± 5.834	76.593 ± 1.838
4	3-Methyl-1-butanol	52.123 ± 1.468	52.68 ± 1.564	55.277 ± 0.689	56.185 ± 1.201	53.353 ± 1.228	54.502 ± 0.174	56.389 ± 1.277	54.137 ± 0.428
5	Isoamyl acetate	7.168 ± 0.146	6.94 ± 0.236	8.044 ± 0.16	8.849 ± 0.241	7.374 ± 0.007	7.953 ± 0.178	9.055 ± 0.168	8.299 ± 0.136
6	Hexanal	16.31 ± 0.913	17.414 ± 0.159	17.552 ± 1.246	18.002 ± 1.449	16.911 ± 0.42	19.817 ± 3.039	20.729 ± 2.196	16.452 ± 0.099
7	2-Methyl-1-propanol	5.165 ± 0.071	5.345 ± 0.09	5.633 ± 0.081	5.769 ± 0.091	5.318 ± 0.09	5.696 ± 0.419	6.204 ± 0.052	5.508 ± 0.054
8	Ethyl butanoate	791.133 ± 19.661	725.792 ± 7.757	641.709 ± 8.416	463.56 ± 13.144	850.006 ± 32.601	838.845 ± 41.953	645.312 ± 7.497	686.617 ± 18.687
9	isobutyl acetate	10.658 ± 0.142	10.204 ± 0.142	11.918 ± 0.385	12.335 ± 0.235	8.979 ± 0.127	13.266 ± 3.644	20.515 ± 0.41	12.092 ± 0.083
10	Pentanal	19.533 ± 1.602	19.185 ± 0.714	19.725 ± 0.556	18.81 ± 0.531	19.572 ± 0.312	17.49 ± 2.186	17.355 ± 0.585	18.974 ± 0.264
11	Ethyl propanoate	90.493 ± 5.042	78.343 ± 2.527	82.719 ± 1.43	76.055 ± 3.016	80.254 ± 1.454	59.23 ± 19.215	46.067 ± 3.208	84.511 ± 1.775
12	Ethyl isobutyrate	12.637 ± 0.381	11.579 ± 0.083	12.313 ± 0.406	11.607 ± 0.436	10.87 ± 0.413	12.232 ± 0.841	15.552 ± 0.537	12.835 ± 0.125
13	Ethyl acetate	2.054 ± 0.016	2.009 ± 0.007	2.014 ± 0.005	1.971 ± 0.008	1.985 ± 0.006	1.984 ± 0.012	1.999 ± 0.013	2.032 ± 0.005
14	Acetone	1.225 ± 0.015	1.188 ± 0.007	1.219 ± 0.013	1.205 ± 0.016	1.204 ± 0.007	1.221 ± 0.011	1.205 ± 0.011	1.244 ± 0.013
15	Acetaldehyde	23.658 ± 0.429	24.429 ± 0.8	25.544 ± 0.299	22.492 ± 0.689	24.566 ± 2.14	24.719 ± 1.073	24.399 ± 2.228	22.116 ± 0.967
16	Propanal	6.074 ± 0.339	6.065 ± 0.292	6.385 ± 0.256	6.5 ± 0.336	5.942 ± 0.145	6.029 ± 0.301	6.257 ± 0.205	6.234 ± 0.141
17	(Z)-3-hexenyl acetate	1.653 ± 0.068	1.622 ± 0.053	1.863 ± 0.144	1.946 ± 0.142	1.645 ± 0.099	1.584 ± 0.059	1.698 ± 0.238	1.686 ± 0.073
18	Ethyl 3-methylbutanoate-M	1199.918 ± 29.439	1132.15 ± 42.03	1234.101 ± 58.948	1213.696 ± 58.676	1141.382 ± 37.058	1041.044 ± 103.454	1288.565 ± 15.688	1192.138 ± 17.852
19	Ethyl 3-methylbutanoate-D	633.563 ± 81.778	668.157 ± 149.962	619.284 ± 92.636	582.841 ± 83.472	707.005 ± 36.459	716.13 ± 41.233	719.029 ± 69.215	671.354 ± 83.818
20	Butanal	5.637 ± 0.254	5.517 ± 0.217	5.609 ± 0.067	5.538 ± 0.429	5.265 ± 0.123	5.451 ± 0.282	5.644 ± 0.129	5.455 ± 0.118
21	3-Methylbutanal	180.374 ± 0.528	144.94 ± 2.482	85.685 ± 9.335	95.195 ± 8.055	184.561 ± 10.337	185.264 ± 11.209	98.137 ± 7.933	110.689 ± 3.628

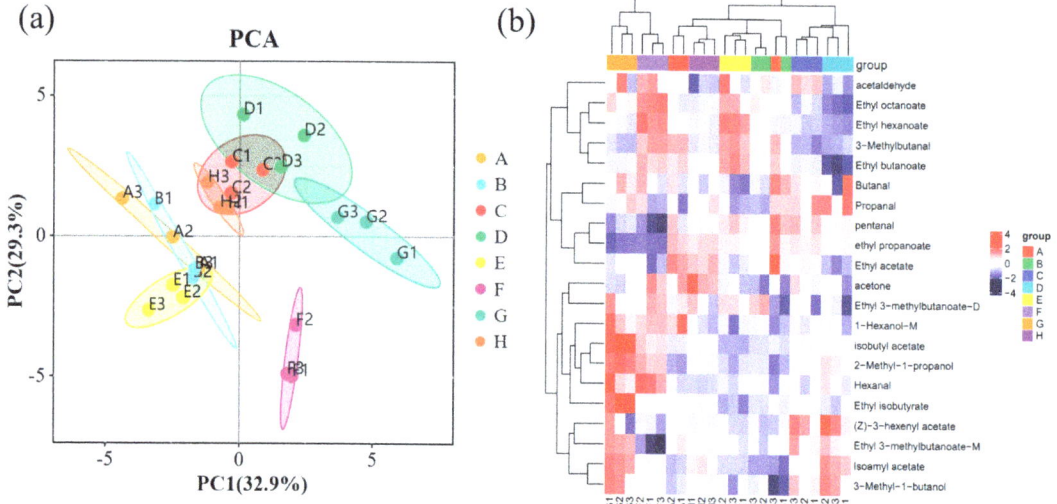

Figure 5. (a) PCA plots and (b) thermograms of volatile aroma compounds with OVA values greater than 1 in ice samples from different treatments of 'Beibinghong'.

Clustering of the concentration of volatile aroma compounds in nine sample wines with OVA values greater than 1. Based on the sample heat map analysis (Figure 5b), it was seen that the red color indicated that the aroma compound component was highly expressed in the sample, and the blue color indicated that the aroma compound was expressed at a lower level in the sample. The concentration of volatile aroma compounds with an OVA value of greater than 1 varied considerably among the samples of each variety. In general, most volatile aroma compounds with high aroma intensity values also have high OAV [57]; moreover, the two methods can be mutually verified. However, a small number of volatile aroma compounds also have low OAV despite high aroma intensity values, or vice versa. In this experiment, screening volatile aroma compounds between different treatment groups using OAV values can more accurately extract the critical volatile aroma compounds that may affect the aroma of 'Beibinghong' ice wine.

2.5.2. Analysis of Volatile Compounds 0PLS-DA in Wine

OPLS-DA is a statistical method for supervised discriminant analysis [54,58,59]. The contribution of each variable to the flavor of the wine was further quantified according to the variable important for the projection (VIP) in the OPLS-DA model [59]. OPLS-DA was validated with 200 permutations and found that R2 and Q2 were more significant than the model after Y replacement (Figure 6b). Thus, the model predictions were reliable, and variables with VIP > 1 could be used as potential biomarkers. This experiment used the OVA values of compounds with an OAV value greater than 1 in the composition of wine samples from different varieties as Y variables for OPLS-DA analysis.

Screening of compounds with VIP values > 1 as marker compounds for wine (Table 7). The results revealed that the compounds that may affect the aroma differences at different SO_2 concentration treatments might be related to ethyl butyrate, ethyl propionate, ethyl 3-methyl butyrate-M, ethyl 3-methyl butyrate-D, and 3-methyl butyraldehyde.

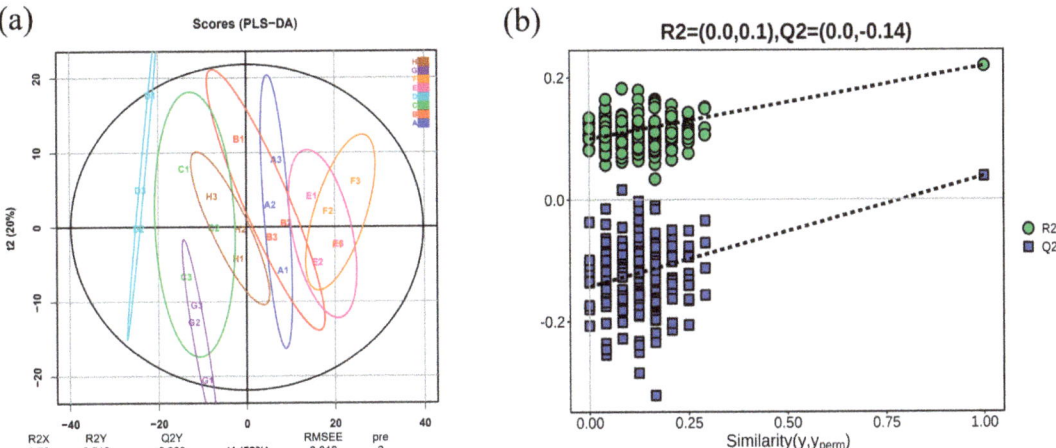

Figure 6. Analysis of different treatments of 'Beibinghong' ice wine samples OPLSA-DA ((**a**) Score chart of the OPLS-DA model of the sample, (**b**) Sample replacement test chart).

Table 7. Analysis of VIP values of aroma compounds in different treatments of 'Beibinghong'.

Substances	VIP Value
Ethyl butanoate	2.741464
ethyl propanoate	1.077886
Ethyl 3-methylbutanoate-M	2.23449
Ethyl 3-methylbutanoate-D	1.761919
3-Methylbutanal	1.550059

2.6. Sensory Evaluation of 'Beibinghong' Ice Wine Brewed with Different Concentrations of SO_2

Sensory evaluation is a crucial way for consumers to assess the quality of a wine [60]. Sensory evaluation of wines influences consumer choice. As seen in the sensory evaluation, it can be found (Table 8) that the color and clarity of wine samples from the eight treatment groups were above 9 points, and their color and clarity scores were the highest in the sixth treatment group, and the aroma and taste scores were the highest in the third treatment group; the typicality of the wines from the third and the sixth treatment groups had the highest scores; the highest total score was obtained from the third treatment group, and the lowest total score was obtained from the first treatment group. In summary, the overall flavor of the 'Beibinghong' ice wines was the highest when SO_2 was added to the ice wines at an added amount of 30 mg/L. Adding too little or too much SO_2 can affect the wine's flavor and lead to certain defects. Pelonnier-Magimel E et al. [30] studied whether the Bordeaux quality wines without added SO_2 have their typicality to assess the organoleptic specificity of the wines without added SO_2. Finally, it was found that wines without added SO_2 had a much higher frequency of defects than wines with added SO_2 (70 percent and 15 percent, respectively). Therefore, the absence of SO_2 in production can significantly impact the wine's flavor. If the wine is to maintain its typicality in production, as little SO_2 as possible can be added, while ensuring the flavor and bactericidal effect [34,61].

Table 8. Sensory score.

Item	Percentage	A	B	C	D	E	F	G	H
Color	10%	9.0 ± 0.06	9.1 ± 0.1	9.7 ± 0.15	9.2 ± 0.06	9.3 ± 0.21	9.9 ± 0.15	9.8 ± 0.10	9.7 ± 0.10
Clarification	10%	9.9 ± 0.06	10 ± 0	10 ± 0	10 ± 0	10 ± 0	10 ± 0	10 ± 0	10 ± 0
Aroma	30%	26.4 ± 1.02	27.2 ± 1.25	29.3 ± 0.67	28.4 ± 1.02	28 ± 0.78	27 ± 0.21	26.7 ± 0.53	26.1 ± 1.02
Taste	40%	36.1 ± 0.17	36.4 ± 2.08	39 ± 0.15	37 ± 1.0	38.1 ± 0.06	37.4 ± 1.03	38 ± 0.57	35.9 ± 0.26
Typicality	10%	9.4 ± 0.15	9.5 ± 1.21	10 ± 0.06	9.8 ± 1.15	9.6 ± 0.27	10 ± 0	9.4 ± 0.21	9.7 ± 0.06
Totals	100%	90.7 ± 1.20	92.2 ± 1.27	98 ± 1.0	94.4 ± 0.57	95 ± 1.0	94.3 ± 2.07	93.9 ± 1.52	91.4 ± 0.70

3. Conclusions

In this study, we investigated the effects of SO_2 addition on the fermentation process and volatile aroma compounds of 'Beibinghong' ice wine. The basic physicochemical properties of ice wine and the types and contents of volatile aroma compounds were analyzed in different SO_2 treatment groups, and the fingerprints of volatile compounds in ice wine brewed with different concentrations of SO_2 were constructed. In this study, it was found that the fermentation time of ice wine was shortest and the total acid content was relatively low at 10 mg/L, but the types and contents of volatile aroma compounds did not increase significantly; the total content of volatile aroma compounds of ice wine was highest at 80 mg/L. OPLS-DA calculated the VIP values, and the joint analysis of OAV and VIP values further identified five compounds: ethyl butyrate, ethyl propionate, ethyl 3-methyl butyrate-M, ethyl 3-methyl butyrate-D, and 3-methyl butyraldehyde, which were significantly different among the groups treated with different concentrations of SO_2. These compounds might be the critical factors of the effect of SO_2 on the volatile aroma compounds of the 'Beibinghong' ice wine. These compounds may be critical factors in the effect of SO_2 on the volatile aroma compounds of 'Beibinghong' ice wine. Tasting of 'Beibinghong' ice wine with different SO_2 additions revealed that the overall flavor of 'Beibinghong' ice wine was the highest at 30 mg/L. In conclusion, the wine's best accumulation of nutrients and flavor was achieved at 30 mg/L of SO_2 additions. To some extent, this experiment reflects the differences in the quality of wines at different concentrations of SO_2 through micro-winemaking, which provided a reference for the development and promotion of wines, and the results of the study provided a basis for optimizing the fermentation process of the 'Beibinghong' ice wine. As SO_2 is a cheap and effective wine preservative, its complete replacement is not feasible. The optimum level of SO_2 addition in the production of 'Beibinghong' ice wine was investigated in order to minimize the excessive use of SO_2 in wine production. The results of this study provide some theoretical basis for optimizing the fermentation process of 'Beibinghong' ice wine and improving the quality of ice wine.

Author Contributions: Conceptualization, B.Z.; Software, W.C.; Validation, W.C.; Formal analysis, H.Q. and S.F.; Investigation, B.Z.; Resources, W.L.; Data curation, C.L., Z.Z. and P.X.; Writing—original draft, B.Z.; Writing—review & editing, Y.L.; Visualization, C.L. and Y.Y.; Supervision, B.Z.; Project administration, W.L.; Funding acquisition, W.L. All authors have read and agreed to the published version of the manuscript.

Funding: This work was supported by the Science and Scientific and Technologic Foundation of Jilin Province (20190301056NY).

Institutional Review Board Statement: Ethics approval was obtained from the Institute of Special Animal and Plant Sciences of Chinese Academy of Agricultural Sciences (Approval Date: 10 May 2020).

Informed Consent Statement: Informed consent was obtained from all subjects involved in the study.

Data Availability Statement: The original contributions presented in the study are included in the article, further inquiries can be directed to the corresponding author.

Conflicts of Interest: The authors declare that they have no competing interests. Compliance with Ethics Requirements: This article does not contain any studies with human or animal subjects.

References

1. Penghui, L.; Yuanlong, J.; Donglin, C.; Xinyuan, W.; Jiahua, L.; Rongchen, Z.; Zhitong, W.; Yang, H.; Liankui, W. Study on the relationship between flavor components and quality of ice wine during freezing and brewing of 'beibinghong' grapes. *Food Chem. X* **2023**, *20*, 101016.
2. Lan, Y.-B.; Qian, X.; Yang, Z.-J.; Xiang, X.-F.; Yang, W.-X.; Liu, T.; Zhu, B.-Q.; Pan, Q.-H.; Duan, C.-Q. Striking changes in volatile profiles at sub-zero temperatures during over-ripening of 'Beibinghong' grapes in Northeastern China. *Food Chem.* **2016**, *212*, 172–182. [CrossRef] [PubMed]
3. Song, R.; Lu, W.; Shen, Y. New Grape Winemaking Variety-Beibinghong. *J. Acta Hortic. Sin.* **2008**, *35*, 1085.
4. Liu, Y.; Fan, S.; Yang, Y.; Zhang, B.; Qin, H.; Lu, W. Effects of Different Fertilization Amounts on Nutrient Elements Accumulation, Yield and Quality of 'Beibinghong' Grape. *North. Hortic.* **2023**, *7*, 46–51.
5. Li, C.; Liu, Y.; Fan, S.; Yang, Y.; Lu, W. Current Situation and Preview of 'Beibinghongs' Cultivation. *Spec. Wild Econ. Anim. Plant* **2019**, *41*, 125–128. [CrossRef]
6. Na, L.; Guanyu, L.; Aihua, L.; Yongsheng, T. Synergy Effect between Fruity Esters and Potential Odorants on the Aroma of Hutai-8 Rose Wine Revealed by Threshold, S-Curve, and σ-τ Plot Methods. *J. Agric. Food Chem.* **2023**, *71*, 13869–13879.
7. Peng, X.; Yang, X.; Aisaiti, A.; Yang, F.; Li, Z.; Li, H. Effects of different concentrations of free SO_2 and dissolved oxygen on wine color and anthocyanins content. *Food Mach.* **2022**, *38*, 11–16+23. [CrossRef]
8. Li, Q.; Zhang, X.; Niu, G.; Dang, Y.; Sun, S. Determination of Wine by Inductively Coupled Plasma Emission Spectrometry (ICP-OES) Study on Sulfur Dioxide Method. *Food Nutr. China* **2023**, *29*, 17–20. [CrossRef]
9. Sara, W.; Pascaline, R.; Soizic, L.; Laura, F.; Georgia, L.; Margaux, C.; Jean-Christophe, B.; Joana, C.; Joana, T.; Isabelle, M.-P. Non-Saccharomyces yeasts as bioprotection in the composition of red wine and in the reduction of sulfur dioxide. *LWT* **2021**, *149*, 111781.
10. Mu, J.; Wang, Y.; Liu, J.; Zhao, X. The Changes of the SO_2 during the Alcohol Fermentation Process in Wine. *Food Res. Dev.* **2011**, *32*, 53–55.
11. Yildirim, H.K. Alternative methods of sulfur dioxide used in wine production. *J. Microbiol. Biotechnol. Food Sci.* **2020**, *9*, 675–687. [CrossRef]
12. The National Standard for Food Safety Packaged Drinking Water (GB19298-2014), the National Standard for Food Safety Use of Food Additives (GB2760-2014) and 37 other national food safety standards were issued. *Beverage Ind.* **2014**, *17*, 46–47.
13. GB 7718-2011; General Principles for the Labelling of Prepackaged Foods Promulgated and Implemented. Ministry of Health: Wellington, New Zealand, 2011.
14. Lin, Z. Analysis and Prospects of Sulfur Dioxide Residues in Wine. *Food Ind.* **2023**, *12*, 104–106.
15. Wang, H.; Tian, X.; Yang, C.; Han, Y.; Shi, X.; Li, H. Wine and health. *China Brew.* **2022**, *41*, 1–5.
16. Du, F.; Huang, Y.; Liu, Z.; Miao, J.; Lai, K. Effects of pH, linoleic acid, and reheating on volatile compounds in glucose-lysine model system. *Food Biosci.* **2024**, *58*, 103631. [CrossRef]
17. Wang, X.; Dang, C.; Liu, Y.; Ge, X.; Suo, R.; Ma, Q.; Wang, J. Effect of indigenous Saccharomyces cerevisiae strains on microbial community successions and volatile compounds changes during Longyan wine fermentation. *Food Biosci.* **2024**, *57*, 103595. [CrossRef]
18. Prezioso, I.; Fioschi, G.; Rustioni, L.; Mascellani, M.; Natrella, G.; Venerito, P.; Gambacorta, G.; Paradiso, V.M. Influence of prolonged maceration on phenolic compounds, volatile profile and sensory properties of wines from Minutolo and Verdeca, two Apulian white grape varieties. *Lebensm.-Wiss. Technol.* **2024**, *192*, 115698. [CrossRef]
19. Ju, Y.L.; Xu, X.L.; Yu, Y.K.; Liu, M.; Wang, W.N.; Wu, J.R.; Liu, B.C.; Zhang, Y.; Fang, Y.L. Effects of winemaking techniques on the phenolics, organic acids, and volatile compounds of Muscat wines. *Food Biosci.* **2023**, *54*, 102937. [CrossRef]
20. Weiyu, C.; Nan, S.; Jinli, W.; Yiming, Y.; Yanli, W.; Wenpeng, L. Widely Targeted Metabolomics Was Used to Reveal the Differences between Non-Volatile Compounds in Different Wines and Their Associations with Sensory Properties. *Foods* **2023**, *12*, 290. [CrossRef]
21. Shu, N. Study on Fermentation Characteristics and Dry Red WineBrewing Technology of New *Vitis amurensis* Cultivar 'Beiguohong'. Master's Thesis, Chinese Academy of Agricultural Sciences, Beijing, China, 2019.
22. Yonghong, Y.; Songyan, Z.; Yuanxing, W. Analysis of aroma components changes in Gannan navel orange at different growth stages by HS-SPME-GC–MS, OAV, and multivariate analysis. *Food Res. Int.* **2024**, *175*, 113622.
23. Qianqian, L.; Bei, L.; Rong, Z.; Shuyan, L.; Shupeng, Y.; Yi, L.; Jianxun, L. Flavoromics Approach in Critical Aroma Compounds Exploration of Peach: Correlation to Origin Based on OAV Combined with Chemometrics. *Foods* **2023**, *12*, 837. [CrossRef] [PubMed]
24. Yijin, Y.; Lianzhong, A.; Zhiyong, M.; Haodong, L.; Xin, Y.; Li, N.; Hui, Z.; Yongjun, X. Flavor compounds with high odor activity values (OAV > 1) dominate the aroma of aged Chinese rice wine (Huangjiu) by molecular association. *Food Chem.* **2022**, *383*, 132370.
25. ISO 8586-2:2008; Sensory analysis. General Guidelines for the Selection, Training and Monitoring of Assessors. Part 2: Professional Sensory Assessors. ISO: Geneva, Switzerland, 2008.
26. ISO 8586:2012; Sensory Analysis. General Guidelines for the Selection, Training and Monitoring of Selected Assessors and Specialised Sensory Assessors. ISO: Geneva, Switzerland, 2012.

27. Jin, Y.; Shu, N.; Xie, S.; Cao, W.; Xiao, J.; Zhang, B.; Lu, W. Comparison of 'Beibinghong' dry red wines from six producing areas based on volatile compounds analysis, mineral content analysis, and sensory evaluation analysis. *Eur. Food Res. Technol.* **2021**, *247*, 1461–1475. [CrossRef]
28. Kim, B.H.; Park, S.K. Volatile aroma and sensory analysis of black raspberry wines fermented by different yeast strains. *J. Inst. Brew.* **2015**, *121*, 87–94. [CrossRef]
29. Sun, H.; Wu, Y. Sulphur dioxide in the wine industry. *Deciduous Fruits* **1996**, *S1*, 45–46. [CrossRef]
30. Pelonnier-Magimel, E.; Mangiorou, P.; Philippe, D.; De Revel, G.; Jourdes, M.; Marchal, A.; Marchand, S.; Pons, A.; Riquier, L.; Tesseidre, P.L.; et al. Sensory characterisation of Bordeaux red wines produced without added sulfites. *Oeno One* **2020**, *54*, 687–697. [CrossRef]
31. Tian, B.; Yuan, M.; Yuan, X. Determination of Dry Leachate in Wine by FOSS Instrument—Specific Gravity Bottle Method. *Sino-Overseas Grapevine Wine* **2016**, *4*, 39–41. [CrossRef]
32. Liu, X.; Xing, J.; Feng, J.; Chen, J.; Jiao, Y.; Yang, B. Research progress on copigmentation of red wine. *China Brew.* **2023**, *42*, 9–14.
33. Margherita, M.; Gianmarco, A.; Anna, M.; Roberto, F.; Serena, F.; Milena, P.; Isabella, T.; Fabio, M.; Andrea, B. Using ethanol as postharvest treatment to increase polyphenols and anthocyanins in wine grape. *Heliyon* **2024**, *10*, e26067. [CrossRef]
34. Ferreira, V.; Carrascon, V.; Bueno, M.; Ugliano, M.; Fernandez-Zurbano, P. Oxygen Consumption by Red Wines. Part I: Consumption Rates, Relationship with Chemical Composition, and Role of SO_2. *J. Agric. Food Chem.* **2015**, *63*, 10928–10937. [CrossRef]
35. Christofi, S.; Malliaris, D.; Katsaros, G.; Panagou, E.; Kallithraka, S. Limit SO_2 content of wines by applying High Hydrostatic Pressure. *Innov. Food Sci. Emerg. Technol.* **2020**, *62*, 102342. [CrossRef]
36. Oliveira, C.M.; Ferreira, A.C.S.; Freitas, V.D.; Silva, A.M.S. Oxidation mechanisms occurring in wines. *Food Res. Int.* **2011**, *44*, 1115–1126. [CrossRef]
37. Ribéreau-Gayon, P.; Dubourdieu, D.; Donèche, B.; Lonvaud, A. *Handbook of Enology, Volume 1: The Microbiology of Wine and Vinifications*; John Wiley & Sons: Hoboken, NJ, USA, 2006; Volume 1.
38. Dallas, C.; Laureano, O. Effects of pH, sulphur dioxide, alcohol content, temperature and storage time on colour composition of a young Portuguese red table wine. *J. Sci. Food Agric.* **1994**, *65*, 477–485. [CrossRef]
39. Sáenz-Navajas, M.-P.; Henschen, C.; Cantu, A.; Watrelot, A.A.; Waterhouse, A.L. Understanding microoxygenation: Effect of viable yeasts and sulfur dioxide levels on the sensory properties of a Merlot red wine. *Food Res. Int.* **2018**, *108*, 505–515. [CrossRef] [PubMed]
40. Bakker, J.; Bridle, P.; Bellworthy, S.; Garcia-Viguera, C.; Reader, H.; Watkins, S. Effect of sulphur dioxide and must extraction on colour, phenolic composition and sensory quality of red table wine. *J. Sci. Food Agric.* **1998**, *78*, 297–307. [CrossRef]
41. Miao, L.; Zhao, X.; Han, A.; Jiang, K. The Main Influencing Factors of Anthocyanins in Red Wine. *Liquor-Mak. Sci. Technol.* **2016**, *2*, 40–46. [CrossRef]
42. Coetzee, C.; Lisjak, K.; Nicolau, L.; Kilmartin, P.; du Toit, W.J. Oxygen and sulfur dioxide additions to Sauvignon blanc must: Effect on must and wine composition. *Flavour Fragr. J.* **2013**, *28*, 155–167. [CrossRef]
43. Liu, P.-T.; Zhang, B.-Q.; Duan, C.-Q.; Yan, G.-L. Pre-fermentative supplementation of unsaturated fatty acids alters the effect of overexpressing ATF1 and EEB1 on esters biosynthesis in red wine. *Lebensm.-Wiss. Technol.* **2020**, *120*, 108925. [CrossRef]
44. Yang, Z.; Zhang, Z.; He, Y.; Zhu, C.; Hu, B. Effect of fermentation with peel on volatile compounds in Kiwi wines based on Headspace Solid Phase Microextraction-Gas Chromatography-Mass Spectrometry and Gas Chromatography-Ion Mobility Spectrometry. *Food Sci.* **2024**, 1–14.
45. Wei, G.; Yang, X.; Zhou, Y.; Zeng, F.; Zhang, H. Review on research progress of esters in wine. *Sci. Technol. Food Ind.* **2015**, *36*, 394–399. [CrossRef]
46. Garde-Cerdán, T.; Ancín-Azpilicueta, C. Effect of SO_2 on the formation and evolution of volatile compounds in wines. *Food Control* **2007**, *18*, 1501–1506. [CrossRef]
47. Liu, J.; Zhang, B.; Zhu, B.; Duan, C.; Yan, G. Effect of pretreatment of sulfur dioxide preservative on aroma and biogenic amines contents in wines. *China Brew.* **2020**, *39*, 32–39.
48. Santos, T.P.M.; Alberti, A.; Judacewski, P.; Zielinski, A.A.F.; Nogueira, A. Effect of sulphur dioxide concentration added at different processing stages on volatile composition of ciders. *J. Inst. Brew.* **2018**, *124*, 261–268. [CrossRef]
49. Daniel, M.A.; Elsey, G.M.; Capone, D.L.; Perkins, M.V.; Sefton, M.A. Fate of damascenone in wine: The role of SO_2. *J. Agric. Food Chem.* **2004**, *52*, 8127–8131. [CrossRef] [PubMed]
50. Zheng, Y.; Li, Y.; Pan, L.; Guan, M.; Yuan, X.; Li, S.; Ren, D.; Gu, Y.; Liang, M.; Yi, L. Aroma and taste analysis of pickled tea from spontaneous and yeast-enhanced fermentation by mass spectrometry and sensory evaluation. *Food Chem.* **2024**, *442*, 138472. [CrossRef] [PubMed]
51. Weiyu, C.; Nan, S.; Jinli, W.; Yiming, Y.; Yuning, J.; Wenpeng, L. Characterization of the Key Aroma Volatile Compounds in Nine Different Grape Varieties Wine by Headspace Gas Chromatography–Ion Mobility Spectrometry (HS-GC-IMS), Odor Activity Values (OAV) and Sensory Analysis. *Foods* **2022**, *11*, 2767. [CrossRef] [PubMed]
52. Yue, M.; Noëlle, B.; Ke, T.; Yuanyi, L.; Marie, S.; Yan, X.; Thierry, T.-D. Assessing the contribution of odor-active compounds in icewine considering odor mixture-induced interactions through gas chromatography–olfactometry and Olfactoscan. *Food Chem.* **2022**, *388*, 132991.

53. Lan, Y.-B.; Xiang, X.-F.; Qian, X.; Wang, J.-M.; Ling, M.-Q.; Zhu, B.-Q.; Liu, T.; Sun, L.-B.; Shi, Y.; Reynolds, A.G.; et al. Characterization and differentiation of key odor-active compounds of 'Beibinghong' icewine and dry wine by gas chromatography-olfactometry and aroma reconstitution. *Food Chem.* **2019**, *287*, 186–196. [CrossRef] [PubMed]
54. Xu, Q.; Mengqi, L.; Yanfeng, S.; Fuliang, H.; Ying, S.; Changqing, D.; Yibin, L. Decoding the aroma characteristics of icewine by partial least-squares regression, aroma reconstitution, and omission studies. *Food Chem.* **2024**, *440*, 138226.
55. Marine, T.; Marina, B.; Warren, A.; Isabelle, M.; Benoit, C.; Philippe, M.; JeanChristophe, B. Impact of Grape Maturity on Ester Composition and Sensory Properties of Merlot and Tempranillo Wines. *J. Agric. Food Chem.* **2022**, *70*, 11520–11530.
56. Sebzalli, Y.M.; Wang, X.Z. Knowledge discovery from process operational data using PCA and fuzzy clustering. *Eng. Appl. Artif. Intell.* **2001**, *14*, 607–616. [CrossRef]
57. Yi, F.; Ma, N.; Zhu, J. Identification of Characteristic Aroma Compounds in Soy Sauce Aroma Type Xi Baijiu Using Gas Chromatography-Olfactometry, Odor Activity Value and Feller's Additive Model. *Food Sci.* **2022**, *43*, 242–256.
58. Xiangwu, H.; Lihong, Z.; Sheng, P.; Yijun, L.; Jianrong, L.; Meiqian, Z. Effects of Varieties, Cultivation Methods, and Origins of Citrus sinensis 'hongjiang' on Volatile Organic Compounds: HS-SPME-GC/MS Analysis Coupled with OPLS-DA. *Agriculture* **2022**, *12*, 1725. [CrossRef]
59. Li, C.; Wang, Y.; Lv, Y.; Qiu, X.; Wu, B.; Ma, T.; Fang, Y.; Sun, X. Physicochemical characterization and antioxidant capacity analysis of commercial Marselan wines from Ningxia and Hebei regions based on OPLS-DA. *Food Ferment. Ind.* **2023**, *49*, 283–292. [CrossRef]
60. Qi, Y.; Wang, M.; Wan, N.; Yin, D.; Wei, M.; Sun, X.; Fang, Y.; Ma, T. Sensory characteristics of "Shine Muscat" grapes based on consumer reviews and human and intelligent sensory evaluation. *Lebensm.-Wiss. Technol.* **2024**, *195*, 115810. [CrossRef]
61. Valásek, P.; Mlcek, J.; Fisera, M.; Fiserová, L.; Sochor, J.; Baron, M.; Juríková, T. Effect of various sulphur dioxide additions on amount of dissolved oxygen, total antioxidant capacity and sensory properties of white wines. *Mitteilungen Klosterneubg.* **2014**, *64*, 193–200.

Disclaimer/Publisher's Note: The statements, opinions and data contained in all publications are solely those of the individual author(s) and contributor(s) and not of MDPI and/or the editor(s). MDPI and/or the editor(s) disclaim responsibility for any injury to people or property resulting from any ideas, methods, instructions or products referred to in the content.

Article

Characterization of Key Compounds of Organic Acids and Aroma Volatiles in Fruits of Different *Actinidia argute* Resources Based on High-Performance Liquid Chromatography (HPLC) and Headspace Gas Chromatography–Ion Mobility Spectrometry (HS-GC-IMS)

Yanli He [1], Hongyan Qin [1], Jinli Wen [1], Weiyu Cao [1], Yiping Yan [1], Yining Sun [1], Pengqiang Yuan [1], Bowei Sun [2], Shutian Fan [1], Wenpeng Lu [1] and Changyu Li [1,*]

[1] Institute of Special Animal and Plant Sciences, Chinese Academy of Agricultural Sciences, Changchun 130112, China; 82101215184@caas.cn (Y.H.); qinhongyan@caas.cn (H.Q.); 82101215188@caas.cn (J.W.); 82101202231@caas.cn (W.C.); 82101225211@caas.cn (Y.Y.); 82101225210@caas.cn (Y.S.); 82101222242@caas.cn (P.Y.); fanshutian@caas.cn (S.F.); luwenpeng@caas.cn (W.L.)
[2] Faculty of Agriculture, Yanbian University, Yanji 136200, China; 2022050841@ybu.edu.cn
* Correspondence: lichangyu@caas.cn

Abstract: *Actinidia arguta*, known for its distinctive flavor and high nutritional value, has seen an increase in cultivation and variety identification. However, the characterization of its volatile aroma compounds remains limited. This study aimed to understand the flavor quality and key volatile aroma compounds of different *A. arguta* fruits. We examined 35 *A. arguta* resource fruits for soluble sugars, titratable acids, and sugar–acid ratios. Their organic acids and volatile aroma compounds were analyzed using high-performance liquid chromatography (HPLC) and headspace gas chromatography–ion mobility spectrometry (HS-GC-IMS). The study found that among the 35 samples tested, S12 had a higher sugar–acid ratio due to its higher sugar content despite having a high titratable acid content, making its fruit flavor superior to other sources. The *A. arguta* resource fruits can be classified into two types: those dominated by citric acid and those dominated by quinic acid. The analysis identified a total of 76 volatile aroma substances in 35 *A. arguta* resource fruits. These included 18 esters, 14 alcohols, 16 ketones, 12 aldehydes, seven terpenes, three pyrazines, two furans, two acids, and two other compounds. Aldehydes had the highest relative content of total volatile compounds. Using the orthogonal partial least squares discriminant method (OPLS-DA) analysis, with the 76 volatile aroma substances as dependent variables and different soft date kiwifruit resources as independent variables, 33 volatile aroma substances with variable importance in projection (VIP) greater than 1 were identified as the main aroma substances of *A. arguta* resource fruits. The volatile aroma compounds with VIP values greater than 1 were analyzed for odor activity value (OAV). The OAV values of isoamyl acetate, 3-methyl-1-butanol, 1-hexanol, and butanal were significantly higher than those of the other compounds. This suggests that these four volatile compounds contribute more to the overall aroma of *A. arguta*. This study is significant for understanding the differences between the fruit aromas of different *A. arguta* resources and for scientifically recognizing the characteristic compounds of the fruit aromas of different *A. arguta* resources.

Keywords: *Actinidia argute* resources; organic acid; volatile compound; orthogonal partial least squares discriminant analysis; odor activity value

1. Introduction

Actinidia arguta [(Sieb. & Zucc) Planch. ex Miq.], also known as soft dates, kiwi berries, kiwi pears, and more, is a large deciduous liana from the kiwifruit family (Actinidiaceae Gilg & Werderm.) and the kiwifruit genus (Actinidia Lindl) [1]. This characteristic berry resource is native to China, with wild resources also found in Japan, the Korean Peninsula, and the Russian Far East [2,3]. Its fruits are tasty and unique in flavor and rich in nutrients, such as proteins, vitamins, amino acids, minerals, dietary fiber [4], polysaccharides, polyphenols, alkaloids, volatile oils, proanthocyanidins, and other active ingredients [5], which have antitumor, antiradiation, antioxidant, antiaging, hypoglycemic, anti-inflammatory, insomnia-inhibiting, immunity-improving, and laxative functions [6–9]. Nowadays, *A. arguta* is popular with the public and the market for its rich nutritional and medicinal value.

Volatile aroma substances are crucial factors that influence fruit quality and consumer enjoyment [10] as well as important indicators of fruit flavor quality. Research on the various aromas of fruits can provide a theoretical basis for screening superior resources and help to better understand and control key flavor quality parameters that may affect fruit processing [11]. Fruit volatile aroma substances are influenced by various factors, such as variety, cultivation conditions, climatic conditions, ripening period, and storage conditions [12,13]. Dozens of compounds, mainly esters, alcohols, aldehydes, alkenes, and ketones, have been identified in the fruits of *A. arguta* varieties [14,15]. However, previous studies on volatile aroma substances of *A. arguta* have mainly focused on varieties and wine products [14,16]. Sun Yang et al. [15] detected 41 compounds from the fruits of different *A. arguta* varieties. There were differences in the types and contents of the aroma components between varieties, with 'Autumn Honey' having the highest number of the kinds of aroma substances. Zhang Baoxiang et al. [17] detected 56 aroma substances in different varieties of *A. arguta*-brewed dry wine, clarified the composition and content of 46 of them, and found that the aroma components of different types of brewed dry wine were the same, but the range varied greatly through analysis. Little research has been performed on the volatile aroma substances of *A. arguta* resource fruits, which should be considered. Meanwhile, the differences between the volatile aroma components of different *A. arguta* resource fruits are not apparent. Therefore, this study aimed to detect their volatile aroma components and to identify the main compounds that affect the volatile aroma components of *A. arguta* resource fruits.

Currently, the commonly used methods for the detection and analysis of fruit aroma substances are gas chromatography–mass spectrometry (GC-MS), gas chromatography–ion mobility chromatography (GC-IMS), and gas chromatography–olfactometry (GC-O-MS) [17–19]. However, GC-MS and GC-O-MS have several disadvantages, including the need for sample pre-treatment, a more complex operation process, a long assay time, and excessive sample consumption [20]. The pre-treatment process may cause damage to the aroma substances present in the models themselves, leading to differences in the types and contents of the detected aroma substances [21]. On the other hand, GC-IMS is an instrumental analytical technique that separates ions of the detected substance according to their ion mobility at atmospheric pressure. It has several advantages, such as simple sample preparation, easy operation, high sensitivity, fast analytical speed, and even trace amounts of volatile compounds can be detected [22–24]. In addition, ion mobility can significantly separate isomers and isobaric compounds [25]. GC-IMS is a recently discovered analytical technique for detecting volatile compounds in mixed analytes [26]. It combines the separation properties of GC with the fast correspondence and high sensitivity of IMS, which allows the detection of alcohols, esters, aldehydes, ketones, and aromatics, including even the most complex and problematic matrices [27], and has been widely used for the study of volatile compounds in food sciences, e.g., in kiwifruit [19], jujube [28], melons [29], wines [30], eggs [31], and honey [32]. Compared with GC-MS, GC-IMS does not require sample pre-processing and preserves the original aroma components of the sample intact. Multivariate statistical methods, such as principal component analysis (PCA) modeling, orthogonal partial least squares discriminant analysis (OPLS-DA) modeling, and cluster analysis, are commonly used when analyzing GC-IMS volatiles. Principal component analysis (PCA) is based on the principle of KL transformation. It uses the idea of dimensionality reduction to transform multiple indicators into a small number of major components that can reflect most of the information of the original variables [33]. Orthogonal partial least squares discriminant analysis (OPLS-DA) is a supervised statistical method of discriminant analysis, and PCA-based OPLS-DA further inputs the transformed score information into the model, identifying the key contributors to the variance-related variables in the model [34,35]. Hierarchical cluster analysis (HCA) calculates the correlation between samples using defined criteria, which are simplified and combined according to the degree of correlation to provide a more intuitive and comprehensive comparison of similar varieties and components [36]. Therefore, HS-GC-IMS mixed multivariate statistical methods have been widely used in metabolomics and flavoromics studies [37,38].

In this study, the sugar and acid contents of 35 *A. arguta* resource fruits were determined. The volatile aroma components were rapidly analyzed and detected by HS-GC-IMS technology. This produced a top view of the differences and established the fingerprints of volatile aroma compounds of different *A. arguta* resource fruits. Furthermore, based on volatile aroma compounds, a quantitative descriptive analysis of the data was performed through multivariate statistical analysis to analyze the differences in volatile aroma compounds between individual resources. In addition, principal component analysis, OPLS-DA analysis, and OAV analysis were combined to screen essential volatile compounds affecting the fruit flavor of *A. arguta* resources. This study provides a theoretical basis for screening *A. arguta* resources with excellent flavor quality, enhancing and improving the flavor quality of *A. arguta* processed products. It also aids in scientifically recognizing the characteristic compounds of the fruit aroma of different *A. arguta* resources and provides a theoretical basis for regulating the flavor quality of processed products.

2. Materials and Methods

2.1. Materials and Reagents

2.1.1. Materials

The 35 resources selected for this study (Table 1) were sampled from the *Actinidia arguta* Resource Nursery of the Institute of Special Animal and Plant Sciences of the Chinese Academy of Agricultural Sciences, Zuojia Town, Jilin City, Jilin Province, China (44°00' N; 126°01' E). The sampling time was September 2022, when the fruits were ripe. Sampling was performed by randomly selecting well-grown, medium-sized vines in the resource nursery, choosing soft date palm kiwifruit with the same degree of exposure to light, the same size, and similar hardness and fruit that was free of pests and diseases. We picked about 300 g of fruit from each resource, placed the samples in separate numbered sampling bags, and transported them back to the lab in an insulated box. We placed the fruit in a −80 °C refrigerator for storage after measuring the relevant indicators on the same day.

2.1.2. Reagents

Analytical purity: anthrone (Sinopharm Chemical Reagent Co., Ltd. Shanghai, China); ethyl acetate, concentrated sulfuric acid, phosphoric acid (Beijing Chemical Factory, Beijing, China).

Chromatographic purity: methanol (TEDIA reagent, Fairfield, OH, USA); oxalic acid, quinic acid, malic acid, shikimic acid, lactic acid, citric acid, ascorbic acid (Shanghai Yuanye Biotechnology Co., Ltd. Shanghai, China); 4-methyl-2-pentanol (Shanghai Lianshuo Biotechnology Co., Ltd. Shanghai, China).

2.2. Instruments and Equipment

High-performance liquid chromatograph (Agilent Technologies, Waldbronn, Germany); FlavourSpec® Flavour Analyzer (G.A.S. It is based on gas chromatography ion mobility spectrometry (GC-IMS), which has both the high separation of gas chromatography and the high sensitivity of ion mobility spectrometry, and can detect trace volatile organic compounds in the samples without enrichment and concentration and other preprocessing to maintain the original flavor of the flavor samples, which is very suitable for the analysis of aroma components. The accompanying software can generate the sample aroma fingerprints, which can easily realize the comparison of sample differences and consistency control); CJJ-931 dual-magnetic heating stirrer (Jiangsu Jintan Jincheng Guosheng Experimental Instrument Factory, Jiangsu); hgs-12 electric thermostatic water bath, KQ-300E ultrasonic cleaner snowflake ice machine (Beijing Changliu Scientific Instrument Co., Ltd. Beijing, China); FA1004B electronic balance (Shanghai Yue Ping Scientific Instrument Co., Ltd. Shanghai, China); IMark enzyme labeling instrument (Biorad, Philadelphia, PA, USA); high-speed freezing centrifuge (Allegra 64R, USA); −80 °C ultra-low-temperature refrigerator (Beijing Chengmaoxing Science and Technology Development Co., Ltd. Beijing, China); WAX columns (RESTEK, Bellefonte, PA, USA).

Table 1. Resources and sources of 35 *A. arguta*.

No.	Name	Source	No.	Name	Source	No.	Name	Source
S1	A020203	Fusong County, Jilin Province, China	S13	A130701	Ji'an County, Jilin Province, China	S25	B080401	Ji'an County, Jilin Province, China
S2	A040103	Ji'an County, Jilin Province, China	S14	A130801	Ji'an County, Jilin Province, China	S26	B080701	Ji'an County, Jilin Province, China
S3	A060902	Zuojia Town, Jilin Province, China	S15	A140101	Zuojia Town, Jilin Province, China	S27	T040501	Fusong County, Jilin Province, China
S4	A100101	Ji'an County, Jilin Province, China	S16	A140301	Zuojia Town, Jilin Province, China	S28	T060203	Ji'an County, Jilin Province, China
S5	A100703	Ji'an County, Jilin Province, China	S17	A140602	Dunhua City, Jilin Province, China	S29	T060301	Fusong County, Jilin Province, China
S6	A100801	Ji'an County, Jilin Province, China	S18	A160701	Zuojia Town, Jilin Province, China	S30	T060503	Ji'an County, Jilin Province, China
S7	A101201	Dunhua City, Jilin Province, China	S19	A170303	Fusong County, Jilin Province, China	S31	SH1	Zuojia Town, Jilin Province, China
S8	A111001	Zuojia Town, Jilin Province, China	S20	A180303	Fusong County, Jilin Province, China	S32	SH2	Zuojia Town, Jilin Province, China
S9	A120403	Dunhua City, Jilin Province, China	S21	A180902	Zuojia Town, Jilin Province, China	S33	SH3	Zuojia Town, Jilin Province, China
S10	A120601	Dunhua City, Jilin Province, China	S22	A191002	Ji'an County, Jilin Province, China	S34	SH4	Zuojia Town, Jilin Province, China
S11	A130101	Ji'an County, Jilin Province, China	S23	B020802	Zuojia Town, Jilin Province, China	S35	SH5	Zuojia Town, Jilin Province, China
S12	A130602	Ji'an County, Jilin Province, China	S24	B070101	Zuojia Town, Jilin Province, China			

2.3. Methods

2.3.1. Determination of Soluble Sugar and Titratable Acid Content

Soluble sugar content was determined by the anthrone reagent method, and titratable acid content was determined by titration method with sodium hydroxide solution, both referring to the *Experiment Guideline of Postharvest Physiology and Biochemistry of Fruits and Vegetables* (1st edition, December 2020). Sugar–acid ratio = soluble sugar content/titratable acid content.

2.3.2. Determination of Organic Acid Content

The organic acid content was determined by high-performance liquid chromatography (HPLC), referring to the previously published literature [39]. Oxalic acid, quinic acid, malic acid, mangiferin acid, lactic acid, and citric acid were analyzed by HPLC using aqueous phosphoric acid at pH = 2.3 as the aqueous phase and methanol as the organic phase. The experimental conditions were as follows: the column temperature of the C18-XT (4.6 mm × 250 mm × 5 mL) column was 25 °C, and the flow rate was set to be 0.3 mL/min, and the injection volume was 10 µL; ascorbic acid was analyzed by the HPLC using aqueous phosphoric acid at pH = 2.3 as the aqueous phase, and methanol as the organic phase. For ascorbic acid, 0.2% aqueous phosphoric acid was used as the aqueous phase, and methanol was used as the organic phase. The test conditions were as follows: the column temperature of the C18-XT (4.6 mm × 250 mm × 5 mL) column was 25 °C, the flow rate was set at 0.5 mL/min, and the injection volume was 10 µL. The standard curves for the seven measured organic acids are shown in Table 2 below.

Table 2. The organic acid standard curve.

Name	Concentration g/L	Standard Curves	R^2
Oxalic acid	1.02	y = 24763x − 735.65	0.9998
Quinic acid	1.01	y = 779.46x − 18.648	0.9962
Malic acid	1.00	y = 1613.5x − 7.0785	0.9999
Shikimic acid	1.00	y = 45865x + 2285.4	0.9977
Lactic acid	1.08	y = 1272.5x − 4.6931	1
Citric acid	1.02	y = 2028.3x − 18.753	0.9999
Ascorbic acid	1.03	y = 24.297x − 41.339	0.9998

2.3.3. HS-GC-IMS Analytical Methods

Headspace gas chromatography–ion mobility spectrometry (HS-GC-IMS) was used for the determination of volatile aroma substances in the soft date kiwifruit resource fruits, and the instrument used in the experiment was a FlavourSpec® Flavour Analyzer. Briefly, 3 g of fruit homogenate was placed in a 20 mL headspace vial, 10 µL of 4-methyl-2-pentanol at 20 ppm was added, and the sample was injected after incubation at 60 °C for 15 min, and three parallel replicates were made for each resource. The chromatographic conditions were as follows (Table 3): the chromatographic column was a WAX column (15 m × 0.53 mm, 1 µm), the column temperature was 60 °C, the carrier gas was N2, and the IMS temperature was 45 °C. The automatic headspace injection conditions were as follows: injection volume was 300 µL, the incubation time was 10 min, the injection needle temperature was 65 °C, the incubation speed was 500 rpm, and the analysis was carried out using 4-methyl-2-pentanol as the internal standard with the concentration of 198 ppb, the signal peak volume of 470.02, and the signal intensity of each signal was about 0.421 ppb. The quantitative calculations were performed according to the following equations.

$$Ci = \frac{Cis * Ai}{Ais}$$

where Ci is the mass concentration of any component used in the calculation, Cis is the mass concentration of the internal standard used, and Ai/Ais is the volume ratio between any signal peak and the signal peak of the internal standard.

Table 3. Gas chromatography conditions.

Time (min: sec)	E1 (Drift Gas)	E2 (Carrier Gas)	Recording
00:00,000	150 mL/min	2 mL/min	rec
02:00,000	150 mL/min	2 mL/min	-
10:00,000	150 mL/min	10 mL/min	-
20:00,000	150 mL/min	100 mL/min	-
30:00,000	150 mL/min	100 mL/min	stop

2.4. Odor Activity Value (OAV) Calculation

The odor activity value (OAV) was used to evaluate the overall aroma contribution of *A. arguta* fruits. The OAV value was calculated by dividing the concentration of volatile aroma compounds by the odor threshold. The odor thresholds are determined by reference to the *Compilations of Odour Threshold Values in Air, Water and Other Media* (Edition 2011). Volatile aroma compounds with OAV > 1 were considered to be aromatically active and contribute significantly to the overall aroma of the samples.

2.5. Data Processing

Excel 2016 was used to organize the experimental data statistically, analysis of variance (ANOVA) was performed by SPSS (version 23.0, IBM, Armonk, NY, USA), and statistical analyses of variance were performed on the experimental data to check for significant differences in the individual results, and all the data were expressed as mean±standard deviation, with $p < 0.05$ indicating significant differences.

The HS-GC-IMS results were analyzed using the Volatile Organic Compounds Analysis Software (VOCal) accompanying the FlavourSpec® Flavour Analyzer, and the volatile aroma compounds were qualitatively analyzed using the retention index database of NIST and the migration time database of IMS built into the GC×IMS Library Search software; the GC-IMS detection was performed by using Savitzky–Golay to perform the smoothing and denoising process, and the migration time normalization method was used by locating the RIP position at position 1, which means that the actual migration time was divided by the peak time of the RIP. The Reporter plug-in was used to compare spectral differences between samples directly, and the Gallery Plot plug-in was used for fingerprinting to visually compare differences in volatile aroma compounds between fruits from different soft date kiwifruit sources. OPLS-DA and VIP values were analyzed using Simca software, and PCA, heatmap, and correlation analyses were performed using the OmicShare tool (https://www.omicshare.com/tools/, accessed on 19 September 2023).

3. Results and Analysis

3.1. Analysis of Soluble Sugar Content, Titratable Acid Content, and Sugar–Acid Ratio of Fruits from Different A. arguta Resources

Analysis of the differential results (Table 4) showed differences in soluble sugar content, titratable acid content and sugar–acid ratio between fruits of different *A. arguta* resources. The variation of soluble sugar content was 2.94–13.97%, the resource with the highest content was S26, which was significantly higher than the other resources, and the lowest resource was S4; the highest titratable acid content was S24 and S26 with 1.59% and 1.51%, respectively, and the lowest content was S35 with 0.32%. Fruit flavor is largely influenced by the levels of sugars and acids in the fruit. A good flavor requires a high sugar content and a suitable sugar–acid ratio. If the acidity is too high, the fruit may not be palatable. If the sugar content is high but the acidity is too low, the flavor may be bland and lack the balance of sweetness and sourness. If both the sugar and acid levels are too low, the fruit may taste watery and insipid [40]. The sugar–acid ratio of 35 *A. arguta* resource fruits was 2.45–28.50, with S35 having the highest sugar–acid ratio but the lowest titratable acid content, resulting in a more homogeneous flavor. In contrast, S12 has a higher sugar–acid ratio with titratable acid content at higher sugar content; therefore, its fruit flavor can be superior to its source.

Table 4. Content of soluble sugar, titratable acid and sugar–acid ratio of different A. arguta resources.

Name	Soluble Sugar %	Titratable Acid %	Sugar–Acid Ratio
S1	5.74 ± 0.34 opq	0.78 ± 0.01 k	7.39 ± 0.39 hij
S2	5.35 ± 0.08 q	1.00 ± 0.06 ij	5.34 ± 0.23 op
S3	6.66 ± 0.17 jkl	0.98 ± 0.08 j	6.80 ± 0.47 ijkl
S4	2.94 ± 0.12 u	1.20 ± 0.04 ef	2.45 ± 0.17 t
S5	5.84 ± 0.24 nop	0.82 ± 0.02 k	7.10 ± 0.38 hijk
S6	5.34 ± 0.09 q	0.97 ± 0.07 j	5.51 ± 0.48 nop
S7	6.49 ± 0.42 klm	0.85 ± 0.04 k	7.63 ± 0.21 hi
S8	7.04 ± 0.16 ij	1.39 ± 0.09 c	5.08 ± 0.30 pq
S9	4.14 ± 0.03 s	0.98 ± 0.01 j	4.24 ± 0.04 r
S10	6.65 ± 0.39 jkl	1.00 ± 0.18 ij	6.65 ± 0.87 jkl
S11	7.14 ± 0.33 hi	1.04 ± 0.03 hij	6.89 ± 0.47 ijkl
S12	9.40 ± 0.41 b	0.77 ± 0.05 k	12.21 ± 0.64 e
S13	4.44 ± 0.09 rs	0.64 ± 0.01 l	6.93 ± 0.20 ijkl
S14	6.03 ± 0.23 no	1.05 ± 0.02 hij	5.72 ± 0.15 mnop
S15	7.20 ± 0.48 ghi	1.11 ± 0.08 fgh	6.47 ± 0.86 klm
S16	6.05 ± 0.21 mno	0.95 ± 0.08 j	6.35 ± 0.71 klm
S17	4.85 ± 0.33 r	0.84 ± 0.04 k	5.73 ± 0.62 mnop
S18	8.25 ± 0.10 cd	1.17 ± 0.02 efg	7.05 ± 0.06 ijk
S19	3.48 ± 0.10 t	1.02 ± 0.03 hij	3.43 ± 0.20 s
S20	6.84 ± 0.32 ijk	1.02 ± 0.01 hij	6.71 ± 0.39 jkl
S21	6.82 ± 0.27 ijk	1.26 ± 0.02 de	5.42 ± 0.23 op
S22	8.31 ± 0.30 cd	1.05 ± 0.03 hij	7.92 ± 0.34 h
S23	5.71 ± 0.09 opq	1.09 ± 0.04 ghi	5.24 ± 0.17 o
S24	8.70 ± 0.40 c	1.59 ± 0.06 a	5.48 ± 0.15 op
S25	9.36 ± 0.54 b	1.47 ± 0.01 b	6.35 ± 0.34 klmn
S26	13.97 ± 0.10 a	1.51 ± 0.05 ab	9.0.26 g
S27	5.43 ± 0.29 pq	1.23 ± 0.02 e	4.40 ± 0.24 qr
S28	7.79 ± 0.17 ef	1.04 ± 0.09 hij	7.47 ± 0.60 hij
S29	8.18 ± 0.36 de	1.33 ± 0.04 cd	6.15 ± 0.41 lmno
S30	7.54 ± 0.06 fgh	1.17 ± 0.04 efg	6.46 ± 0.23 klm
S31	8.30 ± 0.06 cd	0.47 ± 0.03 mn	17.78 ± 1.06 b
S32	6.22 ± 0.06 lmn	0.40 ± 0.02 no	15.67 ± 0.39 c
S33	7.62 ± 0.14 fg	0.53 ± 0.02 m	14.47 ± 0.38 d
S34	3.68 ± 0.09 t	0.37 ± 0.01 o	10.04 ± 0.26 f
S35	9.23 ± 0.07 b	0.32 ± 0.01 o	28.50 ± 1.04 a
CV(%)	30.74	32.03	61.25

Means with different letters in the same column express significant differences (Duncan's test $p < 0.05$).

3.2. Analysis of Organic Acid Content in Fruits of Different A. arguta Resources

The type and content of organic acids affect the acidity of A. arguta fruits and the texture of A. arguta products, and the content of organic acids varies among different resources (Table 5). Organic acid is an essential component of the fruit and an essential factor affecting fruit quality [41]. The highest oxalic acid content was 0.182 g/L for S12, and the lowest was 0.013 g/L for S31. Oxalic acid, as a ubiquitous component in plants, has long been recognized as a metabolic end product with no obvious physiological role, but from the perspective of food nutrition and human health, long-term consumption of oxalic-acid-rich fruits and vegetables not only reduces the effectiveness of calcium and trace elements in the body but also causes the human body to suffer from renal calculi, diseases of the oral and digestive tracts, and so on [42]. The malic acid in fruits inhibits bacterial damage to the pulp and facilitates fruit preservation [43,44], and S8 had the highest malic acid content of 2.868 g/L, while the lowest content of S5 was 0.212 g/L. Quinic acid and shikimic acid will directly affect the bitter taste of the fruit and are intermediate products of the aromatic substance synthesis pathway, thus indirectly affecting the quality of the fruit [45]; the highest content of quinic acid was S2, 11.426 g/L, which was significantly higher than the other resources, and the lowest content was S17, 1.64 g/L. The shikimic acid content was 0.018–0.093 g/L. Lactic acid was detected in the fruits of some A. arguta resources, with the highest level of 0.329 g/L in S1 and the lowest level of 0.015 g/L in S9. Citric acid is characterized by producing acidity quickly and for a sustained period time, and it is capable of causing changes in the threshold of taste substances such as sweetness, sourness, astringency, and bitterness [46]. The citric acid content in the fruits of 35 *A. arguta* resources was 1.987–10.823 g/L, and the resource with the highest content was S2, which was significantly higher than the other resources. Ascorbic acid is widely present in plant tissues and has strong antioxidant properties and a variety of biological functions, such as resistance to stress and disease, but it also can be used for post-harvest storage for horticultural tea growers [47]. The resource with the highest content of ascorbic acid, S5, was 904.739 g/L, which was significantly higher than the other resources, and the content of S7 was the lowest, which was 28.740 g/L. The 35 *A. arguta* resource fruits could be categorized into citric-acid-dominant and quinic-acid-dominant types.

Table 5. Content of organic acids in different *A. arguta* resources.

Name	Oxalic Acid g/L	Quinic Acid g/L	Malic Acid g/L	Shikimic Acid g/L	Lactic Acid g/L	Citric Acid g/L	Ascorbic Acid g/L
S1	0.030 ± 0.003 p	7.714 ± 0.318 c	0.765 ± 0.040 st	0.51 ± 0.002 g	0.329 ± 0.0014 a	8.113 ± 0.051 f	59.617 ± 0.067 x
S2	0.026 ± 0.003 pq	11.426 ± 0.109 a	2.753 ± 0.066 b	0.026 ± 0.002 n	0.208 ± 0.010 b	10.823 ± 0.149 a	67.872 ± 0.063 w
S3	0.133 ± 0.014 efghij	6.085 ± 0.051 g	1.666 ± 0.017 i	0.043 ± 0.004 hi	N.A.	7.890 ± 0.042 g	334.402 ± 15.919 m
S4	0.154 ± 0.014 bcd	5.764 ± 0.039 i	1.591 ± 0.024 j	0.019 ± 0.002 p	0.102 ± 0.003 g	8.642 ± 0.067 d	677.253 ± 0.273 e
S5	0.105 ± 0.011 no	2.872 ± 0.017 v	0.212 ± 0.017 y	0.046 ± 0.002 hi	N.A.	3.479 ± 0.018 x	904.739 ± 0.215 a
S6	0.141 ± 0.013 cedfg	6.682 ± 0.026 e	1.184 ± 0.023 n	0.062 ± 0.002 de	0.046 ± 0.003 m	8.266 ± 0.017 e	82.676 ± 0.195 u
S7	0.156 ± 0.008 bc	5.432 ± 0.018 k	0.684 ± 0.010 u	0.081 ± 0.002 b	0.083 ± 0.002 i	7.267 ± 0.024 j	28.740 ± 0.341 z
S8	0.016 ± 0.001 pq	8.544 ± 0.016 b	2.868 ± 0.014 a	0.073 ± 0.003 c	N.A.	10.547 ± 0.030 b	209.252 ± 0.094 r
S9	0.020 ± 0.001 pq	5.447 ± 0.014 k	1.000 ± 0.011 p	0.043 ± 0.003 hi	0.015 ± 0.002 o	6.735 ± 0.014 l	530.055 ± 0.125 g
S10	0.165 ± 0.010 b	6.378 ± 0.018 f	1.338 ± 0.010 l	0.041 ± 0.004 ij	N.A.	6.793 ± 0.013 k	772.682 ± 0.173 d
S11	0.136 ± 0.017 efghi	4.047 ± 0.014 st	2.174 ± 0.013 e	0.047 ± 0.003 hi	0.060 ± 0.003 kl	5.014 ± 0.019 u	338.561 ± 0.316 m
S12	0.182 ± 0.010 a	5.486 ± 0.013 k	1.924 ± 0.021 f	0.026 ± 0.003 no	0.147 ± 0.008 d	6.837 ± 0.009 k	56.312 ± 0.166 x
S13	0.139 ± 0.013 defgh	5.001 ± 0.013 m	1.697 ± 0.012 hi	0.064 ± 0.002 d	N.A.	3.153 ± 0.012 y	338.813 ± 0.188 m
S14	0.156 ± 0.017 cdef	3.976 ± 0.014 tu	2.428 ± 0.016 c	0.018 ± 0.002 p	N.A.	6.624 ± 0.010 m	393.866 ± 0.133 k
S15	0.020 ± 0.002 pq	6.983 ± 0.009 d	1.744 ± 0.021 g	0.061 ± 0.002 de	0.113 ± 0.004 f	8.076 ± 0.020 f	48.530 ± 0.132 y
S16	0.150 ± 0.015 bcde	4.838 ± 0.010 no	1.230 ± 0.015 m	0.057 ± 0.002 fg	N.A.	5.680 ± 0.024 r	46.768 ± 0.093 v
S17	0.092 ± 0.011 o	1.641 ± 0.013 w	0.302 ± 0.014 x	0.091 ± 0.002 a	0.072 ± 0.002 j	1.987 ± 0.007 z	65.416 ± 0.071 w
S18	0.122 ± 0.008 hijklm	2.871 ± 0.014 v	0.514 ± 0.031 v	0.084 ± 0.003 b	N.A.	3.136 ± 0.021 y	244.467 ± 0.093 p
S19	0.018 ± 0.001 pq	4.615 ± 0.011 q	2.227 ± 0.014 d	0.034 ± 0.003 kl	0.137 ± 0.005 e	8.832 ± 0.022 c	51.441 ± 0.078 y
S20	0.115 ± 0.100 bc	5.901 ± 0.018 h	0.982 ± 0.009 p	0.061 ± 0.005 ef	N.A.	6.486 ± 0.012 n	510.680 ± 0.105 h
S21	0.106 ± 0.007 mno	3.902 ± 0.036 uv	0.920 ± 0.007 q	0.018 ± 0.002 p	0.070 ± 0.003 j	4.362 ± 0.014 w	795.696 ± 0.217 c
S22	0.138 ± 0.027 ijklm	5.828 ± 0.011 hi	1.706 ± 0.013 h	0.035 ± 0.003 kl	N.A.	6.138 ± 0.027 p	125.166 ± 0.182 s
S23	0.016 ± 0.002 pq	7.082 ± 0.005 d	0.903 ± 0.010 q	0.075 ± 0.003 c	N.A.	5.215 ± 0.011 t	109.725 ± 0.074 t
S24	0.142 ± 0.008 cdefg	6.447 ± 0.014 f	0.776 ± 0.010 st	0.054 ± 0.003 fg	0.094 ± 0.003 h	6.291 ± 0.017 o	249.555 ± 0.096 o
S25	0.118 ± 0.010 jklnm	5.647 ± 0.008 j	1.180 ± 0.031 n	0.028 ± 0.002 mn	N.A.	7.695 ± 0.014 i	857.254 ± 0.061 b
S26	0.096 ± 0.010 o	5.312 ± 0.040 l	1.522 ± 0.008 k	0.067 ± 0.003 d	0.054 ± 0.004 l	7.761 ± 0.009 h	457.152 ± 0.089 i
S27	0.173 ± 0.006 pq	4.560 ± 0.007 q	1.500 ± 0.007 k	0.035 ± 0.002 jk	0.061 ± 0.003 k	7.945 ± 0.022 g	582.221 ± 0.123 f
S28	0.146 ± 0.027 fghijk	5.665 ± 0.013 j	0.850 ± 0.012 r	0.049 ± 0.004 h	N.A.	5.176 ± 0.012 t	277.654 ± 0.143 n
S29	0.119 ± 0.007 ijklmn	4.292 ± 0.013 r	1.099 ± 0.027 o	0.046 ± 0.002 hi	0.161 ± 0.002 c	5.595 ± 0.014 s	393.842 ± 0.131 k
S30	0.113 ± 0.007 lmn	4.092 ± 0.009 s	0.744 ± 0.012 t	0.043 ± 0.006 hi	N.A.	3.034 ± 0.013 y	74.301 ± 0.137 v
S31	0.013 ± 0.002 q	4.748 ± 0.024 op	0.295 ± 0.010 x	0.037 ± 0.003 kl	0.027 ± 0.002 n	5.231 ± 0.023 t	237.664 ± 0.058 q
S32	0.115 ± 0.010 klmn	4.602 ± 0.021 q	0.289 ± 0.010 x	0.027 ± 0.002 n	0.033 ± 0.002 n	5.218 ± 0.027 t	66.265 ± 0.064 w
S33	0.128 ± 0.011 ghijkl	4.667 ± 0.040 pq	0.795 ± 0.017 s	0.038 ± 0.003 jk	N.A.	5.619 ± 0.019 s	420.748 ± 0.110 j
S34	0.140 ± 0.013 fghijkl	4.925 ± 0.027 mn	0.463 ± 0.017 w	0.032 ± 0.002 lm	0.083 ± 0.003 i	5.842 ± 0.009 q	456.249 ± 0.090 i
S35	0.142 ± 0.010 defgh	4.560 ± 0.013 q	0.669 ± 0.017 u	0.021 ± 0.002 op	0.044 ± 0.003 m	4.932 ± 0.017 v	352.468 ± 0.095 l
CV(%)	51.7	31.95	62.09	43.17	131.07	32.11	79.72

Means with different letters in the same column express significant differences (Duncan's test $p < 0.05$).

The results of hierarchical clustering analysis (HCA) can better respond to the characteristics of organic acid substances in the fruit samples of different *A. arguta* resources; according to the organic acid cluster analysis of each resource, it can be seen that when the value of the transverse tangent line is taken between 200 and 400 (Figure 1), the 35 *A. arguta* resource fruit samples are divided into six classes: the first class is S5 and S25; the second class is S4, S10, and S21; the third category is S27, S9, and S20; the fourth category is 9 resources, such as S35 and S3; the fifth category is S8, S28, S31, S18, and S214; and the sixth category is 13 resources, such as S22 and S23, which indicates that the samples contained in each category have similarity in organic acids when the value of the transversal line is taken between 200 and 400, and the results of which also show better clustering of fruit samples from different resources of *A. arguta* resources.

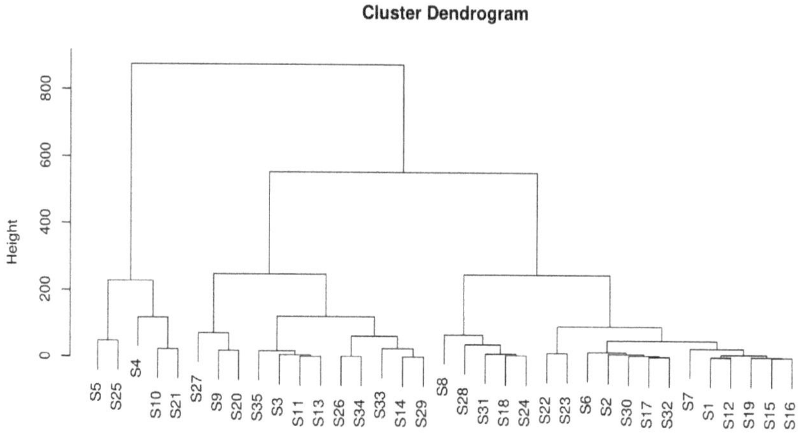

Figure 1. Hierarchical cluster analysis of organic acid content in fruits of different *A. arguta* resources.

3.3. HS-GC-IMS Analysis of Fruits from Different A. arguta Resources

The aroma description of *A. arguta* fruits is one of the critical determinants of their quality, and their flavor is also an essential factor in determining whether they are acceptable to consumers [48]. The type and content of volatile compounds and their interactions are the main factors affecting the quality of *A. arguta* fruits. Gas chromatography–mass spectrometry (HS-GC-IMS) is commonly used to characterize and quantify volatile compounds in food [49].

3.3.1. Two-Dimensional Mapping of Volatile Aroma Substances in Fruits of Different *A. Arguta* Resources

There were differences in the two-dimensional mapping profiles of volatile aroma substances of 35 *A. arguta* resources (Figure 2). The differences were mainly reflected in the content, and the color represented the concentration of the substances, with white representing a low concentration of the substances, red representing a high concentration of the substances, and darker colors representing a higher concentration of the substances. The volatile aroma substances in the 35 *A. arguta* resource fruits were well separated by HS-GC-IMS, and the differences between individual samples could be seen.

3.3.2. Comparative Pattern Spectrum of Differences in Volatile Aroma Substances of Fruits from Different *A. arguta* Resources

HS-GC-IMS was used to obtain full information on the volatiles in the fruits of the *A. arguta* resource, and difference comparison mode spectra represented the differences between the samples. The horizontal and vertical axes of the difference plots represent the ionic migration time of the volatile compounds and the retention time at the ionic peaks of the reactants, respectively, and each point represents the monomer of the volatile compounds extracted from the samples or their dimers [50]. Taking S1 as a reference (Figure 3), the rest of the spectrum subtracts the signal peaks in S1 to obtain the difference comparison mode spectrum between the two. The red area in the graph indicates that the concentration of the substance in this sample is higher than that of S1, and the blue area indicates that the attention of the substance in this sample is lower than that of S1. The white area indicates that the attention of the substance in this sample is comparable to that of S1. Differential mapping analysis showed that S1 contained higher levels of hexyl propanoate, ethyl (E)-hex-2-enoate, ethanol, isobutanol, hexanal, and trans-2-hexenal than some of the resource fruits.

3.3.3. Identification of Substances

For the qualitative analysis of various volatiles in the *A. arguta* resource fruit samples, the drift times and RIs in the IMS were compared to authentic controls. Subsequently, we identified 97 signal peaks (including monomers and dimers) from the two-dimensional profiles, and 76 volatile aroma substances were initially identified, as shown in Table 6. These contain 18 esters, 14 alcohols, 16 ketones, 12 aldehydes, seven terpenoids, three pyrazines, two furans, two acids and two other compounds, which essentially cover the range of aroma compounds found in fruits [19,51–53]. Nineteen of these substances, including methyl butanoate, isoamyl acetate, ethyl hexanoate, ethyl acetate, carveol, 1-hexanol, cineole, and 2-heptanone, formed dimers, which was related to the concentration of the volatile aroma substances and their proton affinity. The transfer of protons from reactants with higher proton affinity than water in highly concentrated substances to substances with higher proton affinity thus contributes to the formation of dimers [54].

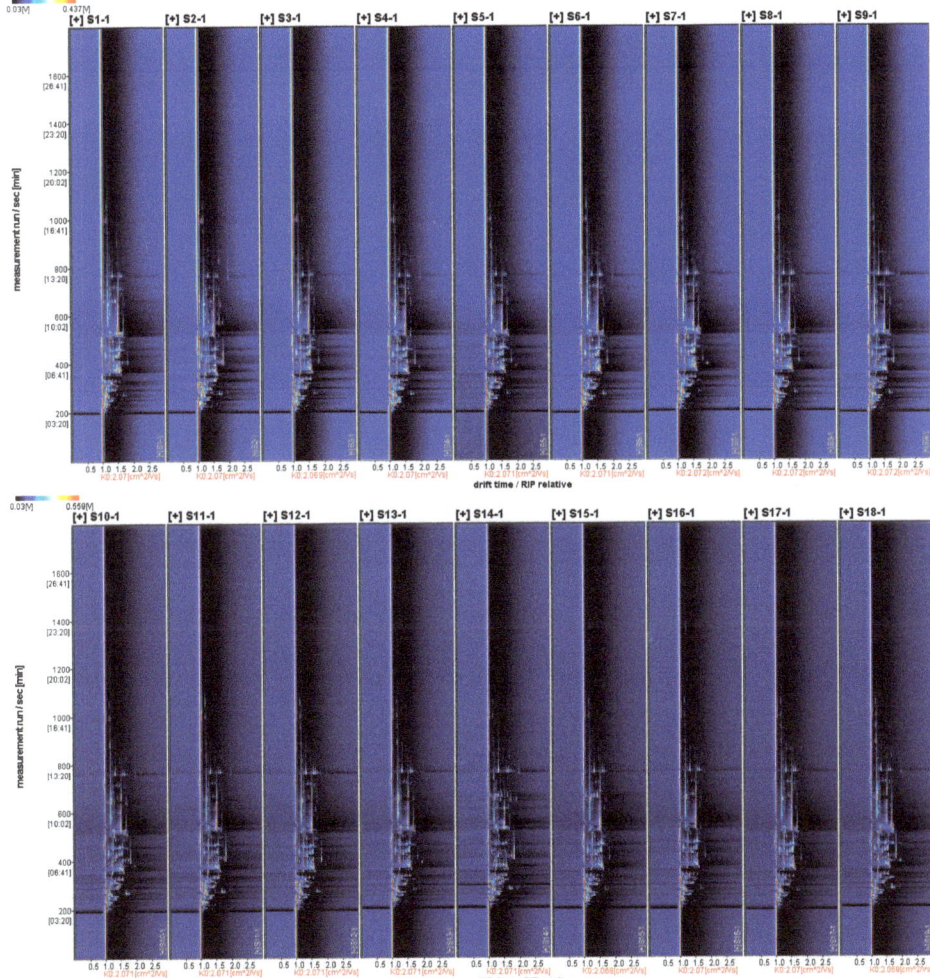

Figure 2. *Cont.*

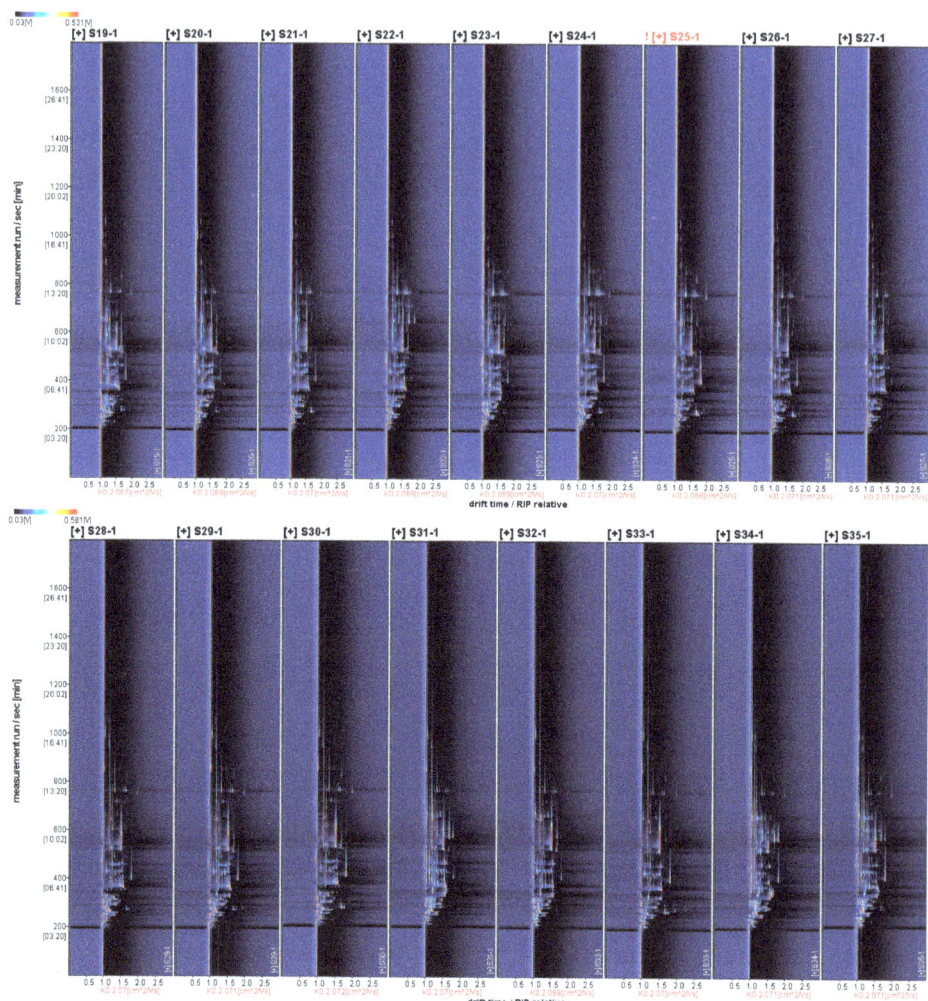

Figure 2. HS-GC-IMS 2D mapping (top view).

3.3.4. Fingerprint Analysis of Volatile Components of Fruits from Different *A. arguta* Resources

Although difference mapping shows overall differences in flammable substances in fruits from different *A. arguta* sources, fingerprinting can more accurately identify differences in the nature and concentration of individual substances. In fingerprint mapping, each row represents the overall signal peak of a sample, and each column represents the same substance in a different model. Color refers to the content of volatile substances; the brighter the color, the higher its content. As shown in Figure 4, the volatile compounds with high variability among the *A. arguta* resource fruit samples were methyl acetate, hexyl propanoate, hexyl acetate, ethyl hexanoate-D, ethyl isovalerate, butyl acetate-D, citronellyl formate, cineole-D, 2-heptanol, 2-octanone, 2-butanone, 3,5-dimethyl-1,2-cyclopentanedione, butanal, isovaleraldehyde, (Z)-4-heptenal, myrcene, and 2-methoxy-3-methylpyrazine.

3.4. Analysis of the Relative Content of Volatile Components

3.4.1. Esters

Ester compounds are the most diverse compounds detected in each resource (Figure 5), which mainly reflect fruity and floral aromas [55]; among the ester compounds detected, methyl butanoate, ethyl acetate, butyl acetate, and ethyl hexanoate have apple and pineapple aromas, ethyl butyrate has a floral aroma, and isoamyl acetate has a sweet aroma. The relative content of esters in 35 resource fruits was 2142.40–6065.74 ppb, accounting for 12.91–30.22% of the total volatiles, of which the relative content of esters in S34 was the highest. The content of ethyl propanoate was the highest among the ester compounds detected in the 35 resource fruits. It best reflected the fruity flavor of *A. arguta* fruits.

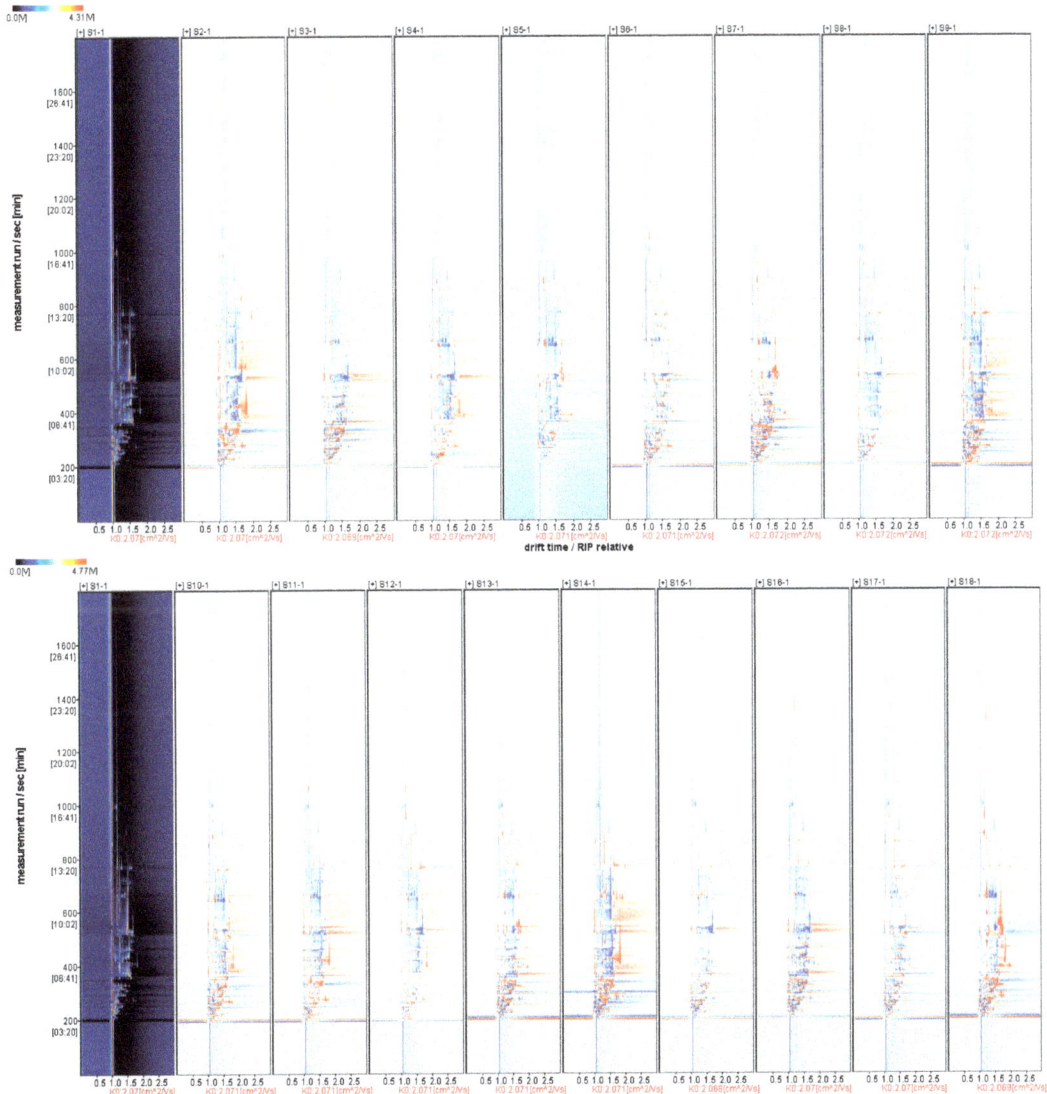

Figure 3. *Cont.*

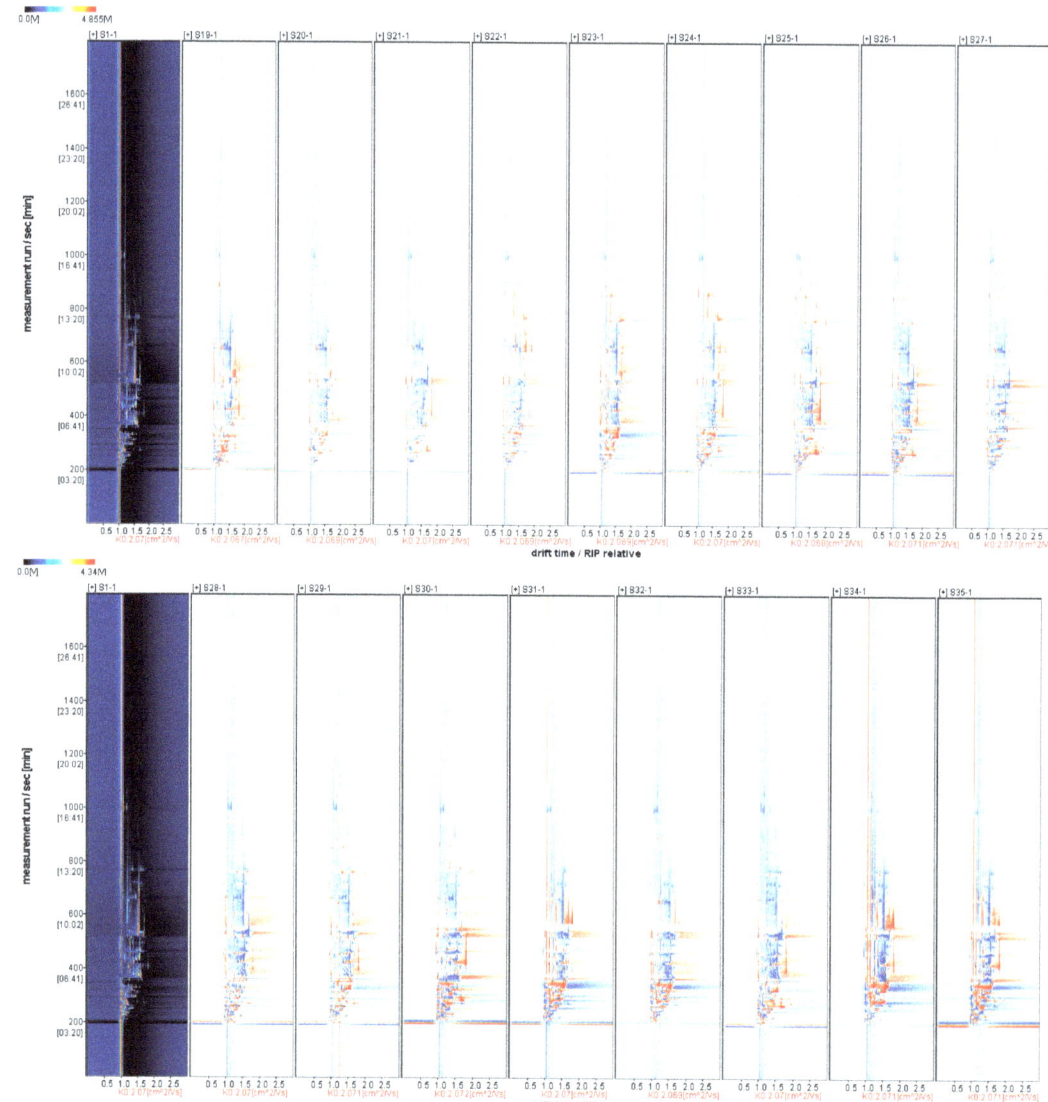

Figure 3. HS-GC-IMS difference comparison mode spectra.

Table 6. Identification of volatile compounds in fruits of different *A. arguta* resources using HS-GC-IMS.

Number	Count	Compound	CAS#	Formula	MW	RI	Rt [sec]	Dt [a.u.]	Comment
1		Methyl butanoate M	623-42-7	C5H10O2	102.1	1018.9	306.187	1.14902	Monomer
2		Methyl butanoate D	623-42-7	C5H10O2	102.1	1010.8	300.593	1.43148	Dimer
3		Methyl acetate	79-20-9	C3H6O2	74.1	890	242.237	1.19625	
4		Isoamyl acetate M	123-92-2	C7H14O2	130.2	1146.5	422.748	1.31005	Monomer
5		Isoamyl acetate D	123-92-2	C7H14O2	130.2	1141.9	417.108	1.75368	Dimer
6		Hexyl propanoate	2445-76-3	C9H18O2	158.2	1300.6	663.746	1.42868	
7		Hexyl acetate	142-92-7	C8H16O2	144.2	1298.1	660.409	1.38933	
8		Ethyl (E)-hex-2-enoate	27829-72-7	C8H14O2	142.2	1044.1	324.256	1.31395	
9		Ethyl propionate M	105-37-3	C5H10O2	102.1	966.4	276.19	1.14517	Monomer
10		Ethyl propionate D	105-37-3	C5H10O2	102.1	984.3	284.817	1.45669	Dimer
11		Ethyl hexanoate M	123-66-0	C8H16O2	144.2	1256.9	585.898	1.34038	Monomer
12		Ethyl hexanoate D	123-66-0	C8H16O2	144.2	1248.9	571.997	1.80357	Dimer
13	Esters	Ethyl formate	109-94-4	C3H6O2	74.1	854.4	227.914	1.0705	
14		Ethyl butyrate	105-54-4	C6H12O2	116.2	1053.1	331.029	1.55657	
15		Ethyl acetate M	141-78-6	C4H8O2	88.1	919.2	254.721	1.10585	Monomer
16		Ethyl acetate D	141-78-6	C4H8O2	88.1	918	254.194	1.33838	Dimer
17		Ethyl isovalerate	108-64-5	C7H14O2	130.2	1077	349.558	1.65689	
18		Butyl propionate	590-01-2	C7H14O2	130.2	1174.4	458.567	1.71886	
19		Butyl acetate M	123-86-4	C6H12O2	116.2	1034.3	317.103	1.23496	Monomer
20		Butyl acetate D	123-86-4	C6H12O2	116.2	1035.3	317.832	1.61627	Dimer
21		Butyl acrylate	141-32-2	C7H12O2	128.2	887	240.999	1.26357	
22		Butyl isovalerate	109-19-3	C9H18O2	158.2	1011.2	300.863	1.3947	
23		1-Methoxy-2-propyl acetate	108-65-6	C6H12O3	132.2	857.5	229.122	1.14191	
24		Citronellyl formate	105-85-1	C11H20O2	184.3	1288.5	643.76	1.8982	
25		Ethanol	64-17-5	C2H6O	46.1	984.1	284.691	1.04754	
26		Cis-2-Penten-1-ol	1576-95-0	C5H10O	86.1	1342.4	721.899	0.94816	
27		1-Penten-3-ol	616-25-1	C5H10O	86.1	1176.3	461.09	0.94578	
28		Isobutanol	78-83-1	C4H10O	74.1	1149.3	426.234	1.36406	
29	Alcohols	Carveol M	99-48-9	C10H16O	152.2	1242.2	560.754	1.29522	
30		Carveol D	99-48-9	C10H16O	152.2	1237.4	552.68	1.68177	
31		3-Methyl-1-butanol	123-51-3	C5H12O	88.1	1223.3	529.961	1.49475	Monomer
32		1-Butanol	71-36-3	C4H10O	74.1	1160.7	440.596	1.18265	Dimer
33		Cyclooctanol	696-71-9	C8H16O	128.2	1164.6	445.668	1.12941	

Table 6. Cont.

Number	Count	Compound	CAS#	Formula	MW	RI	Rt [sec]	Dt [a.u.]	Comment
34		2-Methyl-1-butanol	137-32-6	C5H12O	88.1	1180.1	466.173	1.47668	
35		1-Pentanol	71-41-0	C5H12O	88.1	1272.9	614.561	1.25548	
36		1-Hexanol M	111-27-3	C6H14O	102.2	1375.3	771.107	1.32787	Monomer
37		1-Hexanol D	111-27-3	C6H14O	102.2	1373	767.501	1.64025	Dimer
38		1-Hexanol T	111-27-3	C6H14O	102.2	1367.9	759.689	1.98315	Trimer
39		Cineole M	470-82-6	C10H18O	154.3	1216.4	519.23	1.29225	Monomer
40		Cineole D	470-82-6	C10H18O	154.3	1216.7	519.575	1.72287	Dimer
41		Leaf alcohol	928-96-1	C6H12O	100.2	1383.9	784.497	1.23283	
42		2-Heptanol	543-49-7	C7H16O	116.2	1292.5	651.413	1.71865	
43		2-Octanone	111-13-7	C8H16O	128.2	1304.1	668.411	1.33533	
44		L(-)-Carvone	6485-40-1	C10H14O	150.2	1137	411.188	1.81159	
45		Isomenthone	491-07-6	C10H18O	154.3	1178.9	464.569	1.34028	
46		2-Hexanone	591-78-6	C6H12O	100.2	1064.4	339.629	1.50148	
47		2-Heptanone M	110-43-0	C7H14O	114.2	1194.2	485.826	1.25783	Monomer
48		2-Heptanone D	110-43-0	C7H14O	114.2	1201.1	495.975	1.63226	Dimer
49		Cyclohexanone	108-94-1	C6H10O	98.1	1300.3	663.412	1.15313	
50		2-Butanone M	78-93-3	C4H8O	72.1	894.9	244.296	1.06226	Monomer
51		2-Butanone D	78-93-3	C4H8O	72.1	937.1	262.631	1.2478	Dimer
52	Ketones	5-Methyl-3-heptanone M	541-85-5	C8H16O	128.2	942.3	265.002	1.27861	Monomer
53		5-Methyl-3-heptanone D	541-85-5	C8H16O	128.2	961.6	273.911	1.68433	Dimer
54		Methyl isobutenyl ketone	141-79-7	C6H10O	98.1	1155.1	433.411	1.44875	
55		3-Hydroxy-2-butanone	513-86-0	C4H8O2	88.1	1307.8	673.432	1.05977	
56		3-Hepten-2-one	1119-44-4	C7H12O	112.2	932.2	260.463	1.2265	
57		3,5-Dimethyl-1,2-cyclopentanedione	13494-07-0	C7H10O2	126.2	1066.3	341.109	1.61079	
58		3,4-Dimethyl-1,2-cyclopentanedione	13494-06-9	C7H10O2	126.2	1093.2	362.744	1.62262	
59		1-Penten-3-one	1629-58-9	C5H8O	84.1	1058.9	335.428	1.0793	
60		Hydroxyacetone	116-09-6	C3H6O2	74.1	1277.9	623.753	1.04359	
61		2-Pentanone	107-87-9	C5H10O	86.1	951.4	269.186	1.37493	
62		Hexanal M	66-25-1	C6H12O	100.2	1118.7	389.828	1.25902	Monomer
63		Hexanal D	66-25-1	C6H12O	100.2	1094.5	363.792	1.56769	Dimer
64		Heptanal M	111-71-7	C7H14O	114.2	1202.9	498.672	1.33033	Monomer

Table 6. Cont.

Number	Count	Compound	CAS#	Formula	MW	RI	Rt [sec]	Dt [a.u.]	Comment
65		Heptanal D	111-71-7	C7H14O	114.2	1202.9	498.672	1.69473	Dimer
66		Butanal M	123-72-8	C4H8O	72.1	878.1	237.336	1.11738	Monomer
67		Butanal D	123-72-8	C4H8O	72.1	867	232.889	1.2832	Dimer
68	Aldehydes	Benzaldehyde	100-52-7	C7H6O	106.1	1531.1	1053.979	1.15444	
69		Isovaleraldehyde	590-86-3	C5H10O	86.1	938.6	263.336	1.40951	
70		trans-2-Pentenal	1576-87-0	C5H8O	84.1	1150	427.068	1.10704	
71		2-Methylbutyraldehyde	96-17-3	C5H10O	86.1	875.4	236.261	1.1511	
72		Isobutyraldehyde M	78-84-2	C4H8O	72.1	817.6	213.951	1.09932	Monomer
73		Isobutyraldehyde D	78-84-2	C4H8O	72.1	852.8	227.247	1.28367	Dimer
74		(Z)-4-Heptenal	6728-31-0	C7H12O	112.2	1300.2	663.227	1.61962	
75		trans-2-Pentenal	1576-87-0	C5H8O	84.1	1112	382.209	1.36162	
76		trans-2-Hexena M	6728-26-3	C6H10O	98.1	1251.7	576.747	1.1827	Monomer
77		trans-2-Hexenal D	6728-26-3	C6H10O	98.1	1224.3	531.583	1.51357	Dimer
78		Propionaldehyde	123-38-6	C3H6O	58.1	826.2	217.111	1.04325	
79		Dipentene M	138-86-3	C10H16	136.2	1210.7	510.409	1.21981	Monomer
80		Dipentene D	138-86-3	C10H16	136.2	1215.6	517.85	1.72287	Dimer
81		Camphene	79-92-5	C10H16	136.2	1080.1	352.008	1.20989	
82	Terpenes	β-Pinene	127-91-3	C10H16	136.2	1134.7	408.475	1.21824	
83		Myrcene	123-35-3	C10H16	136.2	1190.1	480.08	1.21772	
84		alpha-Pinene	80-56-8	C10H16	136.2	1033.8	316.769	1.22179	
85		α-Phellandrene	99-83-2	C10H16	136.2	1174.6	458.757	1.21952	
86		Terpinolene	586-62-9	C10H16	136.2	1292.5	651.428	1.21948	
87		Acetic acid M	64-19-7	C2H4O2	60.1	1504.8	999.756	1.05441	Monomer
88	Acids	Acetic acid D	64-19-7	C2H4O2	60.1	1505	1000.243	1.15277	Dimer
89		Isovaleric acid	503-74-2	C5H10O2	102.1	863.4	231.439	1.21454	
90		2-Methoxy-3-methylpyrazine	2847-30-5	C6H8N2O	124.1	985	285.138	1.57071	
91	Pyrazines	2,3,5-Trimethylpyrazine	14667-55-1	C7H10N2	122.2	1445.7	887.951	1.17114	
92		2-Ethyl-3-methylpyrazine	15707-23-0	C7H10N2	122.2	1337.9	715.399	1.59816	
93	Furans	2,5-Dimethylfuran	625-86-5	C6H8O	96.1	930.2	259.546	1.02742	
94		2-Pentylfuran	3777-69-3	C9H14O	138.2	1228.3	537.902	1.24624	
95	Other compounds	Toluene	108-88-3	C7H8	92.1	1033.1	316.26	1.02501	
96		2,4,6-Collidine	108-75-8	C8H11N	121.2	1374.1	769.181	1.5841	

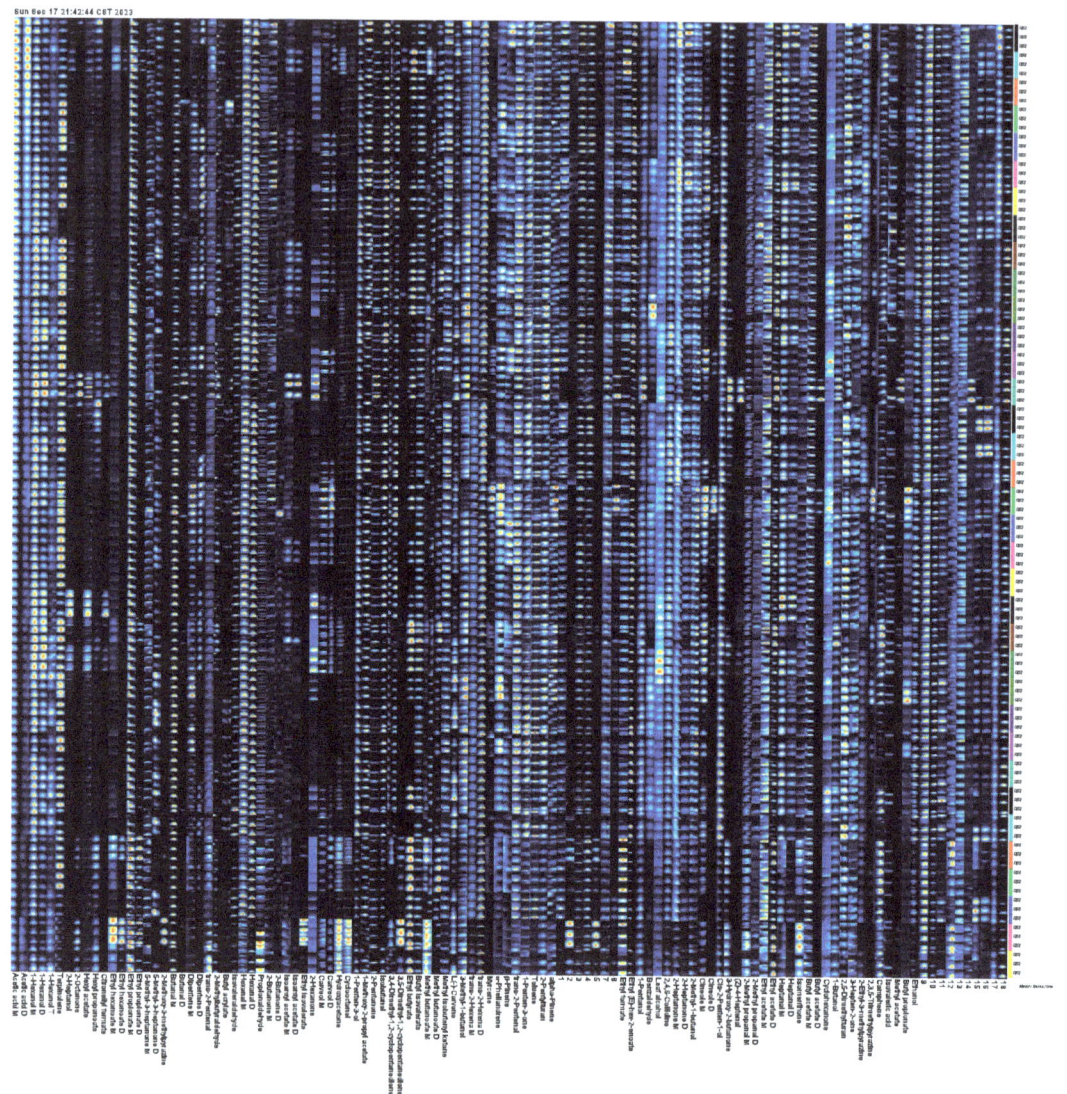

Figure 4. Fingerprints of volatile compounds in fruits of different *A. arguta* resources.

3.4.2. Alcohols

The percentage of alcohols was 8.78–21.45% (Figure 5), and their aroma was mainly grassy and alcoholic. The highest relative content of alcohols was S18, with 4420.72 ppb, followed by S24, with 3126.88 ppb, and the lowest relative content was S33, with 1520.96 ppb. Thirty-five *A. arguta* resource fruits were detected with a higher content of isobutanol and 1-hexanol among the alcohols, which best reflected the grassy aroma of *A. arguta* fruits.

3.4.3. Ketones

The content of ketones was 1581.99–6614.19 ppb, accounting for 8.50–32.95% of the total volatile compounds (Figure 5). The resource with the highest content was S34, and the lowest was S27. The ketones detected in the fruits of 35 *A. arguta* resources were more elevated in 2-heptanone and hydroxyacetone, with 2-heptanone having a banana aroma and slight medicinal flavor.

3.4.4. Aldehydes

Aldehydes were the compounds with the highest relative content detected in the 34 samples except S34, which was similar to the results of Sun Yang [14] et al. at 3480.11–11746.16 ppb, with the highest resource being S6 and the lowest S34, and the content of aldehydes in each sample accounted for 17.34–58.38% of the total volatiles (Figure 5). The highest relative content of aldehydes detected in the fruits of the resources was trans-2-hexenal, which was mainly characterized by grassy, apple, and aldehydic aromas.

3.4.5. Other Compounds

Compounds such as terpenoids, acids, pyrazines, and furans were also detected in the fruits of the *A. arguta* resource, all in low relative amounts, accounting for 2.22–8.51%, 0.65–2.27%, 0.26–1.86%, and 0.64–2.00% of the total volatile compounds, respectively (Figure 5).

3.5. Principal Component Analysis of Fruit Samples from Different A. arguta Resources

In order to better present and differentiate between fruit samples from different *A. arguta* resources, volatile compounds identified by HS-GC-IMS were analyzed by PCA. Unsupervised multidimensional statistics (PCA) were used to determine the samples to distinguish the magnitude of variation among different sample groups, subgroups, and within-group samples of fruits from various *A. arguta* resources. The contribution rate of PC1 was 29.2%, and that of PC2 was 13.1%, with the 35 groups of samples showing a clear tendency to segregate on the two-dimensional plots, and the magnitude of variation of the samples within the groups was obvious. The principal component results (Figure 6) showed significant overall differences in the aroma substances of the 35 groups of samples and differentiated them. As shown in Figure 6, the magnitude of intra-group variation was more significant for S14, S23, and S15, and the distance of the aroma characteristics of S14, S34, S35, S31, S32, S2, and S33 was farther away from each other, indicating that there were significant differences in the aroma characteristics among the different samples.

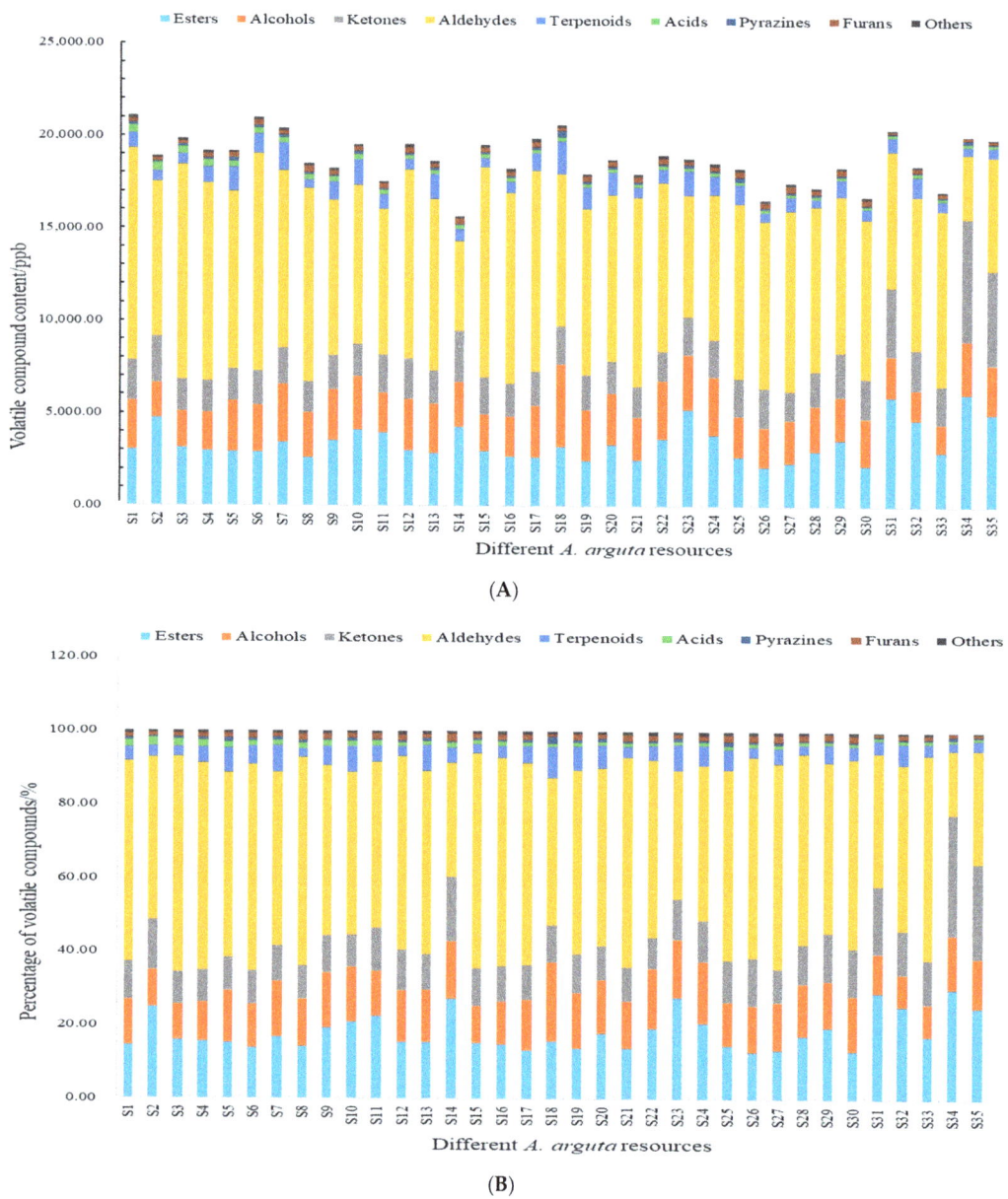

Figure 5. Content (**A**) and percentage (**B**) of volatile compounds in different *A. arguta* resources.

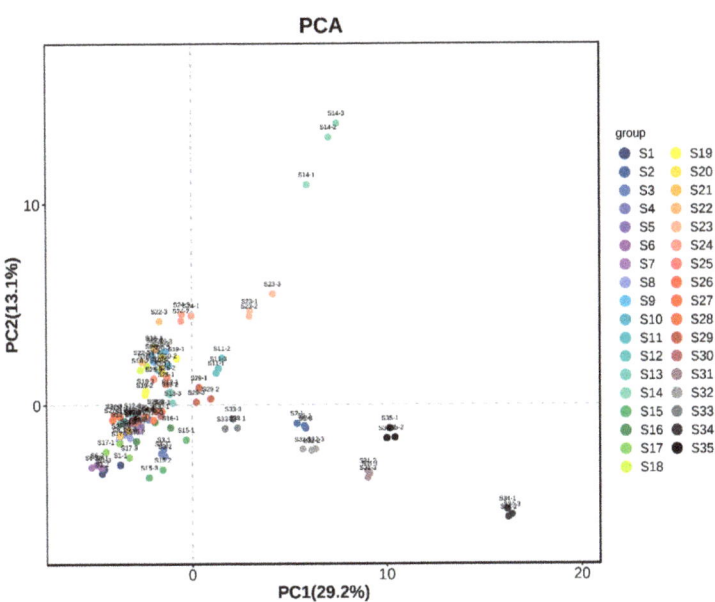

Figure 6. Principal component analysis of fruit samples from different *A. arguta* resources.

3.6. OPLS-DA Analysis and the Model Validation of Volatile Aroma Compounds of A. arguta Resource Fruits

OPLS-DA is a supervised discriminant statistical method that not only realizes the identification of sample differences but also obtains the characteristic markers of sample differences [56]. The contribution of each variable to the aroma of *A. arguta* was further quantified based on the variable importance (VIP) in the OPLA-DA model, and the volatile aroma compounds with VIP values greater than 1 were screened as the main characteristic volatile markers [57]. With 76 volatile aroma substances as dependent variables and different *A. arguta* resources as independent variables, effective differentiation of *A. arguta* fruit samples from 35 resources could be achieved by OPLS-DA (Figure 7A). The fit index (RX2) for the independent variable in this analysis was 0.987, the fit index (RY2) for the dependent variable was 0.793, and the model prediction index (Q2) was 0.554, with R2 and Q2 exceeding 0.5 to indicate acceptable model fit results [58]. After 200 replacement tests, as shown in Figure 7B, the intersection of the Q2 regression line with the vertical axis was less than 0, indicating that there was no overfitting of the model and validating the model, and it was considered that the results could be used for the identification and analysis of volatile aroma compounds in the fruits of different *A. arguta* resources.

The aroma quality of soft date kiwifruit fruit depends on the result of the joint action of several volatile aroma compounds; according to the criteria of $p < 0.05$ and $VIP > 1$, 33 kinds of *A. arguta* resource fruit volatile aroma substances were screened out as the main aroma substances (Figure 8), among which there are eight kinds of esters, five kinds of alcohols, six kinds of ketones, six kinds of aldehydes, two kinds of acids, three kinds of terpenoids, one kind of furan, and two kinds of other compounds.

3.7. OAV Analysis of the Main Aroma Components of Fruit Samples from Different A. arguta Resources

Although HS-GC-IMS characterized and quantified the volatile aroma substances of *A. arguta* resource fruits and OPLS-DA can screen potential characteristic volatile markers of volatile aroma substances of *A. arguta* resource fruits, the level of volatile aroma substance content does not determine the aroma contribution of each substance. Consumers usually judge the acceptability of food by aroma and flavor [59]. The odor activity of volatile compounds in *A. arguta* fruits is one of the main sensory characteristics that determine the quality of the fruit. OAV can reflect the contribution of individual volatile aroma compounds to the characteristic flavor of the sample. The OAV of volatile aroma compounds depends on their concentration and odor threshold. Based on previous studies, it was shown that volatile aroma compounds with OAV>1 contributed more to the overall aroma of the samples, and the larger the OAV value, the greater the contribution of the compound [60]. In this study, the volatile aroma compounds screened by OPLS-DA with VIP values greater than 1 were analyzed for OAV, and a total of 18 volatile aroma compounds with OAV > 1 were detected according to the calculation (Supplementary File S2), among which six types of esters were esters, namely methyl butanoate, isoamyl acetate, hexyl propanoate, butyl acrylate, butyl isovalerate and 1-methoxy-2-propyl acetate; three types of alcohols, namely 3-methyl-1-butanol, 1-hexanol, and leaf alcohol; three types of ketones, namely l(-)-Carvone, 5-methyl-3-heptanone, and 3,4-dimethyl-1,2-cyclopentanedione; three types of aldehydes, namely heptanal, butanal, and isovaleraldehyde; and three types of terpenes, namely dipentene, alpha-pinene, and terpinolene. Although the OAV values of the 35 *A. arguta* samples varied, in comparison, isoamyl acetate, 3-methyl-1-butanol, 1-hexanol, and butanal had higher OAV values than the other compounds, ranging from 183.09 to 1175.54, 10.19 to 6.98, 33.55 to 126.40, and 30.42 to 90.93, respectively, suggesting that the contribution of these four volatile compounds to the overall kiwifruit aroma was greater. Isoamyl acetate had a fruity, sweet, and floral aroma; 3-methyl-1-butanol had an alcoholic and fruity aroma; and 1-hexanol had a grassy, fruity, sweet, and alcoholic aroma, which are essential aromatic characteristics in *A. arguta* fruits.

3.8. Heat Map Analysis, PCA Analysis and Correlation Analysis of Volatile Aroma Compounds with OAV > 1 in Fruits of Different A. arguta Resources

Concentrations of aroma substances with OAV greater than 1 in volatile compounds from 35 *A. arguta* resource fruit samples were clustered using hierarchical analysis, and similarity was calculated using Pearson. Based on the heat map analysis of the samples (Figure 9), the red color indicates the high expression of the volatile aroma compound in the embodiment, and the blue color indicates the low expression of the volatile aroma compound in the selection, which can clearly show the differences between the concentrations of each substance in different *A. arguta* resources.

Volatile aroma compounds with OAV values greater than 1 were analyzed in the PCA of the fruits of *A. arguta* resources (Figure 10). The contribution of PC1 was 20.7% and the contribution of PC2 was 13.6%. The PCA scatters of most of the samples were dispersed, indicating that the similarity between these samples was low. Few samples are distributed in the second quadrant, only S5, S31, and S32, and the distribution is more dispersed. The scatters of the samples distributed in the center of the axes are more clustered, indicating higher similarity between them.

A significant correlation between substances is indicated by a correlation coefficient between 0.8 and 1.0, a strong correlation is indicated by a correlation coefficient between 0.6 and 0.8, a moderate correlation is indicated by a correlation coefficient between 0.4 and 0.6, a weak correlation is indicated by a correlation coefficient between 0.2 and 0.4, and correlation coefficients between 0 and 0.2 indicate that there is no correlation between the substances or that the correlation is very weak. As can be seen in Figure 11, there is a highly significant correlation between 1-hexanol and leaf alcohol and a strong correlation between 1-methoxy-2-propyl acetate and heptanal, alpha-pinene, and terpinolene. Moderate correlations were found between methyl butanoate and hexyl propanoate, 1-methoxy-2-propyl acetate and isovaleraldehyde, 1-hexanol and isovaleraldehyde, and dipentene and alpha-pinene. A strong negative correlation was found between methyl butyrate and heptanal.

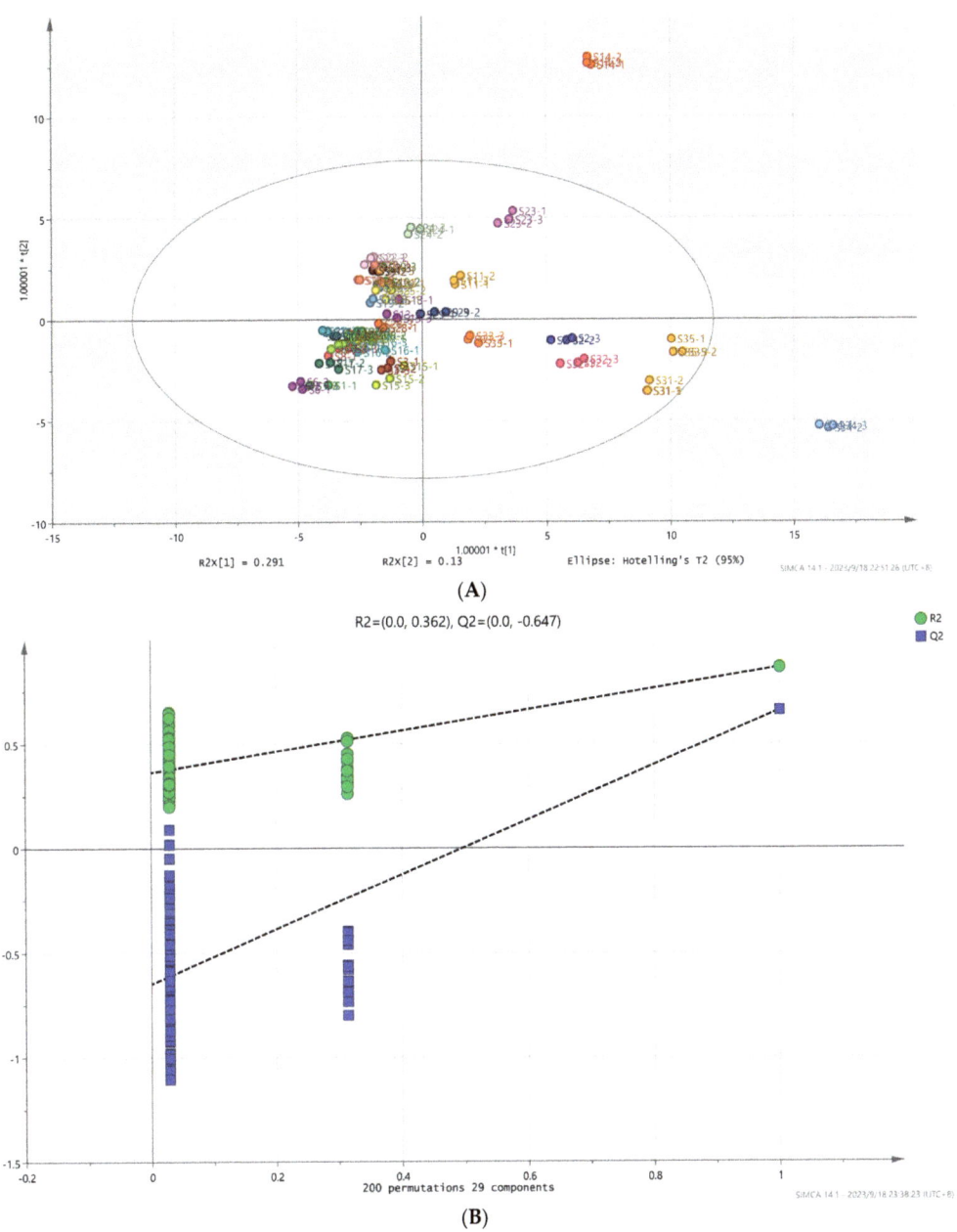

Figure 7. OPLS-DA of volatile aroma compounds in fruits of different *A. arguta* resources (**A**) and model cross-validation results (**B**).

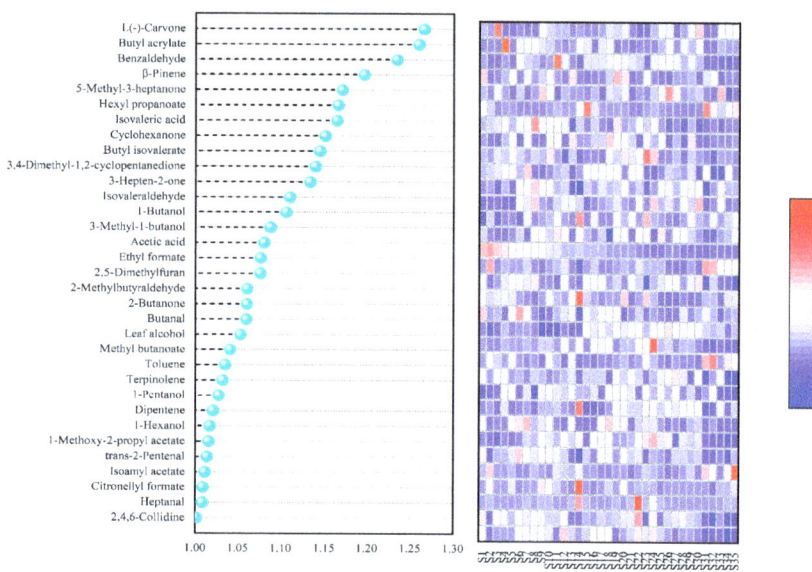

Figure 8. OPLS-DA analysis of VIP values of major volatile aroma substances in fruits of different *A. arguta* resources.

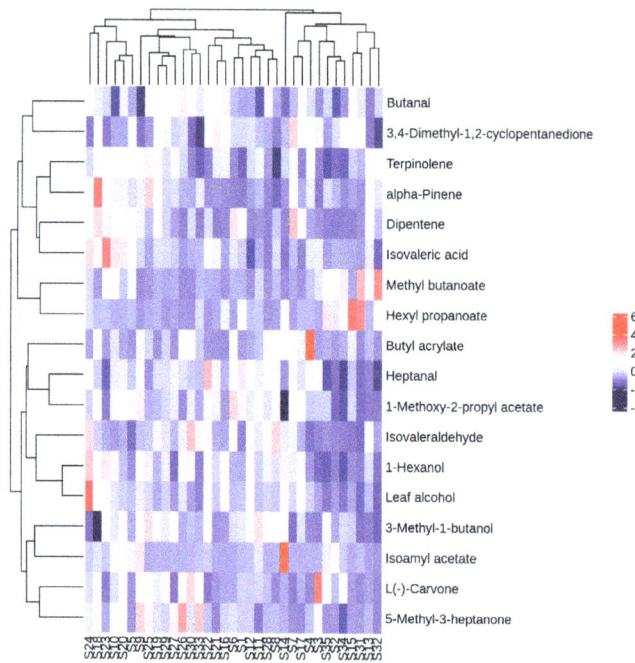

Figure 9. Clustering heat map analysis of volatile aroma compounds with OAV greater than 1 in fruits of different *A. arguta* resources.

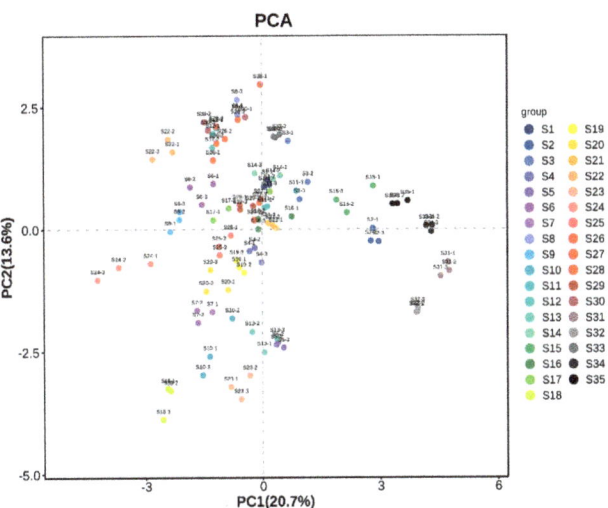

Figure 10. Scatter plot of PCA analysis of volatile aroma compounds with OAV greater than 1 in fruits of different *A. arguta* resources.

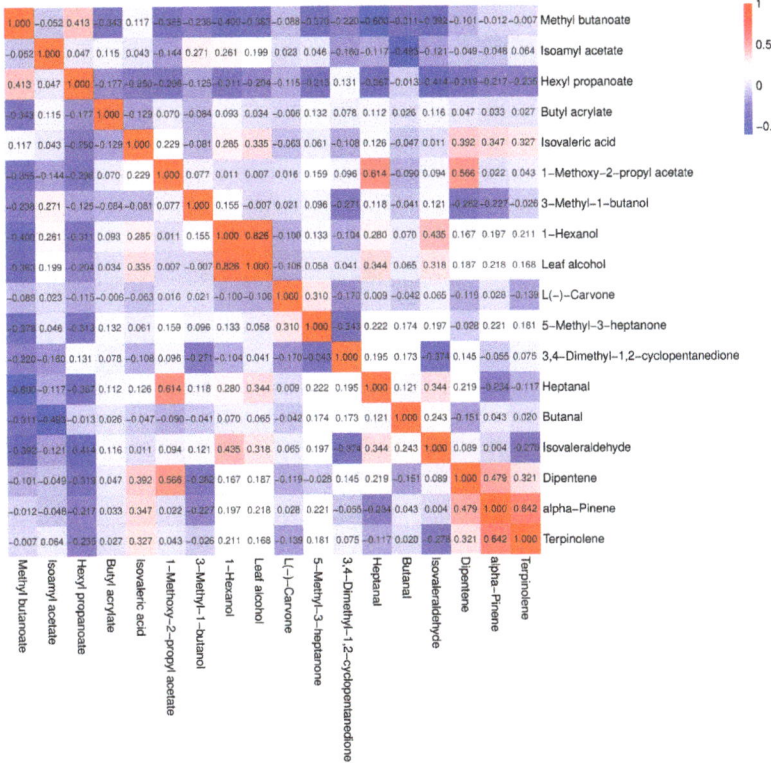

Figure 11. Correlation analysis of volatile aroma compounds with OAV greater than 1 in fruits of different *A. arguta* resources.

4. Conclusions

Actinidia arguta, a type of kiwifruit, has good organoleptic quality and rich nutritional value. Therefore, it is important to study its flavor quality and volatile aroma components. This study used 35 A. arguta resource fruits as materials to measure and analyze their soluble sugar, titratable acid, and sugar–acid ratio. The results showed that the soluble sugar content of 35 *A. arguta* resource fruits was 2.94–13.97%, the content of titratable acid was 0.32–1.59%, and the sugar–acid ratio was 2.45–28.50. In contrast, S12 had a higher sugar–acid ratio with a higher titratable acid content and a higher sugar content, which indicated a superior fruit flavor compared to its source. High-performance liquid chromatography (HPLC) was used to determine the content of organic acids. The results showed that the 35 fruits could be classified into two types: citric-acid-dominant and quinic-acid-dominant. Lactic acid was also detected in some of the fruits.

Headspace gas chromatography–ion mobility spectrometry (HS-GC-IMS) was used to analyze the volatile aroma substances of different *A. arguta* resources, and a total of 76 volatile aroma substances were identified, which contained 18 esters, 14 alcohols, 16 ketones, 12 aldehydes, seven terpenes, three pyrazines, two furans, two acids, and two other compounds, and these compounds basically covered the types of aroma compounds in the fruit. With 76 volatile aroma substances as the dependent variables and different soft date kiwifruit resources as the independent variables, 33 volatile aroma substances with VIP > 1 were screened out as the main aroma substances of *A. arguta* resource fruits by OPLS-DA analysis. The volatile aroma compounds screened by OPLS-DA with VIP values greater than 1 were subjected to OAV analysis, and 18 volatile aroma compounds with OAV>1 were screened based on the calculation of their odor activity values, including six esters, three alcohols, three ketones, three aldehydes, and three terpenoids. Comparison of the OAV values revealed that isoamyl acetate, 3-methyl-1-butanol, 1-hexanol, and butanal had higher OAV values than the other compounds, indicating that these four volatile compounds were the main contributors to the overall aroma of *A. arguta*. Headspace gas chromatography–ion mobility spectrometry can show the commonalities and differences between the samples, which makes up for the perceived inadequacy of sensory evaluation and plays a useful and complementary role in the evaluation of the flavor quality of *A. arguta*. This provides a theoretical basis for screening *A. arguta* resources with excellent flavor quality, enhancing and improving the flavor quality of *A. arguta* processed products, and at the same time, provides a theoretical basis for the scientific understanding of the characteristic compounds of fruit aroma of different *A. arguta*. However, the IMS database is not complete enough, which prevents some compounds isolated by GC from being characterized. Therefore, the gradual enrichment of the IMS database is an important development direction for the detection of volatile aroma compounds in the future. At the same time, it is necessary to further combine the nutritional quality and volatile flavor quality to establish a more detailed evaluation system of *A. arguta* quality to lay a theoretical foundation for the development of excellent *A. arguta* resources.

Supplementary Materials: The following supporting information can be downloaded at: https://www.mdpi.com/article/10.3390/foods12193615/s1, Supplementary File S1: Relative content; Supplementary File S2: OAV values.

Author Contributions: Data curation, Y.H., J.W. and P.Y.; Formal analysis, W.L.; Funding acquisition, C.L.; Investigation, B.S. and S.F.; Methodology, Y.H., H.Q., J.W., Y.Y., W.C. and Y.S.; Project administration, W.L.; Resources, H.Q. and S.F.; Software, Y.H., J.W., W.C. and Y.Y.; Supervision, W.L. and C.L.; Validation, C.L.; Writing—original draft, Y.H.; Writing—review & editing, Y.H. and H.Q. All authors have read and agreed to the published version of the manuscript.

Funding: This work was supported by the by Innovation Project of Chinese Academy of Agricultural Sciences (CAAS-ASTIP-2023-SAPS), Changchun City Science and Technology Development Plan Project (21ZGN09).

Institutional Review Board Statement: Not applicable.

Informed Consent Statement: Not applicable.

Data Availability Statement: All related data and methods are presented in this paper. Additional inquiries should be addressed to the corresponding author.

Conflicts of Interest: The authors declare that they have no competing interests. Compliance with ethics requirements: This article does not contain any studies with human or animal subjects.

References

1. He, Y.L.; Qin, H.Y.; Wen, J.L.; Fan, S.T.; Yang, Y.M.; Zhang, B.X.; Cao, W.Y.; Lu, W.P.; Li, C.Y. Quality analysis and comprehensive evaluation of 35 *Actinidia argute* resources. *J. Fruit. Sci.* **2023**, *40*, 1523–1533. [CrossRef]
2. Piao, Y.L.; Zhao, L.H. Research progress of *Actinidia arguta*. *North. Horticul.* **2008**, *3*, 76–78. Available online: https://kns.cnki.net/kcms2/article/abstract?v=3uoqIhG8C44YLTlOAiTRKgchrJ08w1e7VSL-HJEdEx0PaWcAxm13ZOvEUl25Qn4T8Qm_Vpi0vVHHXSakzk01lbGEvMVwUeUQ&uniplatform=NZKPT (accessed on 19 September 2023).
3. Zhang, M.; Wang, H.X.; Lou, X.; Zhao, L.N.; Ming, D.X. The development status and breeding trend of hardy kiwifruit cultivars in the world. *J. Ecol.* **2017**, *36*, 3289–3297. [CrossRef]
4. Latocha, P. The Nutritional and Health Benefits of Kiwiberry (*Actinidia arguta*)—A Review. *Plant Food Hum. Nutr.* **2017**, *72*, 325–334. [CrossRef] [PubMed]
5. Niu, Q.; Shen, J.; Liu, Y.; Nie, C.Y.; Skripchenko, N.V.; Liu, D.J. Research progress on main active constituents and pharmacological activities of *Actinidia arguta*. *Sci. Technol. Food Ind.* **2019**, *40*, 333–338, 344. [CrossRef]
6. Almeida, D.; Pinto, D.; Santos, J.; Vinha, A.F.; Palmeira, J.; Ferreira, H.N.; Rodrigues, F.; Oliveira, M.B.P.P. Hardy kiwifruit leaves (*Actinidia arguta*): An extraordinary source of value-added compounds for food industry. *Food Chem.* **2018**, *259*, 113–121. [CrossRef] [PubMed]
7. Gong, D.S.; Sharma, K.; Kang, K.W.; Kim, D.W.; Oak, M.H. Endothelium-Dependent Relaxation Effects of *Actinidia arguta* Extracts in Coronary Artery: Involvement of eNOS/Akt Pathway. *J. Nanosci Nanotechnol.* **2020**, *20*, 5381–5384. [CrossRef] [PubMed]
8. Heo, K.H.; Sun, X.; Shim, D.W.; Kim, M.K.; Koppula, S.; Yu, S.H.; Kim, H.B.; Kim, T.J.; Kang, T.B.; Lee, K.H. *Actinidia arguta* extract attenuates inflammasome activation: Potential involvement in NLRP3 ubiquitination. *J. Ethnopharmacol.* **2018**, *213*, 159–165. [CrossRef]
9. Choi, J.J.; Park, B.; Kim, D.H.; Pyo, M.Y.; Choi, S.; Son, M.; Jin, M. Blockade of atopic dermatitis-like skin lesions by DA-9102, a natural medicine isolated from *Actinidia arguta*, in the Mg-deficiency induced dermatitis model of hairless rats. *Exp. Biol. Med.* **2008**, *233*, 1026–1034. [CrossRef]
10. Liu, H.; An, K.; Su, S.; Yu, Y.; Wu, J.; Xiao, G.; Xu, Y. Aromatic Characterization of Mangoes (*Mangifera indica* L.) Using Solid Phase Extraction Coupled with Gas Chromatography-Mass Spectrometry and Olfactometry and Sensory Analyses. *Foods* **2020**, *9*, 75. [CrossRef]
11. Liu, H.; Yu, Y.; Zou, B.; Yu, Y.; Yang, J.; Xu, Y.; Chen, X.; Yang, F. Evaluation of Dynamic Changes and Regularity of Volatile Flavor Compounds for Different Green Plum (*Prunus mume* Sieb. et Zucc) Varieties during the Ripening Process by HS-GC-IMS with PLS-DA. *Foods* **2023**, *12*, 551. [CrossRef] [PubMed]
12. Xu, L.H.; Lai, J.P.; Kuo, J.Q. Research Progress on Factors Influencing Citrus Fruit Aroma and Quality. *J. Gannan Normal Univ.* **2019**, *40*, 85–88. [CrossRef]
13. Huang, X.; Ni, Z.; Shi, T.; Tao, R.; Yang, Q.; Luo, C.; Li, Y.; Li, H.; Gao, H.; Zhou, X.; et al. Novel insights into the dissemination route of Japanese apricot (*Prunus mume* Sieb. et Zucc.) based on genomics. *Plant J.* **2022**, *110*, 1182–1197. [CrossRef] [PubMed]
14. Sun, Y.; Ci, Z.J.; Liu, Z.P.; Lu, L.Y.; Liu, G.P.; Chen, X.Z.; Li, R.H.; You, W.Z. Analysis on fruit quality and aroma components of different *Actinidia arguta* cultivars. *Chin. Fruit Trees* **2021**, *5*, 52–55, 60. [CrossRef]
15. Xin, G.; Zhang, B.; Feng, F.; Li, T.C.; Liu, C.J.; Xu, J.G. Analysis of Aromatic Constituents of *Actinidia arguta* Sieb. *Food Sci.* **2009**, *30*, 230–232.
16. Zhang, B.X.; Qin, H.Y.; Wang, Y.L.; Li, J.Q.; Li, C.Y.; Zhao, Y.; Fan, S.T. Analysis of the aroma composition of dry wine made from different varieties of *Actinidia arguta*. *Spec. Res.* **2023**, *45*, 112–117, 125. [CrossRef]
17. An, K.; Liu, H.; Fu, M.; Qian, M.C.; Yu, Y.; Wu, J.; Xiao, G.; Xu, Y. Identification of the cooked off-flavor in heat-sterilized lychee (*Litchi chinensis* Sonn.) juice by means of molecular sensory science. *Food Chem.* **2019**, *301*, 125282. [CrossRef]
18. Dursun, A.; Caliskan, O.; Guler, Z.; Bayazit, S.; Turkmen, D.; Gunduz, K. Effect of harvest maturity on volatile compounds profiling and eating quality of hawthorn (*Crataegus azarolus* L.) fruit. *Sci. Hortic-amsterdam.* **2021**, *288*. [CrossRef]
19. Tian, Z.; Zhang, M.; Wang, Y.; Liu, Y.P.; Yue, T.L.; Huo, Y.J. Aroma fingerprint characterization of different kiwifruit juices based on headspace-gas chromatography-ion mobility spectrometry. *Food Ferment. Ind.* **2023**, *49*, 279–287. [CrossRef]
20. Li, C.; Xin, M.; Li, L.; He, X.; Yi, P.; Tang, Y.; Li, J.; Zheng, F.; Liu, G.; Sheng, J.; et al. Characterization of the aromatic profile of purple passion fruit (*Passiflora edulis* Sims) during ripening by HS-SPME-GC/MS and RNA sequencing. *Food Chem.* **2021**, *355*, 129685. [CrossRef]
21. Hernandez Mesa, M.; Ropartz, D.; Garcia Campana, A.M.; Rogniaux, H.; Dervilly Pinel, G.; Le Bizec, B. Ion Mobility Spectrometry in Food Analysis: Principles, Current Applications and Future Trends. *Molecules* **2019**, *24*, 2706. [CrossRef] [PubMed]
22. Zhang, L.; Shuai, Q.; Li, P.; Zhang, Q.; Ma, F.; Zhang, W.; Ding, X. Ion mobility spectrometry fingerprints: A rapid detection technology for adulteration of sesame oil. *Food Chem.* **2016**, *192*, 60–66. [CrossRef] [PubMed]

23. Li, M.; Yang, R.; Zhang, H.; Wang, S.; Chen, D.; Lin, S. Development of a flavor fingerprint by HS-GC-IMS with PCA for volatile compounds of Tricholoma matsutake Singer. *Food Chem.* **2019**, *290*, 32–39. [CrossRef] [PubMed]
24. Gerhardt, N.; Birkenmeier, M.; Sanders, D.; Rohn, S.; Weller, P. Resolution-optimized headspace gas chromatography-ion mobility spectrometry (HS-GC-IMS) for non-targeted olive oil profiling. *Anal. Bioanal. Chem.* **2017**, *409*, 3933–3942. [CrossRef] [PubMed]
25. Arce, L.; Gallegos, J.; Garrido-Delgado, R.; Medina, L.M.; Sielemann, S.; Wortelmann, T. Ion Mobility Spectrometry a Versatile Analytical Tool for Metabolomics Applications in Food Science. *Curr. Metabol.* **2014**, *2*, 264–271. [CrossRef]
26. Hernandez Mesa, M.; Escourrou, A.; Monteau, F.; Le Bizec, B.; Dervilly Pinel, G. Current applications and perspectives of ion mobility spectrometry to answer chemical food safety issues. *Trac-Trends Anal. Chem.* **2017**, *94*, 39–53. [CrossRef]
27. Zheng, X.; Smith, R.D.; Baker, E.S. Recent advances in lipid separations and structural elucidation using mass spectrometry combined with ion mobility spectrometry, ion-molecule reactions and fragmentation approaches. *Curr. Opin. Chem. Biol.* **2018**, *42*, 111–118. [CrossRef]
28. Sun, X.; Gu, D.; Fu, Q.; Gao, L.; Shi, C.; Zhang, R.; Qiao, X. Content variations in compositions and volatile component in jujube fruits during the blacking process. *Food Sci. Nutr.* **2019**, *7*, 1387–1395. [CrossRef]
29. Wang, Q.; Chen, X.; Zhang, C.; Li, X.; Yue, N.; Shao, H.; Wang, J.; Jin, F. Discrimination and Characterization of Volatile Flavor Compounds in Fresh Oriental Melon after Forchlorfenuron Application Using Electronic Nose (E-Nose) and Headspace-Gas Chromatography-Ion Mobility Spectrometry (HS-GC-IMS). *Foods* **2023**, *12*, 1272. [CrossRef]
30. Cao, W.; Shu, N.; Wen, J.; Yang, Y.; Jin, Y.; Lu, W. Characterization of the Key Aroma Volatile Compounds in Nine Different Grape Varieties Wine by Headspace Gas Chromatography-Ion Mobility Spectrometry (HS-GC-IMS), Odor Activity Values (OAV) and Sensory Analysis. *Foods* **2022**, *11*, 2767. [CrossRef]
31. Cavanna, D.; Zanardi, S.; Dall'Asta, C.; Suman, M. Ion mobility spectrometry coupled to gas chromatography: A rapid tool to assess eggs freshness. *Food Chem.* **2019**, *271*, 691–696. [CrossRef] [PubMed]
32. Wang, X.; Yang, S.; He, J.; Chen, L.; Zhang, J.; Jin, Y.; Zhou, J.; Zhang, Y. A green triple-locked strategy based on volatile-compound imaging, chemometrics, and markers to discriminate winter honey and sapium honey using headspace gas chromatography-ion mobility spectrometry. *Food Res. Int.* **2019**, *119*, 960–967. [CrossRef]
33. Song, M.T. Evaluation of *Actinidia arguta* Germplasm Based on the Nutrition, Taste and Storge Quality. Master's Thesis, Shenyang Agricultural University, Shenyang, China, 2022. (In Chinese).
34. He, X.; Yangming, H.; Gorska-Horczyczak, E.; Wierzbicka, A.; Jelen, H.H. Rapid analysis of Baijiu volatile compounds fingerprint for their aroma and regional origin authenticity assessment. *Food Chem.* **2021**, *337*, 128002. [CrossRef] [PubMed]
35. Song, X.; Wang, G.; Zhu, L.; Zheng, F.; Ji, J.; Sun, J.; Li, H.; Huang, M.; Zhao, Q.; Zhao, M. Comparison of two cooked vegetable aroma compounds, dimethyl disulfide and methional, in Chinese Baijiu by a sensory-guided approach and chemometrics. *LWT-Food Sci. Technol.* **2021**, *146*, 111427. [CrossRef]
36. Jiao, Y.; Ye, T.; Zhang, J.J.; Guo, X.X.; Luo, G.H. Comprehensive Quality Evaluation of Nostoc commune Vauch. from Gansu Province by Principal Component Analysis and Cluster Analysis. *Food Sci.* **2019**, *40*, 130–135.
37. Pollo, B.J.; Teixeira, C.A.; Belinato, J.R.; Furlan, M.F.; de Matos Cunha, I.C.; Vaz, C.R.; Volpato, G.V.; Augusto, F. Chemometrics, Comprehensive Two-Dimensional gas chromatography and "omics" sciences: Basic tools and recent applications. *TrAC-Trends Anal. Chem.* **2021**, *134*, 116111. [CrossRef]
38. Rocchetti, G.; O'Callaghan, T.F. Application of metabolomics to assess milk quality and traceability. *Curr. Opin. Food Sci.* **2021**, *40*, 168–178. [CrossRef]
39. Shu, N. Study on Fermentation Characteristics and Dry Red Wine Brewing Technology of New Grape New Vitis Amurensis Cultivar 'Beiguohong'. Master's Thesis, Chinese Academy of Agricultural Sciences, Beijing, China, 2019. (In Chinese).
40. Huo, J.Y.; Liu, J.; Feng, H.; Wang, Y.G. Reviews on flavor quality of Tomato. *China Veget.* **2005**, *2*, 38–40. [CrossRef]
41. Zhou, X.Y.; Zhu, C.H.; Li, J.X.; Gao, J.Y.; Gong, Q.; Shen, Z.S.2; Yue, J.Q. Advances in research on organic acid metabolism in fruits. *South. China Fruits* **2015**, *44*, 120–125, 132. [CrossRef]
42. Zhang, Y.P.; Yang, Y.J.; Yang, L.; Li, Y.; Gao, B.M.; Wang, X.J.; Dong, X.M. Hazard of Oxalate in Plants and its Control Manners. *Anhui Agric. Sci. Bull.* **2007**, *10*, 34–39, 92. [CrossRef]
43. Sha, S.F. Pear Organic Acid Components, Content Changes and Genetic Identification. Doctorate's Thesis, Nanjing Agricultural University, Nanjing, China, 2012. (In Chinese).
44. Rovio, S.; Siren, K.; Siren, H. Application of capillary electrophoresis to determine metal cations, anions, organic acids, and carbohydrates in some Pinot Noir red wines. *Food Chem.* **2011**, *124*, 1194–1200. [CrossRef]
45. Hulme, A.C. Quinic and shikimic acids in fruits. *Qual. Plant. Mater. Veget.* **1958**, *3/4*, 468–473. [CrossRef]
46. Zhao, Y.; Zhu, H.Y.; Yang, D.D.; Gong, C.S.; Liu, W.G. Research Progress of Citric Acid Metabolism in the Fruit. *Acta Hortic. Sin.* **2022**, *49*, 2579–2596. [CrossRef]
47. Fan, L.J. Study on Regulating Effect and Mechanism of Asoorbic Acid on Postharvest Persimmon Fruit Softness. Master's Thesis, Guangxi University, Nanning, China, 2016. (In Chinese).
48. Li, M.; Du, H.; Lin, S. Flavor Changes of Tricholoma matsutake Singer under Different Processing Conditions by Using HS-GC-IMS. *Foods* **2021**, *10*, 531. [CrossRef]
49. Zhu, Y.; Lv, H.-P.; Shao, C.-Y.; Kang, S.; Zhang, Y.; Guo, L.; Dai, W.-D.; Tan, J.-F.; Peng, Q.-H.; Lin, Z. Identification of key odorants responsible for chestnut-like aroma quality of green teas. *Food Res. Int.* **2018**, *108*, 74–82. [CrossRef] [PubMed]

50. Chen, Y.; Li, P.; Liao, L.; Qin, Y.; Jiang, L.; Liu, Y. Characteristic fingerprints and volatile flavor compound variations in Liuyang Douchi during fermentation via HS-GC-IMS and HS-SPME-GC-MS. *Food Chem.* **2021**, *361*, 130055. [CrossRef] [PubMed]
51. Du, X.; Rouseff, R. Aroma Active Volatiles in Four Southern Highbush Blueberry Cultivars Determined by Gas Chromatography-Olfactometry (GC-O) and Gas Chromatography-Mass Spectrometry (GC-MS). *J. Agric. Food Chem.* **2014**, *62*, 4537–4543. [CrossRef]
52. Zhu, J.; Wang, L.; Xiao, Z.; Niu, Y. Characterization of the key aroma compounds in mulberry fruits by application of gas chromatography-olfactometry (GC-O), odor activity value (OAV), gas chromatography-mass spectrometry (GC-MS) and flame photometric detection (FPD). *Food Chem.* **2018**, *245*, 775–785. [CrossRef]
53. Guerra Ramirez, D.; Gonzalez Garcia, K.E.; Medrano Hernandez, J.M.; Famiani, F.; Cruz Castillo, J.G. Antioxidants in processed fruit, essential oil, and seed oils of feijoa. *Not. Bot. Horti Agrobot. Cluj-Napoca* **2021**, *49*, 11988. [CrossRef]
54. Wang, S.; Chen, H.; Sun, B. Recent progress in food flavor analysis using gas chromatography-ion mobility spectrometry (GC-IMS). *Food Chem.* **2020**, *315*, 126158. [CrossRef]
55. Wang, J.H.; Ye, X.Y.; Mu, Y.; Ma, L.Z.; Qian, Y.; Ge, Y. Comparative analysis of characteristic aroma components of 3 kinds of representative Actinidia chinensis Planch species in Guizhou. *J. Food Safe Qual.* **2022**, *13*, 6190–6197. [CrossRef]
56. Li, Y.; Zhang, Y.; Peng, Z.; Tan, B.; Lin, H. The difference of quality components of Fuzhuan tea and Qian-liang tea based on the orthogonal partial least squares discriminant analysis model. *J. Food Safe Qual.* **2017**, *8*, 4382–4387.
57. Jin, W.; Zhao, P.; Liu, J.; Geng, J.; Chen, X.; Pei, J.; Jiang, P. Volatile flavor components analysis of giant salamander (*Andrias davidiauns*) meat during roasting process based on gas chromatography-ion mobility spectroscopy and chemometrics. *Food Ferment. Ind.* **2021**, *47*, 231–239. [CrossRef]
58. Yun, J.; Cui, C.; Zhang, S.; Zhu, J.; Peng, C.; Cai, H.; Yang, X.; Hou, R. Use of headspace GC/MS combined with chemometric analysis to identify the geographic origins of black tea. *Food Chem.* **2021**, *360*, 130033. [CrossRef]
59. Heting, Q.; Shenghua, D.; Zhaoping, P.; Xiang, L.; Fuhua, F. Characteristic Volatile Fingerprints and Odor Activity Values in Different Citrus-Tea by HS-GC-IMS and HS-SPME-GC-MS. *Molecules* **2020**, *25*, 6027.
60. Miyazaki, T.; Plotto, A.; Baldwin, E.A.; Reyes-De-Corcuera, J.I.; Gmitter, F.G., Jr. Aroma characterization of tangerine hybrids by gas-chromatography-olfactometry and sensory evaluation. *J. Sci. Food Agric.* **2012**, *92*, 727–735. [CrossRef] [PubMed]

Disclaimer/Publisher's Note: The statements, opinions and data contained in all publications are solely those of the individual author(s) and contributor(s) and not of MDPI and/or the editor(s). MDPI and/or the editor(s) disclaim responsibility for any injury to people or property resulting from any ideas, methods, instructions or products referred to in the content.

Article

Dynamic Changes in Flavor and Microbiota in Traditionally Fermented Bamboo Shoots (*Chimonobambusa szechuanensis* (Rendle) Keng f.)

Zhijian Long [1,2], Shilin Zhao [1], Xiaofeng Xu [1], Wanning Du [1], Qiyang Chen [1] and Shanglian Hu [1,2,*]

[1] School of Life Science and Engineering, Southwest University of Science and Technology, Mianyang 621010, China; long20053182@swust.edu.cn (Z.L.); z740506287@126.com (S.Z.); 18788981570m@sina.cn (X.X.); dwncie@sina.com (W.D.); chenqiyang@swust.edu.cn (Q.C.)
[2] Engineering Research Center for Biomass Resource Utilization and Modification of Sichuan Province, Mianyang 621010, China
* Correspondence: hushanglian@swust.edu.cn

Abstract: Dissecting flavor formation and microbial succession during traditional fermentation help to promote standardized and large-scale production in the sour shoot industry. The principal objective of the present research is to elucidate the interplay between the physicochemical attributes, flavor, and microbial compositions of sour bamboo shoots in the process of fermentation. The findings obtained from the principal component analysis (PCA) indicated notable fluctuations in both the physicochemical parameters and flavor components throughout the 28 day fermentation process. At least 13 volatile compounds (OAV > 1) have been detected as characteristic aroma compounds in sour bamboo shoots. Among these, 2,4-dimethyl Benzaldehyde exhibits the highest OAV (129.73~668.84) and is likely the primary contributor to the sour odor of the bamboo shoots. The analysis of the microbial community in sour bamboo shoots revealed that the most abundant phyla were Firmicutes and Proteobacteria, while the most prevalent genera were *Enterococcus*, *Lactococcus*, and *Serratia*. The results of the correlation analysis revealed that Firmicutes exhibited a positive correlation with various chemical compounds, including 3,6-nonylidene-1-ol, 2,4-dimethyl benzaldehyde, silanediol, dimethyl-, nonanal, and 2,2,4-trimethyl-1,3-pentylenediol diisobutyrate. Similarly, *Lactococcus* was found to be positively correlated with several chemical compounds, such as dimethyl-silanediol, 1-heptanol, 3,6-nonylidene-1-ol, nonanal, 2,2,4-trimethyl-1,3-pentanediol diisobutyrate, dibutyl phthalate, and TA. This study provides a theoretical basis for the standardization of traditional natural fermented sour bamboo production technology, which will help to further improve the flavor and quality of sour bamboo.

Keywords: sour bamboo shoot; microbial compositions; traditional fermentation; volatile flavor compounds; correlation analysis

Citation: Long, Z.; Zhao, S.; Xu, X.; Du, W.; Chen, Q.; Hu, S. Dynamic Changes in Flavor and Microbiota in Traditionally Fermented Bamboo Shoots (*Chimonobambusa szechuanensis* (Rendle) Keng f.). *Foods* **2023**, *12*, 3035. https://doi.org/10.3390/foods12163035

Academic Editor: Thomas Dippong

Received: 7 July 2023
Revised: 8 August 2023
Accepted: 10 August 2023
Published: 12 August 2023

Copyright: © 2023 by the authors. Licensee MDPI, Basel, Switzerland. This article is an open access article distributed under the terms and conditions of the Creative Commons Attribution (CC BY) license (https://creativecommons.org/licenses/by/4.0/).

1. Introduction

Bamboo shoots, the young culms of bamboo plants, are rich in protein, fiber, vitamins, minerals, and phytosterols [1]. Because bamboo shoots are highly moist and lignify quickly, they are hard to store for a long time. The fermentation of bamboo shoots into sour bamboo shoots makes them palatable in flavor, aroma, texture, and appearance. Additionally, fermentation promotes the diversity of bamboo shoot products and extends their shelf life [2].

In China and southeast Asia, fermented bamboo shoots have become increasingly popular due to their unique taste and texture [3]. The critical step in processing sour bamboo shoots is to ferment them by soaking them in cold boiling water or mountain spring water for a certain period (15–30 days) [1]. Under the influence of microorganisms, bamboo shoots gradually turn sour, and the dominant flora and flavor-related flora are diverse among different fermented bamboo shoot varieties. Among sour bamboo shoots in Guangdong, *Pichia*, *Candida*, and *Debaryomyces* dominate, while *Pichia* and *Zygosaccharomyces* dominate in Yunnan [3]. The main genera found in Yunnan sour bamboo shoots (*Dendrocalamus latiflorus* Munro) are *Lactococcus* and *Lactobacillus* [4]. The fermentation of Guanxi sour bamboo shoots (*Dendrocalamus latiflorus* Munro) began with *Weissella*, then *Lactobacillus* increased continuously, finally becoming the dominant bacterial species [5]. When fresh Ma bamboo shoots are fermented for 30 days, *Lactobacillus*, Clostridium_sensu_stricto_1, *Enterobacter*, and *Leuconostoc* exhibit a significant association with sour bamboo shoot flavor. As a characteristic flavor compound of sour bamboo shoots, p-cresol is highly correlated with *Clostridium* [6]. Li, et al. [7] have found that *Lactobacillus*, *Lactococcus*, *Weissella*, and *Cyanobacteria* dominate the bacterial community in Ma bamboo shoots after 35 days of fermentation. According to the research findings of Xia, et al. [8], several types of halophilic bacteria were seen to be the predominant strains during the initial three days of fermentation whereas *Lactococcus lactis* and *Weissella* sp. were seen to take over as the dominant strains after seven days of fermentation. These studies enhance our understanding of the succession of microbial communities throughout the fermentation process. However, there is still a lack of clarity when it comes to identifying the specific compounds that contribute to the notably unpleasant odor of sour bamboo shoots, as well as the microorganisms that are responsible for their fermentation. Additionally, the diverse microbial community present in sour bamboo shoots yields a complex and variable flavor composition [3]. To achieve a comprehensive understanding of fermentation processes and ensure the consistent quality of fermented products, it is essential to investigate the succession and function of the microbes responsible for the formation of characteristic flavor compounds.

Consequently, the current investigation involved acquiring freshly harvested bamboo shoots (*Chimonobambusa szechuanensis* (Rendle) Keng f.) sourced from the southwestern region of China. Following this, the shoots were subjected to a 28 day fermentation process to prepare sour bamboo shoots. During the sour bamboo shoot fermentation process, flavor compounds (organic acids and volatiles), physicochemical properties (pH, titratable acidity, nitrite, and reducing sugar content), and texture properties were analyzed at 1, 7, 14, 21, and 28 days. In addition, high-throughput sequencing of 16S rRNA was used to study microbial communities and observe their dynamic changes. As part of this study, we aim to gain a better understanding of the fermentation mechanisms underlying the production of sour bamboo shoots by examining the associations between physicochemical properties, dominant microbes, and major flavor compounds, and by providing a theoretical basis for the standardization of traditional production methods.

2. Materials and Methods

2.1. Experimental Materials

Fresh bamboo shoots (*Chimonobambusa szechuanensis* (Rendle) Keng f.) were collected from Leshan (Sichuan, China) in October 2022, and delivered to the laboratory premises within 24 h. The bamboo shoots were then carefully cleaned by removing their shells and washing them in clean water. Ultimately, the central edible component of the bamboo shoots was retained for further processing.

2.2. Fermentation Process

Before the bamboo shoots were used, they were washed and drained. The bamboo shoots (shelled) were sliced into julienne strips that were 13 cm long, 0.5 cm wide, and 0.5 cm thick. After placing the strips in a plastic bottle filled with sterile water (450 mL), the bottle was sealed and left to ferment at room temperature (22 ± 2 °C) for 28 days. Preparation and monitoring of 15 bottles were conducted. On days 1, 7, 14, 21, and 28, samples were taken randomly from three unopened bottles each time. In addition to the immediate analysis of the bamboo shoots' texture, fermentation liquids were stored at −80 °C for further analysis.

2.3. Texture Analysis

A texture analyzer (TA-XT. Plus, Stable Micro Systems Ltd., Godalming, UK) was used to measure changes in hardness, friability, and chewiness of sour bamboo shoots during fermentation. A P36R probe was used to test texture properties on consistently sized sour bamboo shoots (30 mm × 5 mm × 5 mm). During the experiment, the parameters were set on a pre-test speed of 2 mm/s, an initial test rate of 1 mm/s, an initial post-test rate of 1 mm/s, a compression rate of 70%, a pause of 5 s, and a trigger force of 5 g.

2.4. Determination of pH and Total Acid (TA)

A pH meter (PHS-25, Greifensee, Switzerland) was used to measure the pH value. TA measurements were conducted following GB/T 12456-2021 [9]. To determine TA, 10 mL of fermentation liquid was titrated with 0.1 N NaOH and 0.5 mL of phenolphthalein was added as an acid–base indicator.

2.5. Determination of Reducing Sugar Content

The reducing sugar content of the fermentation liquids was measured according to GB 5009.7-2021 [10].

2.6. Determination of Nitrite Content

Nitrite content in fermentation liquids was determined using ultraviolet spectrophotometry as described by GB 5009.33-2016 [11]. The nitrite concentration in the fermentation liquid was determined by adding sulfonamide and N-(1-Naphthyl) ethylenediamine dihydrochloride (NED) solutions. At room temperature, the mixtures were incubated for 15 min. Samples were then measured at 538 nm for absorbance. To calculate nitrite, a $NaNO_2$ standard curve was used (0.02–0.25 mg/L), and the results are expressed in mg/kg. A triplicate of each determination was performed.

2.7. Determination of Volatile Compounds

Headspace solid-phase microextraction technique and gas chromatography-mass spectrometry (HS-SPME-GC-MS/MS, Varian450GC, Varian Analytical Instruments, Harbour, CA, USA) were employed to identify the volatile compounds present in the fermentation liquids at various stages of fermentation. As per the methodology, adjusted slightly, of Chen, et al. [4], the sealed headspace vial housed 8 mL of sour bamboo shoot fermentation broth and 3 g of NaCl. The vial was subsequently immersed in a water bath set at a temperature of 60 °C for 10 min. The volatile components were then effectively collected through the use of divinylbenzene/carboxy/polydimethylsiloxane (/CAR/PDMS/DVB, 65 µm) fibers (Supelco Inc., Bellefonte, PA, USA), with the absorption process lasting 30 min. Lastly, the fibers containing the volatiles were subjected to desorption, which was performed on a heated GC jet at a temperature of 250 °C for 5 min. To quantify the volatile compounds, 100 µL n-hexane containing 38.12 µg cyclohexanone was added as an internal standard. Volatile compounds were analyzed using a GC-MS with a flexible capillary column (HP-5MS, 30 m × 0.25 mm × 0.25 µm). Helium gas was used as a carrier gas, with a constant flow rate of 1.0 mL/min and no split flow. The temperature profile for the GC oven was as follows: starting at 40 °C, it was increased at a rate of 12 °C/min to reach 100 °C, followed by a ramp of 5 °C/min to 120 °C. Thereafter, the temperature was increased to 180 °C at the rate of 8 °C/min and held for 5 min. The final temperatures were then reached, in increments of 12 °C/min to 210 °C and 8 °C/min to 250 °C. The mass spectrometer was operated at a source temperature of 230 °C, an ionization potential of 70 eV, and a mass scan range (m/z) of 40–400 amu. These parameters were consistently maintained throughout the experiment. We calculated the odor activity value (OAV) of each compound by dividing the concentration of volatile compounds by the threshold to evaluate the contribution of the compound to the formation of a sour bamboo aroma. Compounds with OAV $\geq$ 1 are believed to be primary contributors to the flavor of sour bamboo shoots [12].

2.8. Microbial Analysis

Genomic DNA from fermentation liquid was isolated by cetyltrimethylammonium bromide (CTAB). Nanodrop and agarose gel electrophoresis were used to determine the purity and concentration of DNA. The obtained DNA was subsequently amplified with 338F (5′-ACTCCTACGGGAGGCAGCA-3′) and 806R (5′-GGACTACHVGGGTWTCTAAT-3′) universal primer sets, targeting the V3–V4 regions of the 16S rRNA gene. Samples exhibiting a brightness reading between 400 and 550 bp were selected for sequencing. A 250 bp double-end sequencing read was obtained using Illumina platform following quality control [13]. By filtering the raw reads obtained from sequencing with Trimmatic v0.33, the results of the information analysis become more accurate and reliable. Subsequently, cutadapt 1.9.1 software was employed to identify and eliminate primer sequences to obtain clean reads devoid of primer sequences. The dada2 method in QIIME2 2020.6 was used for denoising [14,15]; here, the two-ended sequences must be concatenated while the chimeric sequences should be discarded to obtain the final valid data or non-chimeric reads. Sequences showing $\geq 97\%$ similarity were grouped into operational taxonomic units (OTUs). NMDS and PCoA can be used to examine differences in community structure across samples or groups. LDA effect size (LEfSe) was used to quantify the differences in community structure between grouped samples.

2.9. Statistical Analysis

A statistical analysis was performed using SPSS Statistics 25.0 (IBM, Chicago, IL, USA). One-way analysis of variance (ANOVA) with Tukey's honest significant difference (HSD) post-hoc test was utilized to determine significant differences among the groups. In addition, principal component analysis (PCA) and Pearson correlation were used to investigate the interaction between physiochemical characteristics, flavor, and microorganisms. All values are represented as mean $\pm$ standard deviation (SD) with $p < 0.05$ considered statistically significant.

3. Results and Discussion

3.1. Physiochemical Property Analysis

As shown in Table 1, sour bamboo shoots undergo fermentation by gradually lowering and stabilizing pH levels. After 14 days of fermentation, there was no significant variation in the pH value (4.92–4.81) of sour bamboo shoots, which is consistent with the reported trend of fermented bamboo shoots (*Dendrocalamus latiflorus* Munro) [16]. The acidic environment during fermentation may be attributed to the acid or other acidic compounds produced by *Bacillus* and *Lactobacillus*, as suggested by Jeyaram, et al. [17]. Such acidic conditions throughout the fermentation stage are advantageous in enhancing the shelf life and quality of sour bamboo shoots. The content of TA increased steadily from 2.50 ± 0.43 g/L to 8.75 ± 0.29 g/L (Table 1), consistent with that of Ma sour bamboo shoots [6]. The recorded nitrite measurements showed a marked increase from an initial level of 0.12 ± 0.01 mg/kg on day 1 to a peak level of 0.75 ± 0.08 mg/kg. Subsequently, there was a significant decrease, with measurements registering at 0.19 ± 0.03 mg/kg on day 28. A nitrate-reducing bacteria reduces nitrate during the early stages of fermentation, resulting in an increase of nitrite content. Nitrate-reducing bacteria are usually acid-intolerant microorganisms such as *Escherichia coli* and *Pseudomonas* spp. carried by bamboo shoots. The growth of these bacteria is gradually inhibited as fermentation proceeds, and the amount of nitrites decreases. Moreover, under acidic conditions, NO_2^- can combine with H^+ to form HNO_2, and then self-disproportionation occurs to produce nitrogen dioxide and nitric oxide, thus decreasing nitrite. [18]. In compliance with the national health standard level for nitrite (20 mg/kg) in China, the sour bamboo shoots featured in this study are safe for human consumption [19]. In addition, it was recorded that the maximum level of reducing sugar content occurring on day 14 was 1.18 ± 0.02 g/L, followed by a gradual reduction in the reducing sugar content, primarily owing to microbial consumption [6].

Table 1. Physicochemical properties of sour bamboo shoot liquids during the fermentation process.

	1 d	7 d	14 d	21 d	28 d
pH	5.73 ± 0.05 a	5.23 ± 0.14 b	4.92 ± 0.02 c	4.82 ± 0.04 c	4.81 ± 0.01 c
Titratable acidity (g lactic acid/L)	2.50 ± 0.43 e	3.75 ± 0.29 d	5.01 ± 0.29 c	7.5 ± 0.29 b	8.75 ± 0.29 a
Nitrite content (mg/kg)	0.12 ± 0.01 c	0.16 ± 0.01 bc	0.22 ± 0.04 b	0.74 ± 0.08 a	0.19 ± 0.03 bc
Reducing sugar (g glucose/L)	1.05 ± 0.05 b	1.08 ± 0.03 ab	1.18 ± 0.03 a	0.73 ± 0.06 c	0.63 ± 0.03 c

a–e Different lowercase letters in the same row indicate significant differences ($p < 0.05$).

The evaluation of quality in sour bamboo shoots is greatly influenced by their texture, particularly regarding their hardness, fracturability, and chewiness [20]. During the first 7 days of fermentation, there was a significant decrease in hardness, fracturability, and chewiness of fermentation, after which the decline was relatively gradual (Figure 1). A similar softening phenomenon also occurred in Sichuan pickles [21] and pickled chayote [22]. An analysis of physiochemical indexes using Pearson's correlation (Figure S1) revealed a positive correlation between pH and hardness, fracturability, and chewiness. The pH level was lowered, resulting in a decrease in the permeability of cell membranes, which will negatively affect textural characteristics [4]. Additionally, microorganisms involved in the fermentation process have the ability to generate pectinase and cellulose, both of which can break down cell walls and destroy the cellular structure, ultimately impacting the hardness, fracturability, and chewiness of the sample [23].

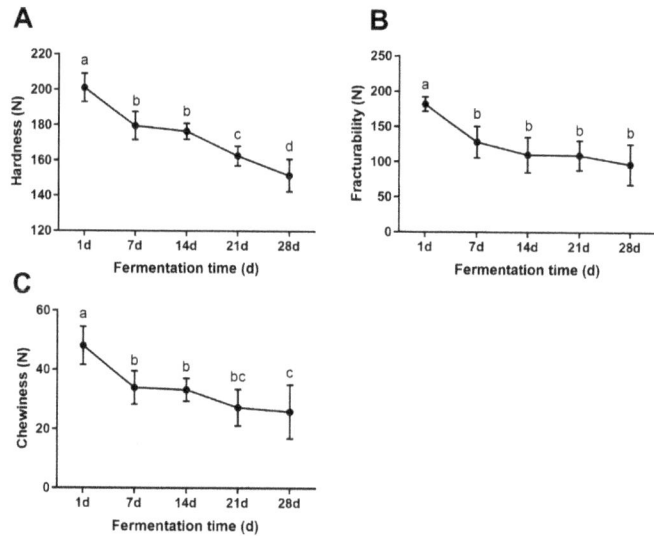

Figure 1. Texture properties of the sour bamboo shoot during the fermentation process. (A) hardness, (B) fracturability, (C) chewiness. Statistically significant differences are indicated by different lowercase letters at $p < 0.05$.

3.2. Volatile Compounds of Sour Bamboo Shoots

During the entire fermentation process of sour bamboo shoots, a total of 43 volatile compounds were detected (Figure S2), among which there were at most 17 types of alcohol volatile compounds, followed by ketones and esters (Figure 2A, and Table S1). In the fermented liquid of sour bamboo shoots, 1-octen-3-ol, 2,4-ditert-butylphenol, and D-limonene were the most dominant volatile compounds on the first day of fermentation. On the 7th day of fermentation, the content of volatile compounds sharply decreased, and some volatile compounds even disappeared, such as 1-octen-3-ol, linalool, 2-ethylhexanol, cyclooctanol, etc. As the fermentation time prolongs, the types and content of volatile substances gradually increase. On the 28th day of fermentation, 35 volatile compounds were detected, including 11 alcohols, 3 phenols, 6 ketones, 5 aldehydes, 6 esters, and 4 other compounds. Among these, 1-octen-3-ol, 2,4-ditert-butylphenol, and dimethylsilanediol had the highest content.

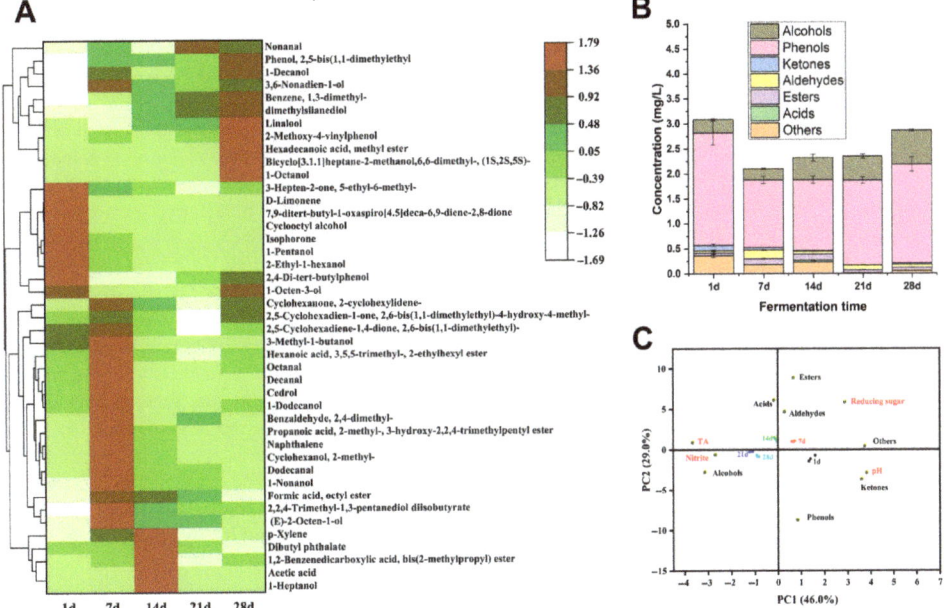

Figure 2. Relative contents of various volatile compounds (**A**), heatmap and dendrogram of the volatile compounds (**B**), and PCA analysis (**C**) in sour bamboo shoots during the 28 days of fermentation. Heatmap plots show the relative abundance of volatile compounds in samples (variables clustered vertically). Color intensity is proportional to volatile compound abundance.

Throughout the process of fermentation, it was observed that three distinct compounds (1-octen-3-ol, 2,4-di-tert-butylphenol, and D-limonene) and their concentration exhibited the most prominent factor (Figure 2B). 1-Octen-3-ol (also known as mushroom alcohol) has a mushroom/earthy aroma [24], 2,4-di-tert-butylphenol has an almond/spicy flavor [25], and D-limonene is a terpene with a citrus-like flavor [26]. 2,4-di-tert-butylphenol is a product obtained from *Lactococcus* sp. with antifungal and antioxidant properties and has demonstrated its potential as a food additive that can improve food safety and promote health [27]. There is no clear understanding of the precursor or pathway that leads to this compound, even though the *Lactobacillus* gene clusters and enzymes for the shikimate and mevalonate pathways have been identified [28]. The latest research shows that 2,4-di-tert-butylphenol is a candidate compound for anti-diabetes-related enzymes [29]. In this study, 2,4-di-tert-butylphenol first decreased and then increased during the fermentation process, which may be related to the dominance of *Lactococcus* in the later stage. It is worth noting that p-cresol is considered a characteristic component in sour bamboo shoots [20], but this substance was not detected in our study, which may be related to factors such as materials and fermentation conditions. The content of alcohol volatile compounds was second only to that of phenols. With the passage of fermentation time, new alcohol volatile compounds were found, including 1-heptanol, 1-decanol, 1-octanol, and dimethylsilanediol. These are produced by Strecker degradation of amino acids through the Ehrlich pathway and by linoleic acid degradation [30]. Ester volatile compounds can exhibit a unique fruit aroma [31], as their lower threshold has a significant impact on flavor. Esters may be generated by the esterification reaction between alcohols and organic acids in sour bamboo shoots [32]. Due to different metabolic types, there are significant differences in the types of volatile aroma compounds produced compared with other studies [4]. Ketone volatile matter has a high threshold value and low content, rendering it minimally impactful to the flavor of the shoots. Maillard reaction, alcohol oxidation, or other acids may generate it under the influence of microorganisms [33]. As fermentation progresses, the content of aldehydes volatile compounds diminishes, with 2,4-dimethyl benzaldehyde exhibiting the highest content of aldehydes, and with volatile acid compounds present in very low concentrations. The results of this study are similar to those of Xu et al. in their study of the fermentation of Jipo bamboo shoots [34].

As shown in Figure 2C, principal component analysis was conducted on the physicochemical properties and volatile compounds of sour bamboo shoots. The analysis showed that PC1 and PC2 accounted for 46.0% and 29.0% of the total variance of the data, with a cumulative value of 75.0%. The distance between the first day of fermentation and other fermentation days is relatively far, indicating significant changes in physicochemical properties and flavor throughout the entire fermentation time. The distance between the 21st and 28th days of fermentation is close, indicating that their flavors are relatively similar. The pH value has a significant impact on flavor on the first day of fermentation, with representative volatile compounds such as ketones and phenols. The typical volatile compounds of sour bamboo shoots on the 7th day of fermentation were aldehydes, esters and others, which were closely related to the content of reducing sugar. The representative volatile compounds on the 14th day of fermentation were acids and closely related to the total acid content. On the 21st and 28th days of fermentation, the representative volatile compounds were alcohols and closely related to nitrite content. Some gram-negative bacteria (such as Enterobacteria) can convert nitrate into nitrite, and the changes in nitrite may be related to the changes in bacteria in the fermentation environment [6].

3.3. Odor Activity Value of Sour Bamboo Shoots

When sour bamboo shoots ferment, we found at least thirteen volatile compounds that we considered to be characteristic volatile compounds, differing from the results obtained by Chen, et al. [4]. In addition, the volatile compounds of sour bamboo shoots included, for the first time, 2,4-dimethylbenzaldehyde and 2,4-di-tert-butylphenol (Table 2). These differences may be caused by factors such as raw material type, raw material growth environment, and fermentation conditions.

The highest OAV (129.73~668.84) of benzaldehyde, 2,4-dimethyl-, which presents an almond/spicy flavor, was considered to be the volatile compound that contributes most to the flavor formation, followed by 2,2,4-trimethyl-1,3-pentanediol diisobutyrate (59.55~197.30), 3,6-nonadien-1-ol (1.29~80.66) and dibutyl phthalate (32.22~96.32). In contrast, phenols have the highest content, but their contribution to flavor is limited because of the high threshold. For example, 2,4-ditert-butylphenol dominates the content, with a high odor threshold (200 µg/kg) and limited contribution to the flavor. Phenol volatile substances typically produce unpleasant and irritating odors [25]. The OAV value of 3,6-nonadien-1-ol, which exhibits a fresh cucumber flavor, is higher on the 7th to 28th days of fermentation, making a critical contribution to the later flavor of sour bamboo shoots. Octanal and nonanal typically exhibit a fresh citrus flavor [35], but due to their low OAV value, they contribute less to the flavor of sour bamboo shoots. Nonanal was also found to be the main component of the sour bamboo shoot flavor by Guo, et al. [36]. Acetic acid, 1-heptanol, and 1-Octen-3-ol were characteristic flavor components, but their relative contents are relatively low, and their contributions to the flavor formation of sour bamboo shoots are limited. Among these, 1-octene-3-ol can be used as an edible essence, which has the aroma of mushroom house, lavender, rose or hay, and is consistent with the main flavor substances reported in other reports [4]. D-limonene and isophorone had high OAV values on the first day of fermentation, which may have contributed significantly to the early flavor formation of sour bamboo shoots.

Table 2. Volatile compounds with OAV >1 in sour bamboo shoots during the 28 days of fermentation.

Compounds	Threshold (µg/kg)	OVA				
		1 d	7 d	14 d	21 d	28 d
1-Heptanol [37]	5.5	0.00	0.00	1.46	0.00	0.00
1-Octen-3-ol [24]	20	5.06	0.00	0.00	0.00	5.88
3,6-Nonadien-1-ol [24]	1.3	1.29	80.66	58.92	53.51	67.30
Silanediol, dimethyl- [24]	21	0.08	0.00	15.12	17.81	20.10
Isophorone [38]	2	35.08	10.53	0.00	0.00	3.91
Octanal [4]	0.7	3.58	7.53	2.23	2.04	3.32
Nonanal [4]	1	9.62	16.30	8.76	19.36	17.26
Benzaldehyde, 2,4-dimethyl- [37]	0.2	129.73	668.84	198.69	346.41	167.06
2,2,4-Trimethyl-1,3-pentanediol diisobutyrate [39]	0.13	59.55	197.30	131.84	113.28	111.13
Dibutyl phthalate [40]	0.26	59.07	50.91	96.32	52.67	32.22
D-Limonene [41]	10	35.42	1.04	0.24	0.00	0.32
2,4-di-tert-butylphenol [42]	200	11.25	6.77	7.09	8.46	9.70
Acetic acid [37]	5.5	0.00	0.00	5.69	0.00	0.00

3.4. Microbial Composition in Sour Bamboo Shoot

Alpha diversity analysis (coverage rate >99%), including Chao1, ACE, Shannon, and Simpson indices, can reflect the diversity and richness of microbial communities. On the first day of fermentation, the richness and diversity of microorganisms were the highest, which may be due to miscellaneous bacteria accompanying bamboo shoots entering the fermentation broth. On the other hand, infiltration may cause some nutrients from bamboo shoots to dissolve into the fermentation broth [43], providing sufficient energy sources for microbial growth and reproduction, and allowing aerobic microorganisms to proliferate in large numbers. The nutrient composition and oxygen decrease as the fermentation process progresses [44], and the growth and reproduction of aerobic microorganisms are inhibited. The dominant bacteria in the fermented liquid of sour bamboo shoots gradually give way to lactic acid bacteria, which rapidly reproduce and multiply, resulting in a higher level of microbial diversity on the 14th day [6]. On the 28th day of fermentation, the richness of microorganisms increased, similarly to the research results of Li, et al. [7] wherein the richness of microbial communities sharply increased on the 28th day of fermentation. PCoA and UPGMA analysis (Figures S3 and S4) found that there were significant differences in microbial structure in the early stages (1d and 7d) of fermentation, while fluctuations were relatively small in the later stages. This result may indicate that microbial succession is intense in the early stage, while dominant bacteria, composed mostly of lactic acid bacteria, form in the later stage [6]. Lactic acid bacteria not only reduce the pH value of the fermentation environment, but also produce bacteriocins to further inhibit the growth of miscellaneous bacteria [45].

3.4.1. The Changes in Microbial Community Structure in Sour Bamboo Shoots during Fermentation

We employed high-quality sequencing reads to examine the bacterial community's structure and dynamic succession at the phylum and genus levels during the fermentation of sour bamboo shoots. At the phylum level of bacteria, a total of 42 phyla were identified, and Firmicutes (20.02%), Proteobacteria (35.18%), and Bacteroidota (17.43%) were the predominant bacteria, accounting for 72.63% of the total bacterial population in the early stage of fermentation (Figure 3A). Then, Firmicutes (20.02–70.05%) increased significantly and dominated the subsequent fermentation stage, attaining the highest relative abundance on day 21 of fermentation. Proteobacteria is also the dominant phyla in the fermentation process, and its abundance varies from 27.93% to 35.18%. The results obtained are in concurrence with the research conducted by Chen, et al. [4], which revealed Firmicutes and Proteobacteria as the predominant phyla. In addition, Firmicutes was also found to be the dominant phyla in numerous traditional fermented vegetables, such as Jiangxi yancai [46], Suancai [47,48], and Kaili red sour soup [49].

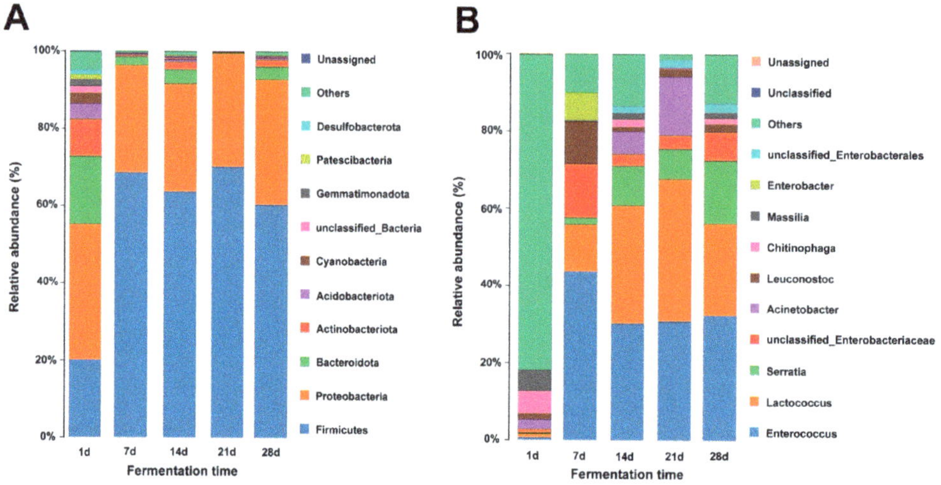

Figure 3. Changes of microbial community at the phylum (**A**) and genus (**B**) levels in sour bamboo shoots during the 28 days fermentation.

At the genus level, the top 10 species at abundance levels in all samples were Enterococcus, Lactococcus, Enterococcus, Serratia, unclassified Enterobacteriaceae, Acinetobacter, Leuconostoc, Chitinophaga, Massilia, Enterobacter, and unclassified Enterobacterales (Figure 3B). Miscellaneous bacteria (81.59%) predominated on day 1 and decreased afterward. According to the findings of Guan, Huang, Peng, Huang, Liu, Peng, Xie and Xiong [6], the initial stage of sour bamboo shoot fermentation is characterized by the presence of five dominant genera, namely Lactococcus, Leuconostoc, Weissella, Enterobacter, and Raoultella. The variation in their abundance may be attributed to the origin of the raw materials used and could serve as an indicator of microbial diversity within the fermentation process. The abundance of Enterococcus in the sample increased significantly from an initial relative abundance of 0.59% to a peak level of 43.62% on day 7. Similarly, Lactococcus also showed a significant increase from the initial relative abundance of 1.07% to a peak level of 36.94% on day 21. According to prior investigations, it has been demonstrated that Lactococcus exhibits a stable function in the process of food fermentation [50,51]. Furthermore, in the Guangxi sour bamboo shoots, dominant genera were identified as Lactobacillus, Serratia, Stenotrichomonas, and Lactococcus [52].

3.4.2. Representative Bacteria in Sour Bamboo Shoots

Linear discriminant analysis effect size (LEfSe) is a statistical method utilized for the identification of biomarkers by analyzing the variations within bacterial communities and recognizing distinctive features based on their abundances and associated categories. LEfSe with a threshold LDA score of 4.0 (Figure 4A,B) was set to distinguish and classify particular microbial taxa present at each stage of the fermentation process [6]. On the first day of fermentation, there were observable biomarkers from four phyla, seven orders, eight families, six genera, and four species. The sample from the 28th day of fermentation exhibited the presence of one order biomarker (Bacillales), one family biomarker (Bacillacea), and one genus biomarker (*Bacillus*), whereas only one genus of biomarkers (*unclassified Lactococcus*) was identified in the 14 day fermentation samples. According to previous studies, *Lactobacillus* is the dominant genus in the mid-to-late stages of fermentation [6]. Such differences may be caused by raw materials, production processes, geographical distributions, or analysis methods, among other things. Further comparative studies from different varieties and regions can be conducted in the future.

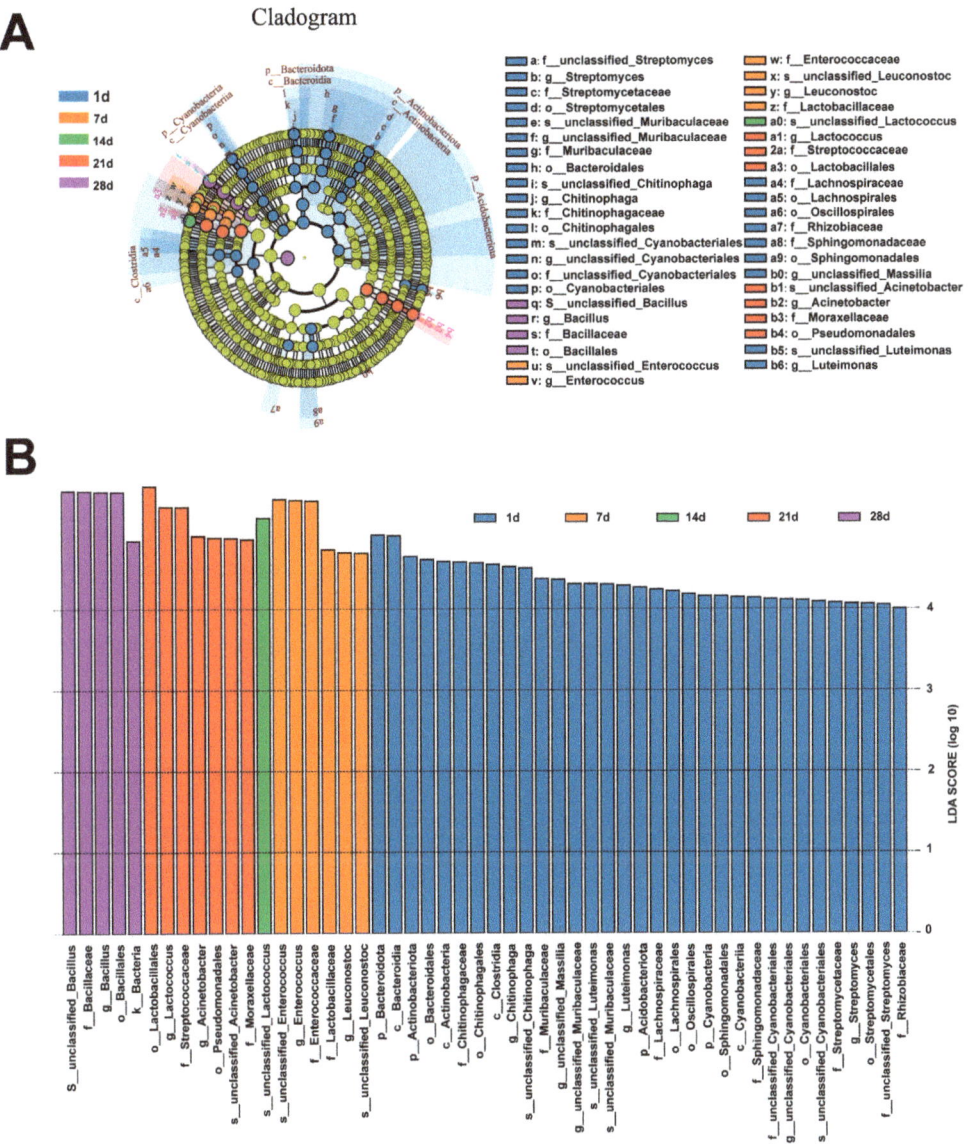

Figure 4. LEfSe of the microbial community during suansun fermentation at the OTU level. (**A**) lefse biomarkers cladogram, (**B**) lefse biomarkers.

3.5. Correlations between Microbial Compositions and Flavor

During the fermentation process, physicochemical properties and main volatile compounds are closely related to the number and type of microbial communities [16]. Pearson correlation analysis was used to study the correlation between the physicochemical properties, characteristic volatile compounds (OAV > 1), and the dominant phyla and genera of the microbial community of sour bamboo shoots (Figure 5).

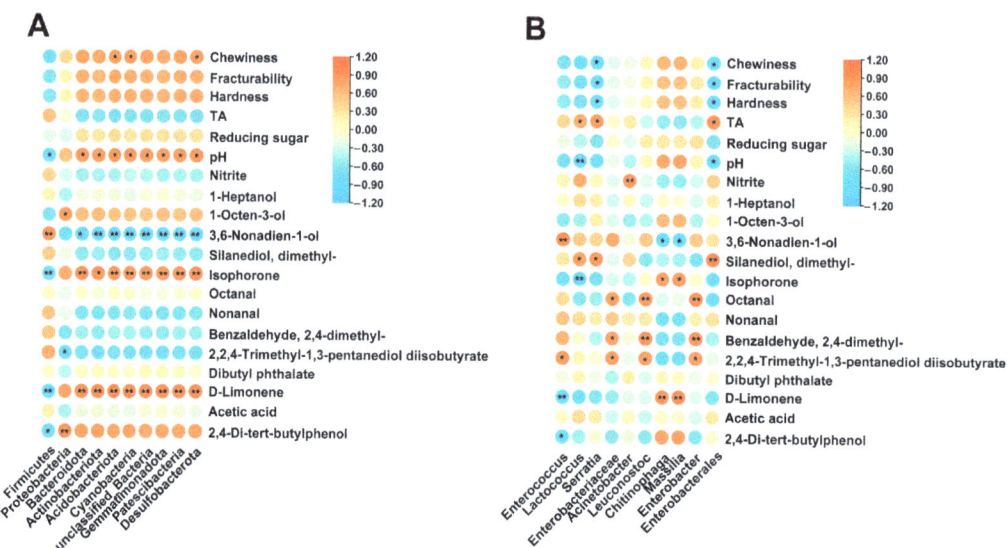

Figure 5. Pearson correlation analysis of the relative abundance of the top ten bacteria phylum levels (**A**) and genus levels (**B**) and variation of volatile compounds (OAV > 1) and physicochemical properties during sour bamboo shoot fermentation. * indicates significance at $p < 0.05$; ** indicates significance at $p < 0.01$.

At the phylum level, Firmicutes was the overwhelmingly advantaged phylum. Firmicutes showed a significant negative correlation with D-limonene (R = −0.98), isophorone (R = −0.93), 2,4-di-tert-butylphenol (R = −0.82), and pH (R = −0.86). It was significantly positively correlated with 3,6-nonylidene-1-ol (R = 0.92), and positively correlated with 2,4-dimethyl benzaldehyde (R = 0.55), silanediol, dimethyl- (R = 0.52), nonanal (R = 0.59), and 2,2,4-trimethyl-1,3-pentylenediol diisobutyrate (R = 0.75). It has been reported by Chen, Li, Cheng, Liu, Yi, Chen, Wang and Cao [20] that Firmicutes can show significant direct correlations with fumaric acid, and geranyl acetone. In addition, a previous study has shown that the thick cell skin of Firmicutes can produce spores to resist extreme environments while being inhibited after the removal of D-limonene [53]. Proteobacteria was significantly positively correlated with 1-Octen-3-ol (R = 0.90), 2,4-di-tert-butylphenol (R = 0.99), and negatively correlated with three characteristic volatile compounds, namely, 2,4-dimethylbenzaldehyde (R = −0.66), 2,2,4-trimethyl-1,3-pentylenediol diisobutyrate (R = −0.84), and dibutyl phthalate (R = −0.40), indicating that proteobacteria was not conducive to the formation of the characteristic flavor of sour bamboo shoots. The majority of the genera attached to raw materials were Proteobacteria, such as *Pseudomonas*, *Vibrio*, and *Acinetobacter*. As a result, these were highly correlated with early fermentation. Fermentation conditions such as falling pH and increasing TA restricted most Proteobacteria genera [22]. Furthermore, it has been observed that Proteobacteria and Cyanobacteria display a positive correlation with textural attributes such as hardness, chewiness, and fracturability, suggesting that the bacteria belonging to the aforementioned phyla may have a crucial role in the alteration of cell wall properties [23].

In general, the genera *Enterobacter* was considered to be a spoilage bacterium in fermented vegetables [54], and exhibited strong positive correlations with two characteristic compounds (3,6-nonylidene-1-ol and 2,2,4-trimethyl-1,3-pentylenediol di-isobutyrate) in our study, suggesting that they may play a major role in suansun's volatile flavor production (Figure 5B). Considering that *Enterobacter* is a foodborne pathogen, its presence in fermented bamboo shoots is undesirable. However, the edible safety of the final products of sour bamboo shoots should not be a concern: sour bamboo shoots are often cooked at a high temperature, which effectively reduces microbial contamination risk. It has also been reported that *Enterobacter* is responsible for the characteristic flavors found in sauerkraut [6], cheeses [55], and yucha [56], among other fermented foods. A common characteristic of lactic acid bacteria is that they produce acids from carbohydrate catabolism [45], while *Lactococcus* was significantly correlated with TA (R = 0.92) and pH (R = −0.94) in this study as well as with seven characteristic flavor substances. Some studies have shown that the fermented vegetable aroma was also produced by lactic acid bacteria, which are responsible for the production of esterases, lipases, and alcohol acetyltransferases, all of which enhance the flavor-forming qualities of fermented vegetables [57]. At the same time, *Serratia* had significantly positive correlations with silanediol, dimethyl- (R = 0.92) and TA (R = 0.89), and significantly negative correlations with hardness (R = −0.91), chewiness (R = −0.93), and fracturability (R = −0.91). Therefore, *Serratia* may affect the texture, a conclusion which is consistent with previously reported results [3].

4. Conclusions

Physicochemical properties, flavor compounds, and microbial composition of sour bamboo shoots produced in southwest China were correlated during the fermentation process (28 days). There was a decreasing trend in pH and TA values of fermentation liquid from sour bamboo shoots, followed by a stabilizing trend. It has been estimated that at least 13 volatile compounds (OAV > 1) are characteristic of sour bamboo shoots. As a result of HTS, it was found that *Enterococcus*, *Lactococcus*, and *Serratia* were the most dominant genera in sour bamboo shoots. Moreover, *Enterococcus* had significantly negative correlations with D-limonene and 2,4-dimethylbenzaldehyde, and significantly positive correlations with 3,6-nonylidene-1-ol and 2,2,4-trimethyl-1,3-pentylenediol di-isobutyrate. There were significant positive correlations between *Lactococcus* and silanediol, dimethyl- and TA, and significant negative correlations between *Lactococcus* and isophorone. *Serratia* had significantly positive correlation with silanediol, dimethyl- and TA, significantly negative correlation with texture properties (hardness, chewiness, and fracturability). A significant correlation was found between 3,6-nonylidene-1-ol and Firmicutes at the phylum level. The findings of this study can be valuable in enhancing the production of starters, which in turn can promote the industrial production of high-quality and safe sour bamboo shoots. However, there is still a lack of key flavor information and fermentation time, and further research is needed to obtain more comprehensive information to comprehensively evaluate the fermentation of sour bamboo shoots.

Supplementary Materials: The following supporting information can be downloaded at: https://www.mdpi.com/article/10.3390/foods12163035/s1, Figure S1: Correlation among reducing sugar, hardness, fracturability, chewiness, TA, pH, and nitrite of sour bamboo shoots conducted by Pearson's correlation analysis; Figure S2: Total ion flow of volatile compounds in sour bamboo shoots during the fermentation process; Figure S3: PCoA analysis of the microbial community in sour bamboo shoots during fermentation process; Figure S4: UPGMA analysis of the microbial community in sour bamboo shoots during fermentation process; Table S1: Volatile compounds of sour bamboo shoots during fermentation measurement by GC-MS; Table S2: Alpha diversity of the microbial community in sour bamboo shoots during the fermentation process.

Author Contributions: Conceptualization, Q.C. and S.H.; investigation, S.Z., X.X. and W.D.; writing—original draft preparation, Z.L.; writing—review and editing, Z.L. and Q.C.; funding acquisition, S.H. All authors have read and agreed to the published version of the manuscript.

Funding: This research was funded by the National Key R&D Program of China (2021YFD2200504_1 and 2021YFD2200505_2), Science and Technology Project of Sichuan Province, China (2023JDRC0130, 2021YFYZ0006 and 2022NSFSC0093), and Ph.D. Foundation (no. 22zx7).

Data Availability Statement: Data is contained within the article or Supplementary Material.

Conflicts of Interest: The authors declare no conflict of interest.

References

1. Wang, Y.; Chen, J.; Wang, D.; Ye, F.; He, Y.; Hu, Z.; Zhao, G. A systematic review on the composition, storage, processing of bamboo shoots: Focusing the nutritional and functional benefits. *J. Funct. Foods* **2020**, *71*, 104015. [CrossRef]
2. Behera, P.; Balaji, S. Health benefits of fermented bamboo shoots: The twenty-first century green gold of Northeast India. *Appl. Biochem. Biotechnol.* **2021**, *193*, 1800–1812. [CrossRef] [PubMed]
3. Guan, Q.; Zheng, W.; Mo, J.; Huang, T.; Xiao, Y.; Liu, Z.; Peng, Z.; Xie, M.; Xiong, T. Evaluation and comparison of the microbial communities and volatile profiles in homemade suansun from Guangdong and Yunnan provinces in China. *J. Sci. Food Agric.* **2020**, *100*, 5197–5206. [CrossRef] [PubMed]
4. Chen, C.; Cheng, G.; Liu, Y.; Yi, Y.; Chen, D.; Zhang, L.; Wang, X.; Cao, J. Correlation between microorganisms and flavor of Chinese fermented sour bamboo shoot: Roles of Lactococcus and Lactobacillus in flavor formation. *Food Biosci.* **2022**, *50*, 101994. [CrossRef]
5. Li, J.; Liu, Y.; Xiao, H.; Huang, H.; Deng, G.; Chen, M.; Jiang, L. Bacterial communities and volatile organic compounds in traditional fermented salt-free bamboo shoots. *Food Biosci.* **2022**, *50*, 102006. [CrossRef]
6. Guan, Q.; Huang, T.; Peng, F.; Huang, J.; Liu, Z.; Peng, Z.; Xie, M.; Xiong, T. The microbial succession and their correlation with the dynamics of flavor compounds involved in the natural fermentation of suansun, a traditional Chinese fermented bamboo shoots. *Food Res. Int.* **2022**, *157*, 111216. [CrossRef]
7. Li, W.; Wu, L.; Suo, H.; Zhang, F.; Zheng, J. Bacterial community dynamic succession during fermentation of pickled Ma bamboo shoots based on high-throughput sequencing. *Food Ferment. Ind.* **2020**, *46*, 9–15. [CrossRef]
8. Xia, X.; Ran, C.; Ye, X.; Li, G.; Kan, J.; Zheng, J. Monitoring of the bacterial communities of bamboo shoots (*Dendrocalamus latiflorus*) during pickling process. *Int. J. Food Sci. Technol.* **2017**, *52*, 1101–1110. [CrossRef]
9. GB/T 12456-2021; Determination of Total Acid in Foods. Standards Press of China: Beijing, China, 2021.
10. GB. 5009.7-2021; Determination of Reducing Sugar in Foods. Standards Press of China: Beijing, China, 2021.
11. GB. 5009.33-2016; Determination of Nitrite and Nitrate in Food. Standards Press of China: Beijing, China, 2016.
12. Zhao, Y.; Wu, Z.; Miyao, S.; Zhang, W. Unraveling the flavor profile and microbial roles during industrial Sichuan radish paocai fermentation by molecular sensory science and metatranscriptomics. *Food Biosci.* **2022**, *48*, 101815. [CrossRef]
13. Ji, X.-G.; Chang, K.-L.; Chen, M.; Zhu, L.-L.; Osman, A.; Yin, H.; Zhao, L.-M. In vitro fermentation of chitooligosaccharides and their effects on human fecal microbial community structure and metabolites. *LWT-Food Sci. Technol.* **2021**, *144*, 111224. [CrossRef]
14. Bolyen, E.; Rideout, J.R.; Dillon, M.R.; Bokulich, N.A.; Abnet, C.C.; Al-Ghalith, G.A.; Alexander, H.; Alm, E.J.; Arumugam, M.; Asnicar, F.; et al. Reproducible, interactive, scalable and extensible microbiome data science using QIIME 2. *Nat. Biotechnol.* **2019**, *37*, 852–857. [CrossRef] [PubMed]
15. Callahan, B.J.; McMurdie, P.J.; Rosen, M.J.; Han, A.W.; Johnson, A.J.; Holmes, S.P. DADA2: High-resolution sample inference from Illumina amplicon data. *Nat. Methods* **2016**, *13*, 581–583. [CrossRef] [PubMed]
16. Chi, H.; Lu, W.; Liu, G.; Qin, Y. Physiochemical property changes and mineral element migration behavior of bamboo shoots during traditional fermentation process. *J. Food Process. Preserv.* **2020**, *44*, e14784. [CrossRef]
17. Jeyaram, K.; Romi, W.; Singh, T.A.; Devi, A.R.; Devi, S.S. Bacterial species associated with traditional starter cultures used for fermented bamboo shoot production in Manipur state of India. *Int. J. Food Microbiol.* **2010**, *143*, 1–8. [CrossRef]
18. Yu, S.M.; Zhang, Y. Effects of lactic acid bacteria on nitrite degradation during pickle fermentation. *Adv. Mater. Res.* **2013**, *781–784*, 1656–1660. [CrossRef]
19. Ye, Z.; Shang, Z.; Li, M.; Qu, Y.; Long, H.; Yi, J. Evaluation of the physiochemical and aromatic qualities of pickled Chinese pepper (Paojiao) and their influence on consumer acceptability by using targeted and untargeted multivariate approaches. *Food Res. Int.* **2020**, *137*, 109535. [CrossRef]
20. Chen, C.; Li, J.; Cheng, G.; Liu, Y.; Yi, Y.; Chen, D.; Wang, X.; Cao, J. Flavor changes and microbial evolution in fermentation liquid of sour bamboo shoots. *J. Food Compos. Anal.* **2023**, *120*, 105273. [CrossRef]
21. Rao, Y.; Qian, Y.; Tao, Y.; She, X.; Li, Y.; Chen, X.; Guo, S.; Xiang, W.; Liu, L.; Du, H.; et al. Characterization of the microbial communities and their correlations with chemical profiles in assorted vegetable Sichuan pickles. *Food Control* **2020**, *113*, 107174. [CrossRef]
22. Shang, Z.; Ye, Z.; Li, M.; Ren, H.; Cai, S.; Hu, X.; Yi, J. Dynamics of microbial communities, flavor, and physicochemical properties of pickled chayote during an industrial-scale natural fermentation: Correlation between microorganisms and metabolites. *Food Chem.* **2022**, *377*, 132004. [CrossRef]
23. Jiménez-Aguilar, D.M.; Grusak, M.A. Minerals, vitamin C, phenolics, flavonoids and antioxidant activity of Amaranthus leafy vegetables. *J. Food Compos. Anal.* **2017**, *58*, 33–39. [CrossRef]
24. Zhang, J.; Sun, Y.; Guan, X.; Qin, W.; Zhang, X.; Ding, Y.; Yang, W.; Zhou, J.; Yu, X. Characterization of key aroma compounds in melon spirits using the sensomics concept. *LWT-Food Sci. Technol.* **2022**, *161*, 113341. [CrossRef]
25. Zhao, Y.; Wei, W.; Tang, L.; Wang, D.; Wang, Y.; Wu, Z.; Zhang, W. Characterization of aroma and bacteria profiles of Sichuan industrial paocai by HS-SPME-GC-O-MS and 16S rRNA amplicon sequencing. *Food Res. Int.* **2021**, *149*, 110667. [CrossRef] [PubMed]
26. Guneser, O.; Demirkol, A.; Yuceer, Y.K.; Togay, S.O.; Hosoglu, M.I.; Elibol, M. Production of flavor compounds from olive mill waste by *Rhizopus oryzae* and *Candida tropicalis*. *Braz. J. Microbiol.* **2017**, *48*, 275–285. [CrossRef] [PubMed]

27. Varsha, K.K.; Devendra, L.; Shilpa, G.; Priya, S.; Pandey, A.; Nampoothiri, K.M. 2,4-Di-tert-butyl phenol as the antifungal, antioxidant bioactive purified from a newly isolated Lactococcus sp. *Int. J. Food Microbiol.* **2015**, *211*, 44–50. [CrossRef]
28. Bolotin, A.; Malarme, K.; Wincker, P.; Weissenbach, J.; Mauger, S.; Ehrlich, S.D.; Jaillon, O.; Sorokin, A. The complete genome sequence of the lactic acid bacterium *Lactococcus lactis* ssp. *lactis* IL1403. *Genome Res.* **2001**, *11*, 731–753. [CrossRef]
29. Sansenya, S.; Payaka, A.; Mansalai, P. Biological activity and inhibition potential against α-glucosidase and α-amylase of 2,4-di-tert-butylphenol from bamboo shoot extract by in vitro and in silico studies. *Process Biochem.* **2023**, *126*, 15–22. [CrossRef]
30. Zhao, D.; Hu, J.; Chen, W. Analysis of the relationship between microorganisms and flavour development in dry-cured grass carp by high-throughput sequencing, volatile flavour analysis and metabolomics. *Food Chem.* **2022**, *368*, 130889. [CrossRef]
31. Yan, J.W.; Ban, Z.J.; Lu, H.Y.; Li, D.; Poverenov, E.; Luo, Z.S.; Li, L. The aroma volatile repertoire in strawberry fruit: A review. *J. Sci. Food Agric.* **2018**, *98*, 4395–4402. [CrossRef]
32. Kaseleht, K.; Paalme, T.; Mihhalevski, A.; Sarand, I. Analysis of volatile compounds produced by different species of lactobacilli in rye sourdough using multiple headspace extraction. *Int. J. Food Sci. Technol.* **2011**, *46*, 1940–1946. [CrossRef]
33. Han, S.; Gao, T.-T.; Liu, Y.-P.; Sun, B.-G. Extraction and analysis of volatile flavor constituents in Laoyipinxiang *Douchi* by SDE-GC-MS. *Food Ferment. Ind.* **2013**, *39*, 192–197. [CrossRef]
34. Xu, X.; Long, Z.; Du, W.; Chen, Q.; Zhang, Y.; Hu, S. Dynamics of physicochemical properties, flavor, and microbial communities of salt-free bamboo shoots during natural fermentation: Correlation between microorganisms and metabolites. *Fermentation* **2023**, *9*, 733. [CrossRef]
35. Hausch, B.J.; Lorjaroenphon, Y.; Cadwallader, K.R. Flavor chemistry of lemon-lime carbonated beverages. *J. Agric. Food Chem.* **2015**, *63*, 112–119. [CrossRef] [PubMed]
36. Guo, R.; Yu, F.; Wang, C.; Jiang, H.; Yu, L.; Zhao, M.; Liu, X. Determination of the volatiles in fermented bamboo shoots by head space—Solid-phase micro extraction (HS-SPME) with gas chromatography—Olfactory—Mass spectrometry (GC-O-MS) and aroma extract dilution analysis (AEDA). *Anal. Lett.* **2020**, *54*, 1162–1179. [CrossRef]
37. Xiao, Z.; Wang, H.; Niu, Y.; Zhu, J.; Ma, N. Analysis of aroma components in four chinese congou black teas by odor active values and aroma extract dilution analysis coupled with partial least squares regression. *Food Sci.* **2018**, *39*, 242–249.
38. Hua, J.; Li, J.; Ouyang, W.; Wang, J.; Yuan, H.; Jiang, Y. Effect of *Strobilanthes tonkinensis* Lindau addition on black tea flavor quality and volatile metabolite content. *Foods* **2022**, *11*, 1678. [CrossRef] [PubMed]
39. Li, X.; Wang, J.; Liu, Y.; Shi, J.; Zhang, X.; Chen, C. Odorant screening and possible origin analysis of odor episodes in one reservoir in northern China. *J. Water Supply Res. Technol.-AQUA* **2015**, *64*, 847–856. [CrossRef]
40. Wenjing, L.; Zhenhan, D.; Dong, L.; Jimenez, L.M.; Yanjun, L.; Hanwen, G.; Hongtao, W. Characterization of odor emission on the working face of landfill and establishing of odorous compounds index. *Waste Manag.* **2015**, *42*, 74–81. [CrossRef]
41. Zhang, Y.; Li, H.; Zhou, S. Analysis of fragrance characteristics in three congou black teas of highly fragrant species using odor active values. *Food Res. Dev.* **2020**, *41*, 184–191.
42. Luo, W.; Du, X.; Xu, Y.; Wu, J.; Yu, Y.; Lu, L. Study on the effect of different lactic acid bacteria fermentation on the quality and volatile flavor of by-products of Chicaixin. *Food Ferment. Ind.* **2023**, 1–11. [CrossRef]
43. Fan, I.; Huang, Q.; Wang, Y. Effects of *Astragalus* extract on physicochemical properties, microbial flora, and sensory quality of low-salt naturally fermented pickles. *Food Ferment. Ind.* **2022**, *48*, 213–218.
44. Duan, X.; Chen, C.; Cao, Y.; Feng, X. Preparation of broccoli's stem pickle and analysis on variance of physicochemical indicators during the fermentation. *Food Ferment. Ind.* **2014**, *40*, 106–111. [CrossRef]
45. Cirlini, M.; Ricci, A.; Galaverna, G.; Lazzi, C. Application of lactic acid fermentation to elderberry juice: Changes in acidic and glucidic fractions. *LWT-Food Sci. Technol.* **2020**, *118*, 108779. [CrossRef]
46. Xiao, M.; Huang, T.; Huang, C.; Hardie, J.; Peng, Z.; Xie, M.; Xiong, T. The microbial communities and flavour compounds of Jiangxi yancai, Sichuan paocai and Dongbei suancai: Three major types of traditional Chinese fermented vegetables. *LWT-Food Sci. Technol.* **2020**, *121*, 108865. [CrossRef]
47. Liang, H.; He, Z.; Wang, X.; Song, G.; Chen, H.; Lin, X.; Ji, C.; Zhang, S. Bacterial profiles and volatile flavor compounds in commercial Suancai with varying salt concentration from Northeastern China. *Food Res. Int.* **2020**, *137*, 109384. [CrossRef] [PubMed]
48. He, Z.; Chen, H.; Wang, X.; Lin, X.; Ji, C.; Li, S.; Liang, H. Effects of different temperatures on bacterial diversity and volatile flavor compounds during the fermentation of suancai, a traditional fermented vegetable food from northeastern China. *LWT-Food Sci. Technol.* **2020**, *118*, 108773. [CrossRef]
49. Li, D.; Duan, F.; Tian, Q.; Zhong, D.; Wang, X.; Jia, L. Physiochemical, microbiological and flavor characteristics of traditional Chinese fermented food Kaili Red Sour Soup. *LWT-Food Sci. Technol.* **2021**, *142*, 110933. [CrossRef]
50. Sharma, N.; Barooah, M. Microbiology of *khorisa*, its proximate composition and probiotic potential of lactic acid bacteria present in *Khorisa*, a traditional fermented Bamboo shoot product of Assam. *Indian J. Nat. Prod. Resour.* **2017**, *88*, 78–88.
51. Singhal, P.; Shukla, L.; Satya, S.; Naik, S.N. Scientific validation and process mechanism of traditional bamboo shoot fermentation by isolation and characterization of lactic acid. *Curr. Nutr. Food Sci.* **2017**, *13*, 176–181. [CrossRef]
52. Guan, Q.; Zheng, W.; Huang, T.; Xiao, Y.; Liu, Z.; Peng, Z.; Gong, D.; Xie, M.; Xiong, T. Comparison of microbial communities and physiochemical characteristics of two traditionally fermented vegetables. *Food Res. Int.* **2020**, *128*, 108755. [CrossRef]
53. Awasthi, M.K.; Lukitawesa, L.; Duan, Y.; Taherzadeh, M.J.; Zhang, Z. Bacterial dynamics during the anaerobic digestion of toxic citrus fruit waste and semi-continues volatile fatty acids production in membrane bioreactors. *Fuel* **2022**, *319*, 123812. [CrossRef]

54. Franco, W.; Pérez-Díaz, I.M. Role of selected oxidative yeasts and bacteria in cucumber secondary fermentation associated with spoilage of the fermented fruit. *Food Microbiol.* **2012**, *32*, 338–344. [CrossRef] [PubMed]
55. Zheng, X.; Liu, F.; Li, K.; Shi, X.; Ni, Y.; Li, B.; Zhuge, B. Evaluating the microbial ecology and metabolite profile in Kazak artisanal cheeses from Xinjiang, China. *Food Res. Int.* **2018**, *111*, 130–136. [CrossRef] [PubMed]
56. Zhang, J.; Wang, X.; Huo, D.; Li, W.; Hu, Q.; Xu, C.; Liu, S.; Li, C. Metagenomic approach reveals microbial diversity and predictive microbial metabolic pathways in Yucha, a traditional Li fermented food. *Sci. Rep.* **2016**, *6*, 32524. [CrossRef] [PubMed]
57. Gammacurta, M.; Lytra, G.; Marchal, A.; Marchand, S.; Christophe Barbe, J.; Moine, V.; de Revel, G. Influence of lactic acid bacteria strains on ester concentrations in red wines: Specific impact on branched hydroxylated compounds. *Food Chem.* **2018**, *239*, 252–259. [CrossRef] [PubMed]

Disclaimer/Publisher's Note: The statements, opinions and data contained in all publications are solely those of the individual author(s) and contributor(s) and not of MDPI and/or the editor(s). MDPI and/or the editor(s) disclaim responsibility for any injury to people or property resulting from any ideas, methods, instructions or products referred to in the content.

Article

The Effects of Sheep Tail Fat, Fat Level, and Cooking Time on the Formation of Nε-(carboxymethyl)lysine and Volatile Compounds in Beef Meatballs

Kübra Öztürk [1], Zeynep Feyza Yılmaz Oral [2], Mükerrem Kaya [1,3] and Güzin Kaban [1,*]

[1] Department of Food Engineering, Faculty of Agriculture, Atatürk University, Erzurum 25240, Türkiye; ozturkkubra1516@gmail.com (K.Ö.); mkaya@atauni.edu.tr (M.K.)
[2] Department of Food Technology, Erzurum Vocational School, Atatürk University, Erzurum 25240, Türkiye; zeynep.yilmaz@atauni.edu.tr
[3] MK Consulting, Ata Teknokent, Erzurum 25240, Türkiye
* Correspondence: gkaban@atauni.edu.tr

Abstract: This study aimed to determine the effects of fat type (sheep tail fat (STF) and beef fat (BF)), fat levels (10, 20, or 30%), and cooking time (0, 2, 4, and 6 min, dry heat cooking at 180 °C) on the carboxymethyl lysine (CML) content in meatballs. pH, thiobarbituric acid reactive substance (TBARS), and volatile compound analyses were also performed on the samples. The use of STF and the fat level had no significant effect on the pH value. The highest TBARS value was observed with the combination of a 30% fat level and STF. CML was not affected by the fat level. The highest CML content was determined in meatballs with STF at a cooking time of 6 min. In the samples cooked for 2 min, no significant difference was observed between STF and BF in terms of the CML content. STF generally increased the abundance of aldehydes. Aldehydes were also affected by the fat level and cooking time. A PCA provided a good distinction between groups containing STF and BF regardless of the fat level or cooking time. Pentanal, octanal, 2,4-decadienal, hexanal, and heptanal were positively correlated with CML.

Keywords: meatball; CML; AGE; sheep tail fat; beef fat; volatile compounds

1. Introduction

The Maillard reaction (MR) is initiated by the condensation of amino groups on proteins, peptides, and amino acids with carbonyl groups on reducing sugars. The main pathways of the MR involve three steps: initial, intermediate or advanced, and final stages [1]. The initial stage of MR consists of a condensation of the reducing saccharide with the amino compound, resulting in the formation of an Amadori product and a Heyns product. In the intermediate/advanced stage of the MR, Amadori and Heyns products can undergo degradation reactions, and several α-dicarbonyl compounds are formed. These compounds are also involved in further reactions with side chains of peptides or proteins, leading to the formation of advanced glycation end products (AGEs). In the final stage of the MR, nitrogen-containing brown polymers or copolymers called melanoidins are formed by the condensation and polymerization reactions of previously formed reactive intermediates [1–3]. The MR is influenced by many factors such as the physical state of the matrix, pH, the type and concentration of reactants, process conditions, the temperature, duration of storage, and water activity [1,4].

The MR contributes significantly to quality characteristics of food products, including flavor, aroma, and color. In addition, mutagenic and toxigenic compounds can also be formed as a result of this reaction. Furthermore, the MR can lead to the loss of the nutritional value of proteins [2]. AGEs formed during the Maillard reaction are heterogeneous compounds, and these compounds have adverse effects on human health and are known to be closely related to many chronic diseases [5–8]. For example, it has been reported that an AGE-rich diet can lead to increased oxidative stress and inflammation linked with type 2 diabetes [9].

AGEs are compounds that occur naturally in foods of animal origin. The AGE level can vary depending on many factors such as the source and composition of meat, the heat treatment process and duration, and the protein and fat contents [5,10–12]. Although AGEs such as N-ε-carboxymethyl-lysine (CML), methyl glyoxal lysine dimer, N-ε-carboxyethyl-lysine (CEL), and pentosidine are commonly found in foods and particularly in processed meat products, CML is established as an indicator of AGEs due to its resistance to acidic food environments [7,11]. The cooking technique applied to meat and meat products, the cooking degree, and the fat content play an important role in CML formation [7].

Minced meat products, such as meatballs, burgers, and meat patties, are widely consumed around the world. They are produced using different methods depending on the type of meat, cost considerations, the shape, the nutritional value, and religious reasons [13]. In these products, fat is a major ingredient in terms of taste, texture, and flavor. In addition to beef fat, sheep tail fat is also used in meatball production. Sheep tail fat contains more unsaturated fatty acids than beef fat (intermuscular fat) [14–16]. Sheep tail fat also contains relatively higher levels of polyunsaturated fatty acids (PUFA) [17–19] and nutraceutical fatty acids (n-3 polyunsaturated fatty acids), which have health-promoting benefits [20], than beef fat. It was reported that the contents of oleic acid (C18:1) and linoleic acid (C18:2) in sheep tail fat varied between 41.51 and 49.7 and between 2.62 and 5.7%, respectively. However, it was stated that the fatty acid composition is affected by breed, age, sex, and nutritional conditions [21]. On the other hand, the fact that sheep tail fat is more susceptible to oxidation is of great importance in determining the effect of this fat on CML formation. In addition, it is important to determine the effect of sheep tail fat on the volatile profile and also to reveal the relationship between volatile compounds and CML in meatballs.

There are limited studies on CML formation in meatballs. In these studies, the effects of adding different amounts of salt to beef patties [22], the storage time of frozen pork patties [23], adding various proportions of wheat, rye, and triticale bran to beef patties [5], and the use of *Kaempferia galanga* L. and kaempferol extracts [24] on CML formation were investigated. However, there is no information on the effects of sheep tail fat and fat levels on CML formation in beef meatballs. The aim of this study is to determine the influences of fat type (beef fat and sheep tail fat), fat level (10, 20, and 30%), and cooking time (0, 2, 4, and 6 min at 180 °C) on CML formation in meatballs. In addition, the effects of these factors on pH, lipid oxidation, and volatile compounds were also investigated.

2. Materials and Methods

2.1. Chemicals, Reagents and Standards

Trichloroacetic acid (TCA), ethylenediaminetetraacetic acid (EDTA), propyl gallate, and sodium borate were purchased from Sigma Aldrich (Steinheim, Germany), and thiobarbituric acid (TBA) and sodium borohydride were purchased from Merck (Darmstadt, Germany). Chromatography-HPLC grade solvents, including chloroform, methanol, and acetonitrile, were purchased from Fisher Scientific (Schwerte, Germany), J.T. Baker (Gliwice, Poland), and Sigma Aldrich (Steinheim, Germany), respectively. The carboxymethyl-lysine standard was obtained from Cayman Chemical (Ann Arbor, MI, USA). The standard substances for volatile compound analysis, ethyl acetate, pentanal, hexanal, heptanal, octanal, nonanal, 2,4-decadienal, 2-heptanone, 1-pentanol, 1-hexanol, and 1-heptanol were purchased from Merck (Hohenbrunn, Germany), and the standard mix was purchased from Supelco (parrafine mix, 44585-U, Bellefonte, PA, USA).

2.2. Material

In the study, lean beef, beef fat (intermuscular fat), and sheep tail fat were used as raw materials. The meat was obtained from the round part of the cattle carcasses by a local butcher. Beef fat and sheep tail fat also came from the same butcher. Since the research was conducted in three replications, meat and fat were procured at three different times.

2.3. Meatball Production and Cooking

Lean beef, beef fat, and sheep tail fat were minced separately through a 3 mm plate using a meat mincer (MADO Typ MEW 717, Dornhan, Germany). The experiment was carried out according to a 2 × 3 × 4 factorial design with two types of fat (beef fat and sheep tail fat), three levels of fat (10%, 20%, and 30%), and four levels of cooking time (0, 2, 4, and 6 min at 180 °C on a hot plate). Six meatball patties were prepared based on fat type and fat level and mixed by hand and processed into meatballs (1.5 cm thick and 7.5 cm diameter) using a metal shaper. The weight of the meatballs was 50 g. Salt was added to the mix at a rate of 1.5%. Three independent experiments were carried out, and thus a total of 18 meatball mixes were prepared.

Each group was divided into four subgroups. The first group of meatballs were evaluated as the control group (raw–uncooked). The second, third, and fourth group of meatballs were cooked for 2 min (1 min per side), 4 min (2 min per side), and 6 min (3 min per side), respectively. The cooking process was carried out on a hot plate preheated at 180 °C and the temperature was controlled using a thermocouple.

2.4. Pysicochemical Analysis

After the cooking process, the samples were cooled down to room temperature and then homogenized with a blender to obtain a uniform sample for the analyses. The homogenized samples were stored at −18 °C until analysis.

2.4.1. pH and TBARSs

For pH measurements, 10 g of homogenized sample was weighed and 100 mL of distilled water was added to it. After homogenizing with an ultra-turrax (IKA Werk T 25, Staufen, Germany) for 1 min, the pH value was determined by a pH meter (Mettler Toledo, Greifensee, Switzerland). The pH meter was calibrated with buffer solutions (pH 4.0 and pH 7.0) before use, and thermal compensation was carried out automatically [25].

In TBARS analyses, 2 g of homogenized sample was mixed with a 12 mL TCA solution (7.5% TCA, 0.1% EDTA, and 0.1% propyl gallate). After homogenization, it was filtered through a Whatman 1 filter paper and 3 mL of filtrate was added to a 0.02 M TBA solution (0.02 M). Afterwards, samples were kept in boiling water bath for 40 min, then centrifugation (Beckman Coulter, Allegra X-30R, Indianapolis, IN, USA) was applied for 5 min at 2000 G. The absorbance of the samples was determined at 530 nm, and the results were expressed as μmol MDA/kg [26].

2.4.2. Nε-(carboxymethyl)lysine (CML)

CML analysis was performed according to the method given by Chen and Smith [11]. After a 0.2 g sample was added to 20 mL of chloroform/methanol (2:1, v/v), it was centrifuged (10,000× g at 4 °C) (Beckman Coulter, Brea, CA, USA) to remove the fat. After incubation for 4 h with 4 mL sodium borohydride (1 M in 0.1 N NaOH) and 8 mL sodium borate buffer (0.2 M, pH 9.4), samples were hydrolyzed with 6 mL 12 M HCl at 110 °C for 20 h. Then, samples were dried using rotary evaporation and were added to 10 mL of water and dissolved in 10 mL of sodium borate buffer, followed by a final filtration. Prior to HPLC analysis, a 50 μL extract was mixed with 200 μL of ortho-phtalaldehyde derivatization reagent for 5 min. The CML amount was determined using HPLC (Agilent 1100, Santa Clara, CA, USA) with a fluorescence detector (Agilent, Santa Clara, CA, USA). The determination was performed with a reverse phase TSK gel ODS-80 TM column (25 cm × 4.6 mm, 5 μm, Tosohass, Montgomeryville, PA, USA) and with the fluorescence settings of 340 nm (excitation) and 455 nm (emission). The flow rate was 1.0 mL/min and the injection volume was 20 μL. Acetate buffer/acetonitrile (90:10, v/v) and acetonitrile were used as mobile phases, and the flow of the mobile phase was gradually changed. The recovery rates were determined by spiking cooked meat with N-ε-(carboxymethyl)-lysine standard at six different levels (1–30 µg/mL). Each treatment was replicated five times. The mean recoveries ranged from 102.20% to 104.46%, with relative standard deviations between 1.10% and 2.18%. The regression line coefficient (R^2) for CML was 0.999. The limit of detection (LOD) and the limit of quantification (LOQ) were calculated using dilutions of the standard solution according to the following formulas: LOD = 3.3 × Sy/s and LOQ = 10 × Sy/s. The LOD and LOQ for CML were 0.64 and 1.95 µg/mL, respectively.

2.4.3. Volatile Compounds

An amount of 5 g of homogenized sample was placed into a 40 mL vial (Supelco, Bellefonte, PA, USA). The extraction of volatile compounds was performed using solid phase microextraction (SPME) with carboxen/polydimethylsiloxane fiber (CAR/PDMS, 75 µm, Supelco, Bellefonte, PA, USA). The sample was kept at 30 °C for 1 h in a thermal block (Supelco, Bellefonte, PA, USA) to collect the volatile compounds. After equilibration, the SPME fiber was exposed to the sample headspace at 30 °C for 2 h. Gas chromatography/mass spectrometry (Agilent, Santa Clara, CA, USA) was used to identify volatile compounds. A DB-624 (J&W Scientific, 30 m × 0.25 mm × 1.4 µm film) was used as the column and the carrier gas was helium. The oven temperature was first set to 40 °C for 6 min, then gradually increased to 210 °C and then held at 210 °C for 12 min. The injector port was in splitless mode. The GC/MS interface was maintained at 280 °C. Mass spectra were obtained by electron impact at 70 eV, and the quadrupole mass spectrometer scan range was 40–400 atomic mass units.

Mass spectrometry libraries (NIST, FLAVOR, and WILEY) and standard substances were used for the identification of compounds, and the Kovats index was determined using the standard mix (Supelco 44585-U). The results are given as AU × 10^6 [27].

2.5. Statistical Analysis

Data were analyzed by an analysis of variance (ANOVA) using a general linear model considering the fat type (beef fat and sheep tail fat), fat level (10, 20, and 30%), and cooking time (0, 2, 4, and 6 min) as the main effects, and the replicates as a random effect for a randomized complete block design. The experiment was repeated 3 times and each experiment was carried out at different times using different raw materials. The differences between the means were determined using Duncan's multiple range tests at the $p < 0.05$ level. All statistical analyses were performed using SPSS version 20 statistical program (SPSS Inc., Chicago, IL, USA). In addition, principal component analysis (PCA) was performed to determine the relationship between fat type, fat level, and cooking time for volatile compounds as well as CML content using Unscrambler software (CAMO version 10.1, Oslo, Norway). The differential profile (cluster heat map) between the factors, volatile compound groups, and CML was also analyzed using heat mapper (http://www.heatmapper.ca, accessed on 1 June 2023).

3. Results and Discussion

3.1. pH and TBARSs

The effects of fat type, fat level, and cooking time on pH, TBARS value, and the CML content of meatballs are given in Table 1. The fat type and fat level had no significant effect on the pH value of samples ($p > 0.05$) (Table 1). In contrast, the cooking time had a very significant impact ($p < 0.01$) on the pH, and the mean pH value increased with increasing cooking times. However, no significant differences in pH value were observed between 4 and 6 min of cooking. Furthermore, the interactions of all factors were insignificant ($p > 0.05$) (Table 1). The increase in pH value with cooking can be explained by the reduction of carboxylic groups on proteins and also by the release of calcium and magnesium ions from proteins [28].

Table 1. The effects of fat type, fat level, and cooking time on pH, TBARSs, and CML of meatballs (mean ± SD).

Factors	N	pH	TBARS (µmol MDA/kg)	CML (µg/g)
Fat type (FT)				
Beef fat	36	5.97 ± 0.12 a	10.36 ± 2.57 b	8.98 ± 2.71 b
Sheep tail fat	36	5.95 ± 0.15 a	12.02 ± 2.07 a	12.31 ± 5.44 a
Significance		ns	**	**
Fat level (FL)				
%10	24	5.94 ± 0.14 a	10.91 ± 1.98 b	10.16 ± 2.94 a
%20	24	5.95 ± 0.15 a	10.68 ± 2.61 b	11.42 ± 4.78 a
%30	24	5.98 ± 0.13 a	11.98 ± 2.64 a	10.36 ± 5.69 a
Significance		ns	*	ns
Cooking time (min)(CT)				
0 (Raw)	18	5.80 ± 0.10 c	9.88 ± 2.20 b	7.27 ± 1.81 c
2 (Rare)	18	5.93 ± 0.09 b	10.55 ± 2.17 b	9.87 ± 2.18 b
4 (Medium)	18	6.03 ± 0.09 a	11.69 ± 2.24 a	11.01 ± 3.54 b
6 (Medium–well)	18	6.06 ± 0.09 a	12.64 ± 2.46 a	14.44 ± 6.26 a
Significance		**	**	**
Interaction				
FT × FL		ns	**	ns
FT × CT		ns	ns	*
FL × CT		ns	ns	ns
FT × FL × CT		ns	ns	ns

a–c: Means marked with different letters in the same column are statistically different ($p < 0.05$), * $p < 0.05$; ** $p < 0.01$. ns: not significant.

Processed meat products are more sensitive to lipid oxidation than fresh meat due to mincing and heat treatment. Malondialdehyde is evaluated to be the most significant degradation product, arising from lipid oxidation, and a TBARS analysis is widely used in its determination [29]. The TBARS value was affected by the fat type ($p < 0.01$), and the highest mean TBARS value was observed in the group with sheep tail fat (Table 1). It is thought that this result is due to the fact that sheep tail fat contains more unsaturated fatty acids than beef fat and therefore is more sensitive to oxidation [14]. In addition, the fat level ($p < 0.05$) and cooking time ($p < 0.01$) had an effect on TBARSs. The highest TBARS value was found for the groups containing 30% fat. In addition, the mean TBARS value increased with increasing cooking time (Table 1). There is no legal limit for the MDA level determined in the TBARS analysis applied to determine the degree of lipid oxidation in meat products [30]. However, it is suggested that when the TBARS value reaches 1 mg MDA/kg, it can create malodor and it can be detected [31]. In this study, in all groups except for 6 min of cooking time, the TBARS value was below 1 mg MDA/kg.

As shown in Figure 1, the fat type × fat level interaction was determined. The TBARS value for beef fat was similar at all fat levels. On the other hand, meatballs prepared with sheep tail fat had the highest TBARS value at 30% fat level, and the lowest TBARS value at 10% fat level. Furthermore, differences between fat types were not significant at 10% and 20% fat levels, and at the 30% fat level, sheep tail fat had a higher mean TBARS value than beef fat. According to these results, 30% sheep tail fat significantly increased lipid oxidation (Figure 1). This result is likely due to the fact that sheep tail fat contains more unsaturated fatty acids than beef fat [14].

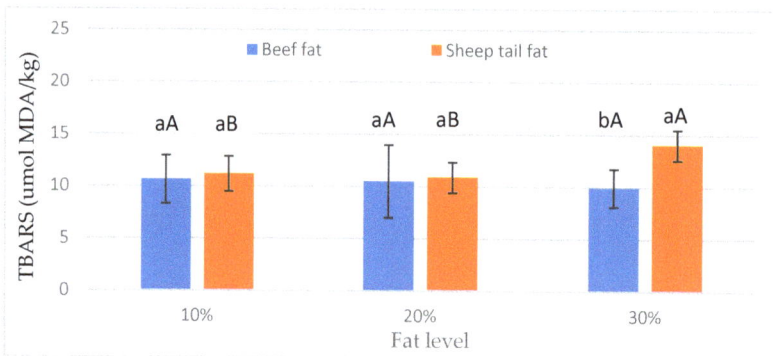

Figure 1. Effect of fat type × fat level interactions on TBARS values of meatballs. a, b: different small letters indicate significant differences between fat types for fat levels. A, B: different capital letters indicate significant differences between fat levels for fat types.

3.2. Nε-(carboxymethyl)lysine (CML)

Fat type had a very significant ($p < 0.01$) effect on CML content. Cooking time also showed an effect at the $p < 0.01$ level on CML content (Table 1). Sheep tail fat has a higher CML content than beef fat. At the same time, sheep tail fat showed a higher TBARS value, an indicator of lipid oxidation, than beef fat (Table 1). These results show a positive relationship between CML formation and lipid oxidation. It has also been determined in some other studies conducted on meat products that lipid oxidation increases CML formation [7,32]. In addition, in a study performed on fresh cooked meats, it was reported that irradiation accelerated lipid oxidation and thus enhanced CML formation [33]. Yu et al. [34], in their study on raw and heat-treated meats, reported that there was a positive correlation between the TBARS value and CML and that high temperatures promoted AGE formation. In the present study, sheep tail fat, which is more susceptible to autoxidation than beef fat, caused a significant increase in the TBARS value and therefore accelerated the formation of CML.

In the present study, the CML content increased with the duration of cooking time. However, no statistically significant difference was found between the samples cooked for 2 min and 4 min. The highest mean CML content was observed after 6 min (Table 1). The cooking method, temperature, and time are important factors in the formation of CML in meat products. In addition, in a study conducted on ground beef, it was revealed that the CML content increased with the heat treatment time and temperature (65–100 °C and 0 to 60 min) [35]. Additionally, it was stated that in cooked meat products, factors such as the source and composition of the meat, protein, and fat content are effective in the formation of AGEs [10–12]. In the present study, the CML content of the meatballs had no effect depending on the fat level. However, Bayrak Kul et al. [36] reported that the fat level is an important factor in the formation of CML, and the lowest CML content was found when using 10% fat. It is believed that this difference is probably due to the product type and cooking conditions.

The CML content of meatballs was significantly ($p < 0.05$) affected by the interaction of fat type and cooking time (Figure 2). The CML level increased in meatballs prepared using beef fat after 2 min of cooking, and a prolonged cooking time did not affect the CML levels (Figure 2). On the other hand, the CML content increased with increasing cooking time in meatballs with sheep tail fat. However, there was no statistical difference in CML contents between 2 and 4 min of cooking time. In addition, it was observed that the fat type was not important in the raw and cooked samples after 2 min. In the case of prolonged cooking times, the groups containing sheep tail fat had a higher CML content (Figure 2). This result shows that sheep tail fat is more effective in CML formation when the heat treatment time is increased.

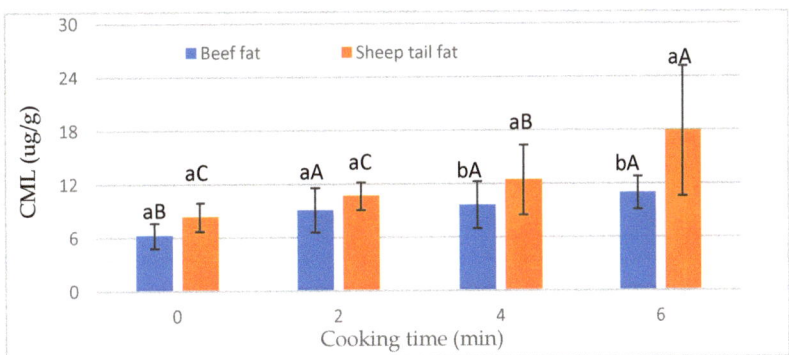

Figure 2. Effect of interaction of fat type and cooking time on the CML value of meatballs. a, b: different small letters indicate significant differences between fat type for cooking time. A–C: different capital letters indicate significant differences between cooking time for fat type.

Although the mechanism of the Maillard reaction has not yet been fully explained [37,38], it is known that the reaction rate may vary depending on the amount of reducing sugar and free amino groups in the environment and the temperature [37]. In our study, it was also revealed that the CML level increased as the time increased in dry heat cooking at 180 °C.

3.3. Volatile Compounds

A total of 25 compounds including aliphatic hydrocarbons, esters, aldehydes, ketones, alcohols, and furans were identified (Table 2). The fat type was found to have a significant impact on ethyl acetate, heptanal, 2-nonenal, 2,4-decadienal, 2-butanone, 2-octanone, 2-nonanone, 1-pentanol, 1-hexanol, 1-octene-3-ol, and furan. In addition, butyl propionate, pentanal, hexanal, and 4-decenal were affected by the fat type at a level of $p < 0.05$. The fat level caused a statistical difference at the $p < 0.05$ or $p < 0.01$ levels for five compounds (pentanal, hexanal, heptanal, octanal, and 2-butanone). The cooking time, another factor, had a very significant or significant effects on ethyl acetate, pentanal, hexanal, heptanal, octanal, 2,4-decadienal, and 2-octanone ($p < 0.05$ or $p < 0.01$) (Table 2).

Ethyl acetate, butyl propionate, 2-butanone, 1-octen-3-ol, 1-hexanol, and 1-pentanol gave higher mean values in the groups containing beef fat compared to the groups containing sheep tail fat. On the other hand, the highest abundances were observed in meatballs with sheep tail fat for pentanal, hexanal, heptanal, 2-nonenal, 4-decenal, 2,4-decadienal, 2-octanone, 2-nonanone, and furan (Table 2). As can be seen from Table 2, aldehydes were higher in the group containing sheep tail fat than the group with beef fat. This result is due to the high content of polyunsaturated fatty acids in sheep tail fat [14] and thus to faster lipid oxidation. In fact, compounds such as pentanal, hexanal, heptanal, octanal, and nonanal result from lipid oxidation. At the same time, most of the volatile compounds in cooked meat are formed by lipid reactions [39]. Pentanal, hexanal, heptanal, octanal, and 2-butanone compounds, which were found to be statistically significant, gave the highest average values in the groups containing 30% fat. However, pentanal, heptanal, and octanal compounds did not differ statistically between the groups containing 20% fat and the groups containing 30% fat. There was no statistical difference between the groups containing 10% and 20% fat in terms of hexanal, heptanal, octanal, and 2-butanone compounds ($p > 0.05$) (Table 2).

Table 2. The effects of fat type, fat level, and cooking time on volatile compounds of meatballs (arbitrary units × 10^6).

Compounds	KI	R	Fat Type		Fat Level (%)			Cooking Time (min)			
			Beef fat	Sheep Tail Fat	10%	20%	30%	0	2	4	6
Hexane	600	a	3.67 ± 3.03	3.24 ± 2.46	3.44 ± 2.77	3.54 ± 3.00	3.39 ± 2.58	3.82 ± 2.67	4.66 ± 2.96	2.40 ± 2.74	2.93 ± 2.22
Octane	800	a	1.03 ± 1.51	0.67 ± 0.98	0.53 ± 0.96	0.73 ± 1.11	1.29 ± 1.61	0.74 ± 1.06	0.87 ± 1.11	0.70 ± 1.37	1.11 ± 1.58
Decane	1000	a	2.11 ± 2.91	1.18 ± 2.26	1.51 ± 2.47	1.40 ± 2.42	2.03 ± 3.02	2.01 ± 3.05	1.64 ± 2.62	1.28 ± 2.35	1.66 ± 2.62
Dodecane	1200	a	1.74 ± 2.32	2.06 ± 2.34	1.66 ± 1.80	1.21 ± 2.01	2.82 ± 2.81	2.80 ± 1.98	1.48 ± 2.43	1.15 ± 1.85	2.16 ± 2.74
Ethyl acetate	648	a	3.82 ± 1.68 a	1.26 ± 2.30 b	2.22 ± 1.65	2.18 ± 2.22	3.21 ± 3.01	1.39 ± 1.26 b	2.76 ± 2.52 a	3.21 ± 2.85 a	2.80 ± 2.34 a
Butyl propionate	952	a	1.08 ± 2.37 a	0.27 ± 0.69 b	0.58 ± 1.29	1.19 ± 0.53	1.26 ± 2.69	0.56 ± 1.31	0.26 ± 0.63	0.63 ± 2.18	1.25 ± 2.41
Pentanal	742	a	2.25 ± 1.84 b	2.92 ± 1.65 a	1.62 ± 1.36 c	2.68 ± 1.61 b	3.46 ± 1.85 a	1.00 ± 1.08 c	2.32 ± 1.53 b	3.45 ± 1.62 a	3.58 ± 1.54 a
Hexanal	849	a	2.88 ± 1.32 b	3.30 ± 1.26 a	2.20 ± 0.90 b	3.34 ± 1.22 a	3.74 ± 1.26 a	1.88 ± 0.70 c	3.05 ± 1.06 b	3.40 ± 1.17 b	4.03 ± 1.21 a
Heptanal	955	a	2.33 ± 0.91 b	3.36 ± 1.10 a	2.45 ± 0.82 b	2.86 ± 1.22 ab	3.22 ± 1.22 a	2.53 ± 0.97 b	2.53 ± 1.37 b	3.07 ± 0.97 ab	3.25 ± 1.07 a
Octanal	1044	a	3.27 ± 1.03	3.56 ± 1.02	3.20 ± 0.87 b	3.26 ± 1.06 ab	3.77 ± 1.09 a	2.56 ± 0.82 c	3.43 ± 0.82 b	3.57 ± 0.81 ab	4.08 ± 1.08 a
Nonanal	1146	a	1.88 ± 2.86	2.67 ± 2.40	2.58 ± 2.83	2.23 ± 2.85	2.01 ± 2.33	2.14 ± 2.32	2.60 ± 2.77	2.76 ± 3.02	1.59 ± 2.52
2-Nonenal	1219	b	1.13 ± 2.04 b	3.65 ± 2.29 a	2.07 ± 2.37	2.19 ± 2.36	2.91 ± 2.76	1.70 ± 2.62	1.92 ± 2.03	2.41 ± 2.30	3.52 ± 2.78
4-Decenal	1246	b	2.53 ± 3.56 b	4.47 ± 3.62 a	4.24 ± 3.83	3.00 ± 3.60	3.26 ± 3.69	3.57 ± 3.61	3.56 ± 3.77	3.73 ± 3.96	3.14 ± 3.72
2,4-Decadienal	1363	a	1.00 ± 1.79 b	3.72 ± 1.08 a	2.01 ± 1.81	2.34 ± 1.97	2.71 ± 2.23	1.69 ± 1.83 b	1.58 ± 1.79 b	2.81 ± 2.21 a	3.34 ± 1.73 a
2-Butanone	780	b	2.36 ± 2.07 a	1.43 ± 2.38 b	1.50 ± 1.90 b	1.66 ± 1.97 b	2.53 ± 2.77 a	1.26 ± 1.81	2.08 ± 2.42	1.99 ± 2.42	2.25 ± 2.41
2-Heptanone	948	b	2.28 ± 2.27	1.82 ± 2.63	2.27 ± 2.52	1.44 ± 2.16	2.44 ± 2.63	1.63 ± 2.07	2.09 ± 2.87	1.95 ± 2.15	2.52 ± 2.74
2-Octanone	1035	b	0.00 ± 0.00 b	2.50 ± 2.91 a	1.16 ± 2.60	0.94 ± 2.15	1.65 ± 3.54	2.39 ± 3.54 a	0.93 ± 1.87 b	0.94 ± 1.84 b	0.76 ± 1.68 b
2-Nonanone	1079	b	0.00 ± 0.00 b	2.34 ± 2.56 a	1.22 ± 2.29	0.69 ± 1.51	1.60 ± 2.51	0.91 ± 1.04	1.20 ± 2.07	1.36 ± 2.33	1.21 ± 2.90
1-Pentanol	835	a	2.58 ± 2.22 a	0.00 ± 0.00 b	0.85 ± 1.36	1.17 ± 1.76	1.86 ± 2.68	0.58 ± 2.16	1.35 ± 1.98	1.55 ± 1.92	1.68 ± 2.04
1-Hexanol	935	a	2.67 ± 2.10 a	1.09 ± 2.03 b	1.79 ± 1.85	1.85 ± 2.33	2.00 ± 2.47	1.13 ± 2.08	2.11 ± 2.12	2.69 ± 2.15	1.58 ± 2.32
1-Heptanol	1033	a	1.37 ± 1.63	1.00 ± 2.04	1.40 ± 2.15	1.22 ± 1.95	0.93 ± 1.38	1.05 ± 1.93	1.63 ± 1.80	0.99 ± 1.83	1.06 ± 1.90
1-Octen-3-ol	1041	b	2.16 ± 2.83 a	0.85 ± 1.60 b	1.37 ± 2.18	1.42 ± 2.07	1.71 ± 2.89	2.04 ± 2.92	1.16 ± 1.89	0.97 ± 1.82	1.85 ± 2.71
2-Ethyl-1-hexanol	1084	b	1.95 ± 2.70	2.65 ± 2.79	2.36 ± 2.81	1.94 ± 2.88	2.60 ± 2.63	1.70 ± 2.35	1.94 ± 2.53	1.63 ± 2.16	3.92 ± 3.36
2-Pentyl furan	1022	b	0.85 ± 2.36	2.12 ± 3.14	2.02 ± 3.36	0.82 ± 2.05	1.61 ± 2.91	0.88 ± 2.06	1.78 ± 2.93	1.70 ± 3.33	1.59 ± 2.99
Furan	1107	b	0.50 ± 1.20 b	2.74 ± 3.18 a	1.11 ± 1.36	2.23 ± 3.51	1.52 ± 2.58	1.50 ± 2.64	1.08 ± 1.29	2.23 ± 3.57	1.68 ± 2.64

Results are expressed in arbitrary area units (×10^6); KI: Kovats index calculated for a DB-624 capillary column (30 m × 0.25 mm × 1.4 μm film) installed on a gas chromatograph equipped with a mass selective detector. R: reliability of identification, a: mass spectrum and retention time identical to an authentic sample; b: mass spectrum and Kovats index from the literature. a–c: means marked with different letters in the same row in same section are statistically different ($p < 0.05$).

Ethyl acetate increased with the cooking time of 2 min, but increases in the cooking times to 4 and 6 min did not cause a statistical difference. Pentanal, hexanal, octanal, and 2,4-decadienal generally increased with the prolongation of the cooking time, but no statistical difference was found between the 4 and 6 min durations of the other compounds except for hexanal ($p > 0.05$). On the other hand, the level of 2-octanone, which is statistically significant among the defined ketone compounds, decreased with cooking (Table 2).

A principal component analysis (PCA) was applied to evaluate the relationships between factors (fat type, fat level, and cooking time) and volatile compounds (Figure 3). The PCA provided a good distinction between groups containing sheep tail fat (STF) and beef fat (BF) regardless of the fat content and cooking time. The groups with STF, which have different fat levels and cooking times, were on the positive side of PC1, while all BF groups were on the negative side of PC1. This result indicated that the fat type is very important for the volatile component profile of meatballs. Lipids play an important role in the development of odor and flavor of foods due to them being precursors of odor and flavor compounds or modifying the odor and flavor of other components [40].

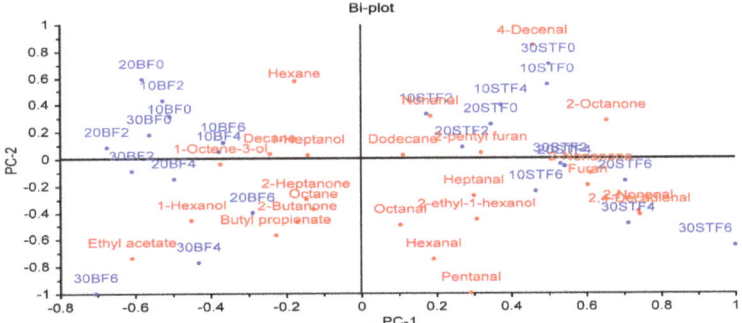

Figure 3. The biplot result of a principal component analysis of the relationships between factors and volatile compounds (the first number indicates the % fat level, BF: beef fat, SFT: sheep tail fat, and the last number indicates the cooking time (min)).

Among the volatile compounds, aldehydes showed a higher positive correlation with groups containing STF. Furthermore, meatballs with 20% and 30% BF cooked for 4 and 6 min, and meatballs containing 10, 20, and 30% STF cooked for 6 min were placed on the negative side of PC2 and showed a positive correlation with each other (Figure 3).

3.4. Evaluation of the Relationship between Fat Type, Fat Level, Cooking Time, CML, and Volatile Compounds

A PCA was applied to determine the relationships between fat type, fat level, cooking time, volatile compounds, and CML (Figure 4). The first PC explained 54% of the variation. PC2 accounted for 26% of the total variance. The first two principal components explained 80% of the total variance. CML showed a positive correlation with sheep tail fat, 20 and 30% fat levels, and a cooking time of 6 min. These factors were on the positive side of PC1. In contrast, beef fat, a 10% fat level, and 0 (raw) and 2 min cooking times were on the negative side of PC1, showing a negative correlation with CML. Although CML was more affected by cooking time, it was determined that fat level and fat type also had an effect on CML formation.

In this study, the correlation between factors, volatiles, and CML was also evaluated using a heat map (Figure 5). Average linkage was used as the clustering method and Pearson correlation was used as the distance measure to illustrate the heat map. The X and Y axes represent the factors (fat type, fat level, and cooking time), and volatile compounds groups and CML, respectively. Yellow and blue were used for higher and lower correlation coefficients, respectively.

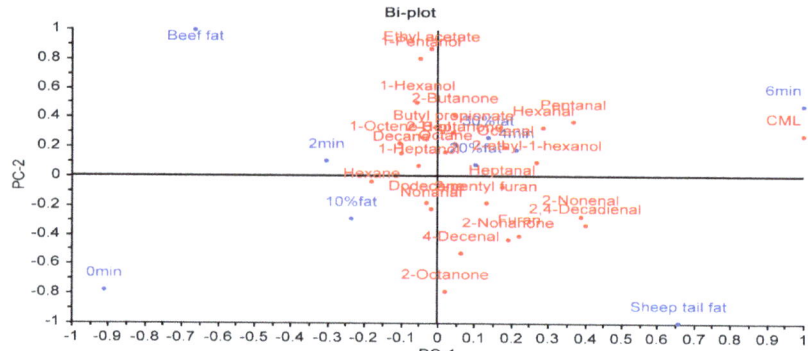

Figure 4. The biplot result of a principal component analysis of the relationships between factors, volatile compounds, and CML.

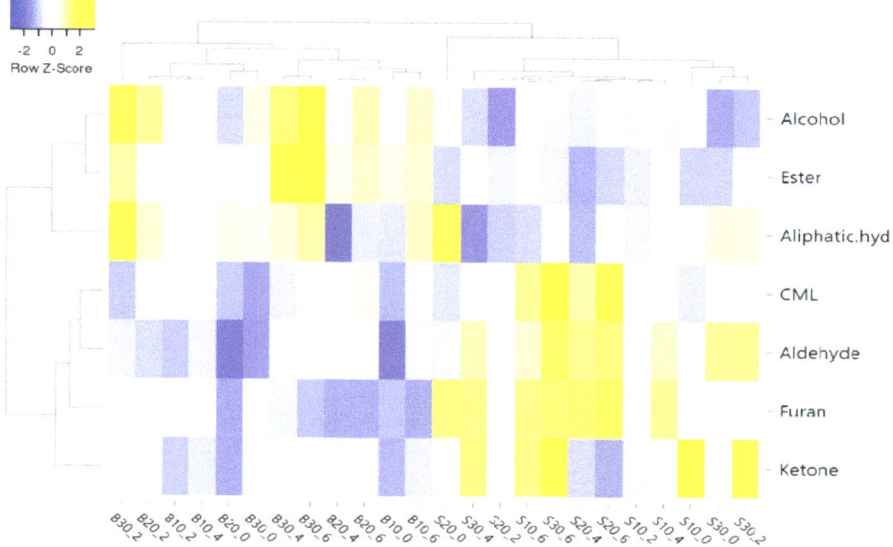

Figure 5. Cluster analysis of a heat map showing the relationship between factors, volatile compound groups, and CML (B: beef fat; S: sheep tail fat; 10, 20, and 30: fat level %; 0, 2, 4, and 6: cooking time (min)).

Figure 5 shows the two main clusters, and the first cluster includes ketones, furans, aldehydes, and CML, while the second cluster includes aliphatic hydrocarbons, esters, and alcohols. The first cluster was separated two subclusters. CML was generally more correlated with STF than BF. In addition, groups containing 10, 20, and 30% STF and cooked for 6 min were more correlated with CML. CML had a more positive correlation with aldehydes (Figure 5). Among the aldehydes, pentanal, octanal, 2,4 decadienal, hexanal, and heptanal were positively correlated with CML. In terms of volatile compounds, all groups containing beef fat and sheep tail fat were divided into two different clusters, regardless of the fat level and cooking time (Figure 6).

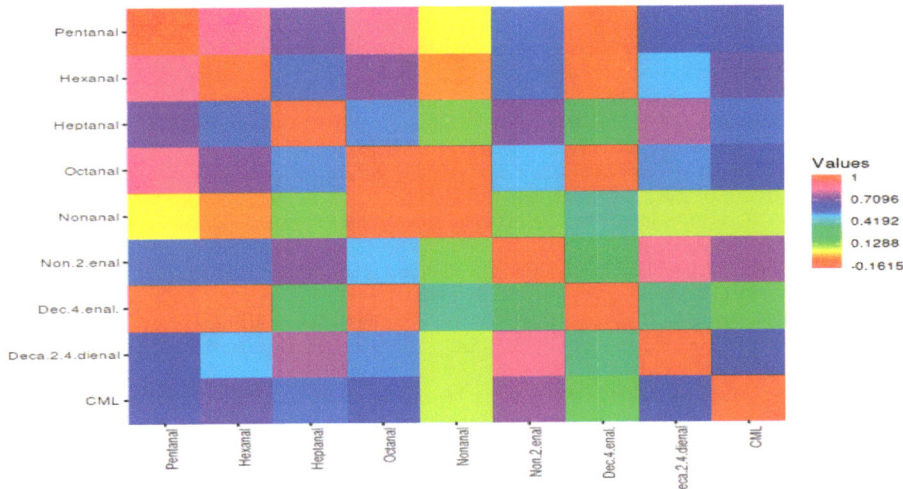

Figure 6. Heat map correlation matrix of CML and aldehyde compounds.

4. Conclusions

The use of STF at the level of 30% in meatball production significantly increased lipid oxidation. On the other hand, the BF level did not affect lipid oxidation. In contrast, the cooking time had an effect on TBARSs. Likewise, the CML content increased with increasing cooking times. However, in meatballs prepared with BF, no significant differences in the CML content were observed between cooked samples. In the presence of STF, the highest CML value was determined at the 30% fat level. In addition, fat type did not have a significant effect on CML in both raw samples and samples cooked for 2 min. When the cooking time was increased to 4 or 6 min, STF gave a higher CML content than BF. Among the volatile compounds, aldehydes were more affected by the factors examined, and these compounds exhibited a close relationship with CML. The use of sheep tail fat in meatball production, especially at the 30% level, significantly increased both CML formation and lipid oxidation.

The results showed that cooking time and the type of fat used in production are important factors for the formation of CML in meatballs. An increase in the fat level in meatballs had a positive effect on the formation of CML if the fat contained a high amount of unsaturated fatty acids (especially PUFA). These findings suggest that lipid oxidation plays an important role in the formation of CML during the cooking of beef meatballs. On the other hand, the results of this study provide new ideas for future studies on reducing CML in cooked meat products.

Author Contributions: Methodology, K.Ö. and G.K.; validation, G.K.; formal analysis, K.Ö. and Z.F.Y.O.; investigation, K.Ö. and G.K.; writing—original draft preparation, K.Ö. and Z.F.Y.O.; writing—review and editing, M.K. and G.K.; supervision, G.K.; project administration, G.K. All authors have read and agreed to the published version of the manuscript.

Funding: This research received no external funding.

Data Availability Statement: The data presented in this study are available on request from the corresponding author.

Conflicts of Interest: The authors declare no conflict of interest.

References

1. Etxabide, A.; Kilmartin, P.A.; Maté, J.I.; Prabakar, S.; Brimble, M.; Naffa, R. Analysis of advanced glycation end products in ribose-, glucose-and lactose-crosslinked gelatin to correlate the physical changes induced by Maillard reaction in films. *Food Hydrocoll.* **2021**, *117*, 106736. [CrossRef]
2. Şen, D.; Gökmen, V. Kinetic modeling of Maillard and caramelization reactions in sucrose-rich and low moisture foods applied for roasted nuts and seeds. *Food Chem.* **2022**, *395*, 133583. [CrossRef] [PubMed]
3. Jia, W.; Guo, A.; Zhang, R.; Shi, L. Mechanism of natural antioxidants regulating advanced glycosylation end products of Maillard reaction. *Food Chem.* **2022**, *404*, 134541. [CrossRef] [PubMed]
4. Lund, M.N.; Ray, C.A. Control of Maillard reactions in foods: Strategies and chemical mechanisms. *J. Agric. Food Chem.* **2017**, *65*, 4537–4552. [CrossRef] [PubMed]
5. Chen, G.; Madl, R.L.; Smith, J.S. Inhibition of advanced glycation end products in cooked beef patties by cereal bran addition. *Food Cont.* **2017**, *73*, 847–853. [CrossRef]
6. Arena, S.; Salzano, A.M.; Renzone, G.; Dambrosio, C.; Scaloni, A. Non-enzymatic glycation and glycoxidation protein products in foods and diseases: An interconnected. complex scenario fully open to innovative proteomic studies. *Mass Spectro. Rev.* **2013**, *33*, 49–77. [CrossRef]
7. Zhu, Z.; Huang, M.; Cheng, Y.; Khan, A.I.; Huang, J. A comprehensive review of Nε-carboxymethyllysine and Nε- carboxyethyllysine in thermal processed meat products. *Trends Food Sci. Technol.* **2020**, *98*, 30–40. [CrossRef]
8. Uribarri, J.; Woodruff, S.; Goodman, S.; Cai, W.; Chen, X.; Pyzik, R.; Yong, A.; Forvet, A.E.; Vlassara, H. Advanced glycation end products in foods and a practical guide to their reduction in the diet. *J. Am. Diet. Assoc.* **2010**, *110*, 911–916. [CrossRef]
9. Dong, L.; Li, Y.; Chen, Q.; Liu, Y.; Wu, Z.; Pan, D.; Yan, N.; Liu, L. Cereal polyphenols inhibition mechanisms on advanced glycation end products and regulation on type 2 diabetes. *Crit. Rev. Food Sci. Nutr.* **2023**, *24*, 1–19. [CrossRef]
10. Poulsen, W.M.; Hedegaard, V.R.; Andersen, M.J.; Courten, B.; Bügel, S.; Nielsen, J.; Skibsted, H.; Dragsted, O.L. Advanced glycation end products in food and their effects on health. *Food Chem. Toxicol.* **2013**, *60*, 10–37. [CrossRef]
11. Chen, G.; Scott Smith, J. Determination of advanced glycation endproducts in cooked meat products. *Food Chem.* **2015**, *168*, 190–195. [CrossRef]
12. Sun, X.; Tang, X.; Wang, J.; Rasco, A.B.; Lai, K.; Huang, Y. Formation of free and protein-bound carboxymethyllysine and carboxyethyllysine in meats during commercial sterilization. *Meat Sci.* **2016**, *116*, 1–7. [CrossRef]
13. Turgut, S.S.; Işıkçı, F.; Soyer, A. Antioxidant activity of pomegranate peel extract on lipid and protein oxidation in beef meatballs during frozen storage. *Meat Sci.* **2017**, *129*, 111–119. [CrossRef]
14. Şişik Oğraş, Ş.; Akköse, A.; Kaban, G.; Kaya, M. Volatile profile and fatty acid composition of kavurma (a cooked uncured meat product) produced with animal fat combinations. *Int. J. Food Prop.* **2018**, *21*, 364–373. [CrossRef]
15. Akköse, A.; Oğraş, Ş.Ş.; Kaya, M.; Kaban, G. Microbiological, physicochemical and sensorial changes during the ripening of sucuk, a traditional Turkish dry-fermented sausage: Effects of autochthonous strains, sheep tail fat and ripening rate. *Fermentation* **2023**, *9*, 558. [CrossRef]
16. Sallan, S. Influence of sheep tail fat and autochthonous starter culture on the formation of volatile nitrosamines in sucuk. *Kafkas Univ. Vet. Fak. Derg.* **2023**, *29*, 171–176. [CrossRef]
17. Ghafari, H.; Rezaeian, M.; Sharifi, S.D.; Khadem, A.A.; Afzalzadeh, A. Effects of dietary sesame oil on growth performance and fatty acid composition of muscle and tail fat in fattening Chaal lambs. *Anim. Feed Sci. Technol.* **2016**, *220*, 216–225. [CrossRef]
18. Jiang, T.; Busboom, J.R.; Nelson, M.L.; O'Fallon, J.; Ringkob, T.P.; Joos, D.; Piper, K. Effect of sampling fat location and cooking on fatty acid composition of beef steaks. *Meat Sci.* **2010**, *84*, 86–92. [CrossRef]
19. Jiang, T.; Mueller, C.J.; Busboom, J.R.; Nelson, M.L.; O'Fallon, J.; Tschida, G. Fatty acid composition of adipose tissue and muscle from Jersey steers was affected by finishing diet and tissue location. *Meat Sci.* **2013**, *93*, 153–161. [CrossRef] [PubMed]
20. Li, Y.; Li, Y.B.; Liu, C.J. Changes in lipid oxidation and fatty acids in altay sheep fat during a long time of low temperature storage. *J. Oleo. Sci.* **2017**, *66*, 321–327. [CrossRef] [PubMed]
21. Ünsal, M.; Aktaş, N. Fractionation and characterization of edible sheep tail fat. *Meat Sci.* **2003**, *63*, 235–239. [CrossRef] [PubMed]
22. Li, Y.; Xue, C.; Quan, W.; Qin, F.; Wang, Z.; He, Z.; Zeng, M.; Chen, J. Assessment the influence of salt and polyphosphate on protein oxidation and Nε-(carboxymethyl)lysine and Nε-(carboxyethyl)lysine formation in roasted beef patties. *Meat Sci.* **2021**, *177*, 108489. [CrossRef]
23. Yu, L.; Li, Q.; Li, Y.; Yang, Y.; Guo, C.; Li, M. Impact of frozen storage duration of raw pork on the formation of advanced glycation end-products in meatballs. *LWT Food Sci. Technol.* **2021**, *146*, 111481. [CrossRef]
24. Xue, C.; Quan, W.; Li, Y.; He, Z.; Qin, F.; Wang, Z.; Zeng, M. Mitigative capacity of Kaempferia galanga L. and kaempferol on heterocyclic amines and advanced glycation end products in roasted beef patties and related mechanistic analysis by density functional theory. *Food Chem.* **2022**, *385*, 132660. [CrossRef]
25. Kaban, G.; Sallan, S.; Çinar Topçu, K.; Sayın Börekçi, B.; Kaya, M. Assessment of technological attributes of autochthonous starter cultures in Turkish dry fermented sausage (sucuk). *Int. J. Food Sci. Technol.* **2022**, *57*, 4392–4399. [CrossRef]
26. Lemon, D.W. *An Improved TBA Test for Rancidity New Series Circular*; No:51; Halifax-Laboratory: Halifax, NS, Canada, 1975; Available online: https://waves-vagues.dfo-mpo.gc.ca/library-bibliotheque/15127.pdf (accessed on 5 June 2023).
27. Kaban, G. Changes in the composition of volatile compounds and in microbiological and physicochemical parameters during pastırma processing. *Meat Sci.* **2009**, *82*, 17–23. [CrossRef] [PubMed]

28. Medyński, A.; Pospiech, E.; Kniat, R. Effect of various concentrations of lactic acid and sodium chloride on selected physico-chemical meat traits. *Meat Sci.* **2000**, *55*, 285–290. [CrossRef]
29. Lazárková, Z.; Kratochvílová, A.; Salek, R.N.; Polášek, Z.; Šiška, L.; Petová, M.; Bunka, F. Influence of heat treatment on the chemical. physical. microbiological and sensorial properties of pork liver pâté as affected by fat content. *Foods* **2023**, *12*, 2423. [CrossRef]
30. Warriss, P.D. *Meat Science: An Introductory Text*; CABI International: New York, NY, USA, 2000; ISBN 0851994245.
31. Djenane, D.; Sanchez-Escalante, A.; Beltrán, J.A.; Roncalés, P. Ability of α-tocopherol. taurine and rosemary. in combination with vitamin C. to increase the oxidative stability of beef steaks packaged in modified atmospheres. *Food Chem.* **2002**, *76*, 407–415. [CrossRef]
32. Yu, L.; Chai, M.; Zeng, M.; He, Z.; Chen, J. Effect lipid oxidation on the formation of Nε-carboxymethyl-lysine and Nε-carboxyethyl-lysine in Chinese-style sausage during storage. *Food Chem.* **2018**, *269*, 466–472. [CrossRef]
33. Yu, L.; He, Z.; Zeng, M.; Zheng, Z.; Chen, J. Effect of irradiation on Nε—Carboxymethyl-lysine and Nε–carboxyethyl-lysine formation in cooked meat products during storage. *Radiat. Phys. Chem.* **2016**, *120*, 73–80. [CrossRef]
34. Yu, L.; He, Z.; Zeng, M.; Zheng, Z.; Chen, J. Effects of raw meat and process procedure on Nε-carboxymethyllysine and Nε-carboxyethyl-lysine formation in meat products. *Food Sci. Biotechnol.* **2016**, *25*, 1163–1168. [CrossRef] [PubMed]
35. Sun, X.; Tang, J.; Wang, J.; Rasco, A.B.; Lai, K.; Huang, Y. Formation of advanced glycation endproducts in ground beef under pasteurisation conditions. *Food Chem.* **2015**, *172*, 201–207. [CrossRef] [PubMed]
36. Bayrak Kul, D.; Anlar, P.; Yılmaz Oral, Z.F.; Kaya, M.; Kaban, G. Furosine and Nε-carboxymethyl-lysine in cooked meat product (kavurma): Effects of salt and fat levels during storage. *J. Stored Prod. Res.* **2021**, *93*, 101856. [CrossRef]
37. Chen, G. Formation and Inhibition of Advanced Glycation End Products in Meat and Model Systems. Ph.D. Thesis, Kansas State University, Manhattan, KS, USA, 2006.
38. Zhang, O.; Ames, J.M.; Smith, R.D.; Baynes, J.W.; Metz, T.O. A perspective on the Maillard reaction and the analysis of protein glycation by mass spectrometry: Probing the pathogenesis of chronic disease. *J. Proteome Res.* **2009**, *8*, 754–769. [CrossRef] [PubMed]
39. Domínguez, R.; Gómez, M.; Fonseca, S.; Lorenzo, J.M. Influence of thermal treatment on formation of volatile compounds. cooking loss and lipid oxidation in foal meat. *LWT Food Sci. Technol.* **2014**, *58*, 439–445. [CrossRef]
40. Shahidi, F.; Hossain, A. Role of lipids in food flavor generation. *Molecules* **2022**, *27*, 5014. [CrossRef]

Disclaimer/Publisher's Note: The statements, opinions and data contained in all publications are solely those of the individual author(s) and contributor(s) and not of MDPI and/or the editor(s). MDPI and/or the editor(s) disclaim responsibility for any injury to people or property resulting from any ideas, methods, instructions or products referred to in the content.

Article

Study on the Changes in Volatile Flavor Compounds in Whole Highland Barley Flour during Accelerated Storage after Different Processing Methods

Wengang Zhang [1,2,3], Xijuan Yang [1,2,3], Jie Zhang [1,2,3], Yongli Lan [4] and Bin Dang [1,2,3,*]

1. Academy of Agriculture and Forestry Sciences, Qinghai University, Xining 810016, China; 2017990098@qhu.edu.cn (W.Z.); 2007990025@qhu.edu.cn (X.Y.); 2015990070@qhu.edu.cn (J.Z.)
2. Key Laboratory of Qinghai Province Tibetan Plateau Agric-Product Processing, Qinghai University, Xining 810016, China
3. Laboratory for Research and Utilization of Qinghai Tibet Plateau Germplasm Resources, Qinghai University, Xining 810016, China
4. College of Food Science and Engineering, Northwest A & F University, Yangling 712100, China; yonglilan@nwsuaf.edu.cn
* Correspondence: 2008990019@qhu.edu.cn; Tel.: +86-15897185340

Abstract: The effect of heat processing on the flavor characteristics of highland barley flour (HBF) in storage was revealed by analyzing differences in volatile compounds associated with flavor deterioration in HBF using GC-MS identification and relative odor activity values (ROAVs). Hydrocarbons were the most abundant in untreated and extrusion puffed HBFs, while heterocycles were found to be the most abundant in explosion puffed, baked, and fried HBFs. The major contributors to the deterioration of flavor in different HBFs were hexanal, hexanoic acid, 2-pentylfuran, 1-pentanol, pentanal, 1-octen-3-ol, octanal, 2-butyl-2-octanal, and (E,E)-2,4-decadienal. Amino acid and fatty acid metabolism was ascribed to the main formation pathways of these compounds. Baking slowed down the flavor deterioration in HBF, while extrusion puffing accelerated the flavor deterioration in HBF. The screened key compounds could predict the quality of HBF. This study provides a theoretical basis for the regulation of the flavor quality of barley and its products.

Keywords: highland barley; heat processing; volatile compounds; flavor deterioration; gas chromatography-mass spectrometry (GC-MS)

Citation: Zhang, W.; Yang, X.; Zhang, J.; Lan, Y.; Dang, B. Study on the Changes in Volatile Flavor Compounds in Whole Highland Barley Flour during Accelerated Storage after Different Processing Methods. *Foods* **2023**, *12*, 2137. https://doi.org/10.3390/foods12112137

Academic Editor: Thomas Dippong

Received: 27 April 2023
Revised: 22 May 2023
Accepted: 23 May 2023
Published: 25 May 2023

Copyright: © 2023 by the authors. Licensee MDPI, Basel, Switzerland. This article is an open access article distributed under the terms and conditions of the Creative Commons Attribution (CC BY) license (https://creativecommons.org/licenses/by/4.0/).

1. Introduction

Highland barley (*Hordeum vulgare* L. var. *nudum* Hook. f.) is a gramineous barley crop of the genus barley. It is also known as highland barley due to the separation of its lemma from its caryopsis. Highland barley is rich in starch, protein, dietary fiber, β-glucan, vitamins, minerals, and other nutrients, and it exhibits nutritional qualities of "three highs and two lows" [1,2]. In recent years, there has been increasing interest in the health benefits of highland barley, and its use in the food industry is growing [3]. Long-term intake of highland barley has been shown to positively affect the prevention and alleviation of certain diseases, such as obesity, type II diabetes, hypertension, atherosclerosis, cardiovascular disease, and colon cancer [4,5]. In the food production industry, highland barley is mainly added to various food products in the form of barley flour, and the amount can reach 30% to 60% [5]. However, poor storage stability, a short storage life, and a tendency towards rancidity have limited the development of the highland barley industry. Thus, approaches to extend the shelf life of highland barley through processing are key issues that need to be addressed.

Different heat processing treatments can effectively improve the texture, aroma, and storage stability of grain flours [6]. Changes in freshness during the storage of highland barley flour are most evident as changes in flavor, which often determine the overall sensory characteristics of the analyzed product [7]. The volatile compounds in fresh highland barley flour, mainly from lipid oxidation and the degradation of proteins, carbohydrates, and amino acids, include alcohols, esters, aldehydes, ketones, and alkanes [8]. These compounds vary in aroma type and activity and synergistically constitute the unique flavor of highland barley flour [9]. During heat processing, highland barley is further subjected to various reactions such as lipid oxidation, Maillard, and caramelization, which usually results in a greater abundance of heterocycles and aldehydes and a significant reduction in acids, alcohols, and ketones [10]. Concurrently, heat processing also partially passivates antinutritional factors and lipases, reduces the formation of undesirable flavor compounds due to the oxidation of unsaturated fatty acids, improves pregelatinization, and gives the flour a pleasant roasted aroma, which are important for improving the processing quality and extending the shelf life of highland barley flour [11,12]. Wang et al. found that hot steam treatment could inactivate lipase, lipoxygenase, and peroxidase in buckwheat, thus reducing the production of lipid oxidation markers such as 3-methylbutane and hexanal and thereby delaying the flavor deterioration of stored buckwheat noodles [13]. Extrusion puffing treatment was found to slow down oxidative flavor production during the storage of millet compared to cooking [14]. Jiao et al. showed that roasting treatments based on hot air-assisted RF heating technology could improve the quality and extend the shelf life of nut products such as peanuts [15]. These studies provide new ideas for improving the flavor quality and shelf life of highland barley flour.

At present, the identification of volatile flavor compounds in highland barley and their changes during processing have not been studied much. Tatsu et al. identified 22 and 23 odor-active compounds in highland barley and hulled barley tea, respectively, which mainly included 2-methoxyphenol, trans-isobutanol, 2-acetyl pyrazine, 2-acetyl-1-pyrroline, and 3-methylbutyraldehyde [10]. The main undesirable flavor compounds in cooking barley were reported to be hexanal, (E)-nonenal, (E,E)-2,4-nonadienal, and (E,E)-2,4-decadienal [16]. The sensory quality of barley could be improved by using hot steam treatment to reduce aldehydes (hexanal and (E,E)-2,4-dodecenal) and acids (acetic and hexanoic acids) [9]. However, there is still uncertainty regarding the effects of different heat processing methods on the characteristic flavor and the mechanisms of flavor deterioration in highland barley flour during storage. This study used GC-MS detection and relative odor activity values (ROAVs) to investigate the effects of different processing methods (extrusion puffing, explosion puffing, baking, and frying) on the volatile flavor compounds in highland barley flour. By identifying the key volatile compounds involved in flavor deterioration and the processing methods that can help extend the shelf life of highland barley flour, the study provides a theoretical basis for regulating the flavor quality of highland barley products.

2. Materials and Methods

2.1. Chemicals and Materials

Kunlun 15 white highland barley was provided by the Qinghai Academy of Agriculture and Forestry Sciences. The test material was planted in 2021 in the experimental field of Qinghai Academy of Agriculture and Forestry Sciences (Xining, Qinghai) (36°67' N 101°77' E, 2300 m above sea level). 2-octanol ($\geq$99.5% purity) was purchased from TCI. The test water was deionized.

2.2. Preparation of Highland Barley Flour by Different Processing Methods

Highland barley seeds were cleaned, de-mixed, and then processed in different ways. The processing parameters for different treatments in this study were determined through pre-experimental optimization of the preparation conditions for mature highland barley flour. A sample of untreated highland barley flour (Y) was obtained by crushing highland barley seeds using a XL-10B buckling and swinging small crusher (Tianjin Xinhua Instrument Factory, Tianjin, China) and passing them through a 60-mesh sieve. The highland barley flour was extruded using a DZ65-II twin-screw extrusion puffing machine (Jinan

Saixin Machinery Co., Ltd., Jinan, China) at a moisture content of 34%, a feed frequency of 22 Hz, a screw speed of 800 r/min, a zone I temperature of 55 °C, a zone II temperature of 180 °C, and a zone III temperature of 160 °C. The extruded material was cooled to room temperature and further powdered through a 60-mesh sieve to obtain the extrusion puffed sample (J). To obtain explosion puffed highland barley flour (Q), 1000 g of highland barley seeds with a moisture content of 10% was added to a XSS-QPD explosion puffing machine (Wuhan Xinshishang Food Machinery Co., Ltd., Wuhan, China) with rotary heating to a pressure of 1.25 MPa after about 7 min, and the valve was opened thereafter. The puffed material was collected and powdered through a 60-mesh sieve to obtain the sample. To obtain baking barley flour (H), highland barley seeds were first soaked for 6 h in water to reach a moisture content of 40%, then evenly spread on a baking tray with a thickness of 0.5 cm and baked in a CK-2 far-infrared food baking oven (Guangzhou Maisheng Baking Equipment Co., Ltd., Guangzhou, China) for 20 min at 150 ± 5 °C for the primer and 170 ± 5 °C for the surface. Finally, the baked highland barley was powdered through a 60-mesh sieve to obtain the sample. For fried barley flour (C), highland barley seeds were first soaked for 6 h in water to reach a moisture content of 40%, then added to a CH50 constant temperature automatic stir fryer (Henan Hui'an Machinery Equipment Co., Ltd., Zhengzhou, China) and fried at 105 ± 5 °C for 10 min. After cooling to room temperature, the fried highland barley was powdered through a 60-mesh sieve to obtain the sample.

2.3. Accelerated Storage Tests

Lipid oxidation is an important cause of flavor deterioration in whole grain flour. Thus, we chose accelerated storage conditions for this study by referring to a relevant research report on lipid oxidation during the storage of matured whole grain flour [17]. The accelerated storage test conditions were as follows: 500 g of highland barley flour from the different treatments was stored in woven flour bags (polyethylene) at $50 \pm 1\%$ relative humidity and 50 ± 1 °C for 0–84 d to simulate accelerated storage conditions. Appropriate samples were removed every 14 d for the differently treated highland barley flours, repackaged, and stored at −80 °C until gas chromatography–mass spectrometry (GC-MS) test was performed. The control samples were fresh highland barley flour that had not undergone accelerated storage and were directly stored at −80 °C after preparation. Based on a preliminary analysis of the type and relative content of volatile flavor compounds in the analyzed samples, samples stored for 0, 14, and 84 d were selected as representative of the stages of flavor change for subsequent analysis.

2.4. The Gas Chromatography–Mass Spectrometry Detection of Volatile Flavor Compounds of Highland Barley Flour

Different samples of highland barley flour (300 mg) were placed in 20 mL headspace bottles. Ten microliters of 2-octanol was added as an internal standard. The volatile flavor compounds were analyzed using a 7890B gas chromatograph, 5977B mass spectrometer, and DB-Wax chromatographic column (30 m × 250 μm × 0.25 μm; Agilent Technologies Inc., Santa Clara, CA, USA). The GC conditions were: extraction temperature, 60 °C; preheating time, 15 min; extraction time, 30 min; resolution time, 4 min; splitless mode; and carrier gas, helium. The column flow rate was 1 mL/min, the sample inlet temperature was 250 °C, and the quadrupole temperature was 150 °C. The temperature rise program of the column box was 40 °C for 4 min, followed by 5 °C/min at 245 °C for 5 min. The mass spectrometry conditions included: ionization voltage, −70 eV; ion source temperature, 230 °C; transmission line temperature, 250 °C; mass scan range, m/z 20–400; and solvent delay, 0 min.

2.5. Analysis of Key Volatile Flavor Compounds

The ROAV method was used to evaluate the contribution of different volatile compounds to the overall flavor of the samples [18]. The greater the ROAV, the greater the contribution of the compound to the overall flavor of the sample. The components that contributed most to the overall flavor of the sample were defined as having ROAVs = 100; herein, all components in the sample had ROAVs ≤ 100. The components with ROAV ≥ 1 were the key flavor compounds in the sample. The components with $0.1 \leq$ ROAV < 1 had a modifying effect on the overall flavor of the sample.

The ROAV was calculated according to Equation (1):

$$\text{ROAVi} \approx 100 \times \frac{C_i}{C_s} \times \frac{T_s}{T_i} \qquad (1)$$

where Ci is the relative content of volatile compound i (%); Ti is the sensory threshold of volatile compound i (μg/kg); Cs is the relative content of the component contributing most to the overall flavor of the sample (%); and Ts is the sensory threshold of the component contributing most to the overall flavor of the sample (μg/kg).

2.6. Data Processing

The MS data were subjected to peak extraction, baseline correction, deconvolution, peak alignment, and MS matching using Chroma TOF software (V4.3x, LECO) and the National Institute of Standards and Technology (NIST) library for quantitative analysis based on relative peak areas. Data were statistically analyzed using Microsoft Excel 2007, Origin 2019b was used for component analysis and graphing, and Simca 14.1 was used for orthogonal partial least squares-discriminant analysis (OPLS-DA). Values were expressed as the mean ± standard deviation (SD) resulting from replicate samples. The annotation and metabolic pathways of differential metabolic volatile compounds in highland barley flour were performed used the KEGG database (http://www.kegg.jp (accessed on 23 September 2021)).

3. Results and Discussion

3.1. Volatiles in Different Highland Barley Flour

A total of 186 volatile metabolites, including alcohols, esters, aldehydes, ketones, acids, heterocycles, phenols, and hydrocarbons, were identified in different samples by GC-MS analysis. The distribution of these compounds is shown in Figure 1, according to which the untreated group was most enriched with hydrocarbons, which were more abundant after extrusion puffing. The heterocyclic compounds were most abundant in the explosion puffed, baked, and fried groups. A significant increase in ketones was observed in the explosion puffed group. Zheng et al. studied the effects of different thermal treatment methods on lipid oxidation and flavor compounds in highland barley and found that aldehydes were the most abundant in untreated, infrared-treated, and steamed highland barley, while esters were the most abundant in roasted, microwaved, and fried highland barley [19]. These results differed from the results of this study, indicating that the variety and treatment methods as well as the conditions affect the characteristic volatile compounds of highland barley flour. During heat processing, the hydrocarbons formed mainly through the oxidative decomposition of unsaturated fatty acids, free radical reactions (cyclization and conjugation), and heat decarboxylation of saturated fatty acids have a high flavor threshold and are generally not considered to be an important factor affecting food flavor [10,15,20]. Ketones are formed mainly through autoxidation, β-oxidation, and decarboxylation of fatty acids and generally have a pleasant aroma. However, their flavor activity is relatively low, and their contribution to the overall flavor of highland barley flours is likely to be small [6]. Heterocyclic compounds, which are mainly produced by meladation, caramelization, and Strecker degradation reactions between reduced sugars and amino acids, have nutty, roasted, and cocoa aromas and are considered to be "good" flavoring substances for the samples [15]. With the increase in storage time, the number of highly expressed volatile compounds tended to decrease in the untreated, explosion puffed, and fried groups, whereas their numbers increased in the extrusion puffed and baked groups. At the end of storage, new low-threshold and highly expressed compounds appeared in different groups. These results suggested significant differences in the volatile flavor compounds of highland barley flour after different treatments or at different storage stages and increased flavor deterioration in different highland barley flours at longer storage times.

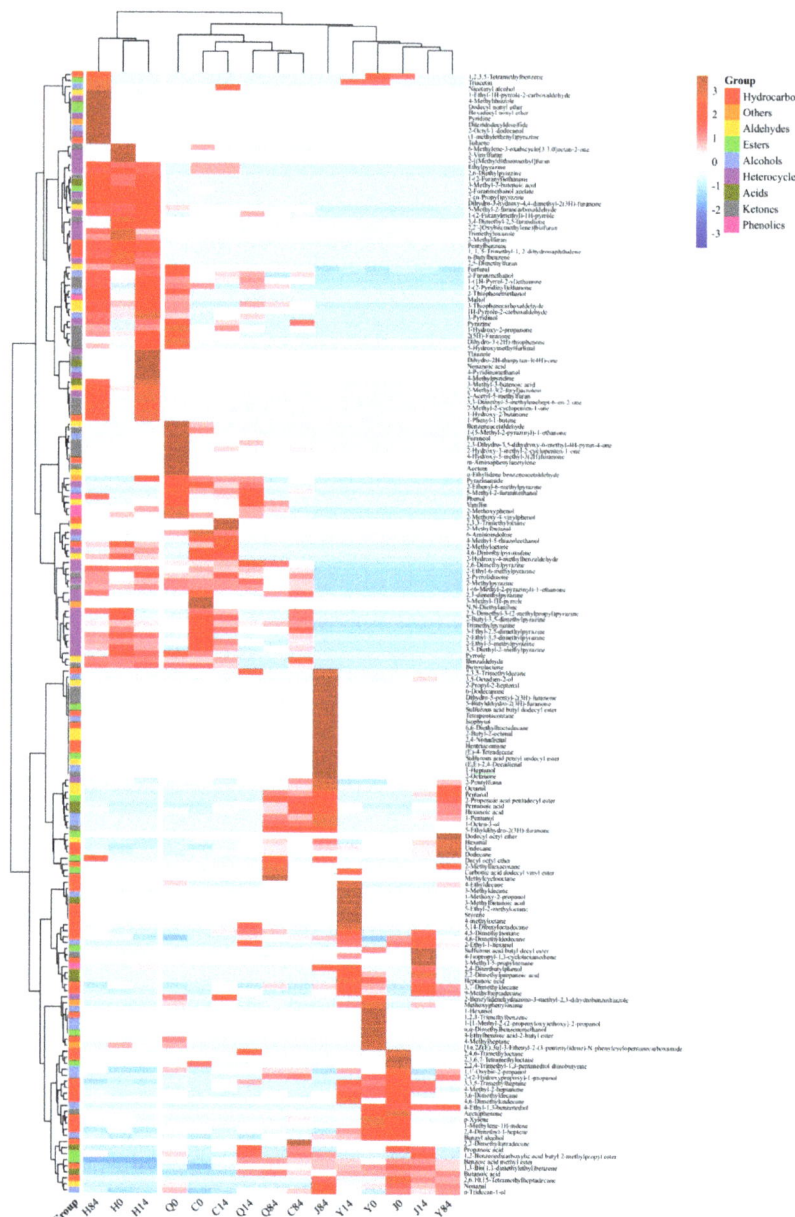

Figure 1. Heatmap of volatile flavor compounds in stored highland barley flour after different heat processing methods. Y0, Y14, Y84: unprocessed samples stored for 0, 14, and 84 d, respectively. Q0, Q14, Q84: explosion puffed samples stored for 0, 14, and 84 d, respectively. J0, J14, J84: extrusion puffed samples stored for 0, 14, and 84 d, respectively. H0, H14, H84: baked samples stored for 0, 14, and 84 d, respectively. C0, C14, C84: fried samples stored for 0, 14, and 84 d, respectively.

3.2. Differential Volatile Compounds in Different Highland Barley Flour

Figure 2 showed that 22, 24, 14, 19, and 20 differential metabolic volatile compounds were detected in untreated, explosion puffed, extrusion puffed, baked and fried highland barley flours, respectively, during storage. As storage time increased, the composition of the volatile flavor compounds changed significantly in the groups. In the untreated group, hydrocarbons and alcohols were highly expressed within 14 d of storage, while the low-threshold values of hexanoic acid, 1-pentanol, and 2-pentylfuran were significantly up-regulated at the end of storage. 1-Pentanol and 2-pentylfuran have been reported to be important indicators reflecting lipid oxidation in grains, and hexanoic acid has been associated with undesirable flavor formation in highland barley [9,21]; these compounds may be an important cause of deteriorated flavor in untreated highland barley flour. In the explosion puffed group, the levels of heat-processed characteristic aroma compounds such as pyrazine, 2-methoxy-4-vinylphenol, and 5-methyl-2-furancarboxaldehyde were reduced during storage, while those of low-threshold compounds such as hexanal, pentanal, hexanoic acid, 1-pentanol, and 1-octen-3-ol were increased [6,21]. The significantly upregulated, low-threshold compounds in the explosion puffed group were closely related to lipid oxidation, where hexanal had a grassy and fatty flavor and 1-octen-3-ol had an 'odor'; these compounds may be the main source of flavor deterioration in explosion puffed highland barley flour [9,14]. In the extrusion puffed group, the highly expressed compounds were mainly hydrocarbons at the beginning of storage. After 14 d of storage, pentanal, hexanal, and 4-methyldecane were significantly upregulated, whereas after 84 d of storage, hexanoic acid, 2-pentylfuran, 1-heptanol, 1-pentanol, 1-octen-3-ol, 3,5-octadien-2-ol, and octanal were significantly upregulated, suggesting that these compounds may be the primary cause of flavor deterioration in extruded highland barley flour. In the baked group, the relative contents of 2-methylfuran, 2-pentylfuran, and 4,6-dimethylpyridine were significantly decreased during storage, while those of hexanoic acid, hexadecyl nonyl ether, 2,3-dimethylpyrazine, and dodecyl nonyl ether were significantly enhanced, suggesting that the loss of some of the characteristic roasted flavors in baked highland barley flour was accompanied by an increase in hexanoic acid content after long-term storage. This significant increase in hexanoic acid and hexanal contents may have contributed more to the deterioration of flavor in baked highland barley flour [9]. At the beginning of storage, the highly expressed compounds in the fried group were mainly heterocyclic, with significant upregulation of 1-hydroxy-2-propanone and hexanal after 14 d of storage and 1-octen-3-ol, propionic acid, pentanoic acid, 2-pentylfuran, and 1-pentanol after 84 d of storage. The low-threshold compounds upregulated at the end of storage may have contributed more to the deterioration of flavor in fried highland barley flour [21]. The changes in pyrazines in the different heat-processed samples were generally small, suggesting that the "rancid flavor" may be mainly the result of masking the "good" aromas by low-molecular-weight aldehydes, ketones, acids, and alcohols produced by lipid oxidation rather than the polymerization or degradation of pyrazines. This finding was consistent with the findings of Warner et al. on flavor deterioration in peanuts [22].

In summary, different processing methods and storage times had significant effects on the differential flavor compounds in highland barley flour. Hexanal, pentanal, octanal, 1-pentanol, 1-octen-3-ol, 1-heptanol, 3,5-octadien-2-ol, hexanoic acid, propanoic acid, pentanoic acid, and 2-pentylfuran may be closely related to the deterioration of flavor in highland barley flour. Variations were observed in the distribution and expression of these compounds in different samples.

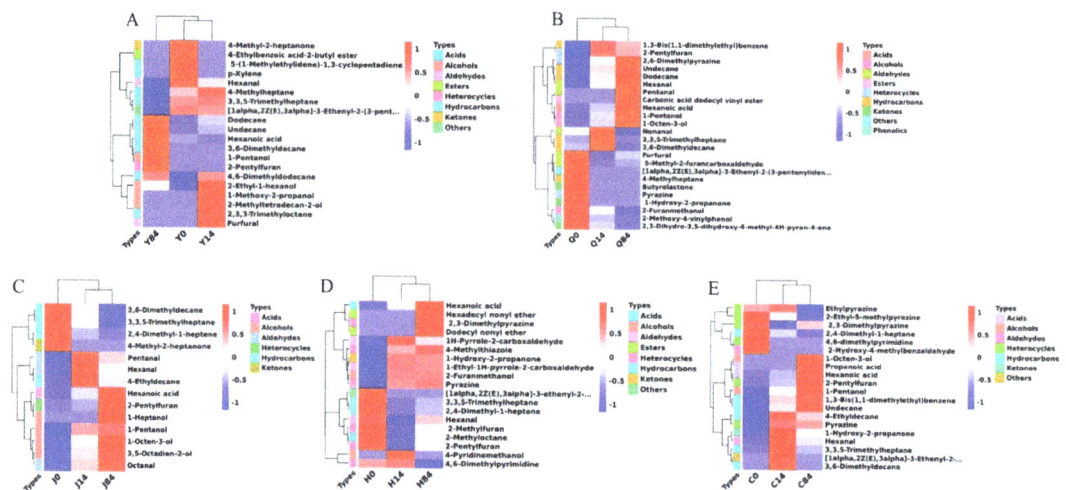

Figure 2. Heatmap of differential metabolic flavor compounds in stored highland barley flour after different processing methods. (**A–E**) represent the differential metabolic compounds in unprocessed samples (Y), explosion puffed samples (Q), extrusion puffed samples (J), baked samples (H), and fried samples (C) stored for different times (0, 14, and 84 d), respectively.

3.3. Analysis of Key Volatile Compounds

As shown in Table 1, 34 volatile compounds with ROAV ≥ 0.1 were detected in different highland barley flours, including 5 alcohols, 1 ester, 9 aldehydes, 1 ketone, 3 acids, 12 heterocycles, and 3 phenols. The distribution of these compound species differed from their ROAVs in different highland barley flours. The explosion puffed, baked, and fried groups had the highest number of key volatile compounds at 0 d of storage (all 10) and the lowest number in the untreated group (4). The largest contributor to flavor in untreated highland barley flour was hexanal, followed by hexanoic acid, mainly expressed as a green aroma [9]. In the explosion puffed, baked, and fried groups, 2,6-dimethylpyrazine was the largest contributor to flavor, followed by 3-ethyl-2,5-dimethylpyrazine, 2-methoxyphenol, and 2-pentylfuran, which mainly conferred the samples nutty, roasted, smoky, and mung bean-like aromas [23,24]. In the extrusion puffed group, hexanal still contributed the most to the flavor, followed by 2-ethyl-1-hexanol, while the contribution of hexanoic acid was lower than in the untreated group but higher than in the other groups. The order of the effect of the different processing methods on the roasted highland barley flavor profile was baked > fried > explosion puffing > extrusion puffing.

At 14 d of storage, the explosion puffed and fried groups had the most key volatile compounds (9), while the untreated group had the fewest (4). Types of the key volatile flavor compounds decreased after storage of the heat-processed highland barley flours. The compound that contributed most to flavor in the untreated group changed to 2-ethyl-1-hexanol and the ROAV of hexanoic acid increased to 50.71, both with a flowery aroma and sweaty flavor, respectively [25]. The biggest contributor to flavor in the explosion puffed, extrusion puffed, baked, and fried groups continued to be 2,6-dimethylpyrazine, accompanied by the appearance of new key volatile compounds. Among these volatile compounds, 3-methylbutyric acid (ROAV = 2.18) appeared in the untreated group; phenol and 2-ethyl-1-hexanol appeared in the explosion puffed group; 1-pentanol (ROAV = 6.53) and pentanal appeared in the extrusion puffed group; propionic acid and benzoic acid methyl ester appeared in the baked group; and phenol and 2-methylbutanal appeared in the fried group. These compounds, of which 3-methylbutyric acid has a sweat and perspiration flavor and 1-pentanol has a grassy flavor, were found to contribute somewhat to the poor flavor of highland barley flour [9,24]. Pentanal, 2-methylbutanal, and propionic acid had fermented, malty, and sweaty flavors, respectively, but their ROAVs were all less than 1 and they had a secondary role in the overall flavor [10,26]. The other compounds in the different groups showed an overall reduction in ROAV. Overall, the degree of flavor deterioration in the different highland barley flours increased with storage time, which was most significant in the extrusion puffed group, followed by the untreated group, and to a lesser extent in the explosion puffed, baked, and fried groups.

After 84 d of storage, different groups had significant increases in the key volatile flavor compounds; the fried group had the greatest increase and the untreated group had the least increase, indicating further flavor deterioration in the highland barley flour. The largest contributor to flavor was again hexanal in the untreated group and 1-octen-3-ol in the extrusion puffed group, while 2,6-dimethylpyrazine remained the largest contributor to flavor in the explosion puffed, baked, and fried groups. In the explosion puffed, baked, and fried groups, 2-ethyl-6-methylpyrazine, 2,3,5-trimethylpyrazine, and 3-ethyl-2,5-dimethylpyrazine showed a variable increase in ROAV, but they were accompanied by a decrease in ROAV for some other pyrazine compounds. The upregulation of some pyrazines may have been generated again in the accelerated storage test. The key volatile compounds of hexanal (4.53–100.00), hexanoic acid (16.06–95.22), 2-pentylfuran (7.12–40.12), 1-pentanol (9.61–33.49), and pentanal (0.74–1.92) in different samples contributed more significantly after 14 d of storage. In addition, six new key volatile compounds were observed, including 1-octen-3-ol (10.94–100.00), octanal (22.50–69.94), pyridine (1.54), 4-methylthiazole (0.89), 2-butyl-2-octenal (2.69), and (E,E)-2,4-decadienal (1.27). Apart from pyridine and 4-methylthiazole, the remaining nine of these compounds were the main constituents contributing to the undesirable flavor of the different stored highland barley flours, and the formation of these compounds in large amounts may be the result of lipid rancidity [14].

The above-mentioned nine key volatile compounds were most abundant in the extrusion puffed group (9), and their ROAVs was generally high, followed by the fried (7) and untreated groups (7), the explosion puffed group (6), and the baked group, which had the least number of species (3) and their ROAVs were the lowest. Wilkins and Scholl used GC-MS to analyze the volatile metabolites of molds growing on barley and found that the main compounds were 1-octen-3-ol, styrene, and 3-methyl-1-butanol, with smaller amounts of 2-pentylfuran, 3-methylmethyl ether, 2-(2-furyl)pentanal, and 2-ethyl-5-methylphenol [27]. Later, Wang et al. found that 1-octen-3-ol, (E)-2-octen-1-ol, 3-nonen-1-ol, (E)-2-nonenal, decanal, 2-undecanone, and 2-methylnaphthalene appeared or increased with longer storage time in steamed millet flour, in which the main cause of the bad odor of the samples was attributed to 1-octen-3-ol and (E)-2-nonenal [14]. Kaneko et al. reported that hexanal, (E)-nonenal, (E,E)-2,4-nonadienal, and (E,E)-2,4-decadienal caused undesirable odors in cooked barley [16]. The volatile components of aldehydes increased during the storage of buckwheat, and hexanal and 3-methylbutanal could be used as markers of lipid oxidation and flavor changes [13]. Our study findings were congruent with these above-mentioned reports. In summary, nine compounds, including aldehydes, alcohols, acids, and furans, were the main causes of flavor deterioration in stored highland barley flour after different processing methods, and the degree of flavor deterioration in the different highland barley flours was in the order of extrusion puffed group > fried group > untreated group > explosion puffed group > baked group.

Table 1. Analysis results of key volatile compounds in stored highland barley flour after different processing methods.

No.	Compounds	OT (μg/kg^{-1})	Odorant Description	Y0	Q0	J0	H0	C0	Y14	Q14	J14	H14	C14	Y84	Q84	J84	H84	C84
1	2-Methylfuran	9	Chocolate, vanilla				3.05					0.95					0.97	
2	2-Methylbutanal	20	Musty, cocoa, pungent, sweet										0.55					
3	Pentanal	20	Fermented, bready, woodsy, fruity								0.63			1.55	0.74	1.92		1.45
4	Hexanal	4.5	Grass-like, green, fatty	100.00	4.53	100.00	6.74	8.18		5.41	100.00		8.13	100.00	27.96	68.44	4.53	15.88
5	Pyridine	7.9	Sour, fishy															1.54
6	2-Pentylfuran	6	Green bean, butter, vegetable		5.66	29.06	12.83	7.70		3.66	5.17	4.22	2.43	12.10	5.67	40.12	7.12	27.03
7	1-Pentanol	5	Grassy, fruity								6.53			9.76	9.61	33.49		17.03
8	2-Methylpyrazine	60	Nutty, cocoa, roasted		4.34		3.49	3.75		1.79		3.93	3.49		1.24		4.61	3.98
9	4-Methylthiazole	20	Nutty, green, vegetable														0.89	
10	Acetoin	55	Sweet, milky, buttery		0.43													
11	Octanal	0.7	Fatty, soapy, green											32.87		69.94		22.50
12	2,6-Dimethylpyrazine	1.5	Chocolate, nutty, roasted		100.00		100.00	100.00		100.00		100.00	100.00		100.00		100.00	100.00
13	2,3- Dimethyl pyrazine	400	Nutty, peanut, butter					0.13										0.11
14	2-Ethyl-6-methylpyrazine	40	Earthy, roasted		1.25		3.37	2.00		0.71		2.80	1.42		0.80		3.06	1.94
15	2-Ethyl-5-methylpyrazine	100	Coffee bean, nutty, roasted		0.21		0.88	1.09		0.11		0.66	0.59				0.70	0.57
16	Nonanal	260	Citrus-like, rose, grassy	0.43	0.12	0.81			100.00		0.10			0.14		0.31		
17	2,3,5-Trimethylpyrazine	400	Nutty, baked potato, roasted peanut		0.17		0.36	0.48				0.23	0.22		0.14		0.25	0.39
18	3-Ethyl-2,5-dimethylpyrazine	8.6	Roasted, nutty, earthy		5.99		31.44	29.87		4.28		14.46	20.59		5.33		15.41	36.30
19	1-Octene-3-ol	1	Mushroom											10.94	33.33	100.00		71.60
20	2-Ethyl-3,5-dimethylpyrazine	40	Roasted, chocolate, cocoa-like				1.47	1.23				0.97					0.87	0.69
21	2-Ethyl-1-hexanol	0.8	Floral, citrus, sweet			85.96			100.00	12.05	19.03			8.28				
22	Benzaldehyde	300	Almond, burnt sugar, cherry				0.15					0.13					0.13	0.20

Table 1. Cont.

No.	Compounds	OT (μg·kg⁻¹)	Odorant Description	ROAV														
				Y0	Q0	J0	H0	C0	Y14	Q14	J14	H14	C14	Y84	Q84	J84	H84	C84
23	Propionic acid	57	Pungent, acidic, cheesy	0.69	0.27	2.23			0.70	0.27		0.39	0.14					0.64
24	Benzoic acid methyl ester	73	Floral, fruity	1.37	0.39	1.91		0.17	1.92	0.21		0.37	0.26	0.33	0.30	0.46		0.61
25	Benzeneacetaldehyde	9	Flora, honey-like		1.29			0.32										
26	2-Butyl-2-octenal	20	Lemony, fruity													2.69		
27	3-Methylbutyric acid	12	Sweaty						2.18									
28	(E,E)-2,4-Decadienal	1.81	Fatty, green													1.27		
29	Hexanoic acid	2.52	Sweaty, sour, fatty	29.00	14.15	16.18	6.54	3.03	50.71	3.08	4.88	6.21	9.80	38.87	20.34	91.37	16.06	95.22
30	2-Methoxyphenol	0.84	Gammon-like, smoky		24.99		9.84	5.93		4.07		12.10	3.92		5.12		8.40	
31	Benzyl alcohol	100	Sweet, flowery	0.58		0.57												
32	Phenol	0.65	Rubber, phenolic	9.46						3.18			2.91				8.51	
33	Furaneol	27.4	Caramel-like, sweet, maltol-like		0.96			0.14					0.10					
34	2-Methoxy-4-vinylphenol	21	Smoky, clove-like		2.03		1.00	0.76		0.31		0.59	0.24		0.07		0.18	0.15

Note: OT is the abbreviation of 'odor threshold'. Blank means that the compound was not detected or the aroma type or flavor threshold of the substance was not found. The types of aroma and flavor thresholds were mainly obtained from http://www.thegoodscentscompany.com (accessed on 18 March 2022) and the literature [6,9,10,13,25,28–30].

3.4. Changes in Key Volatile Compounds during Storage

The relative contents of the nine key volatile compounds were monitored throughout the storage period. The results are shown in Figure 3, according to which the relative content of the nine compounds in the different highland barley flours showed an overall increasing trend with the increase in storage time. The relative content of hexanal in the untreated group increased rapidly and was significantly higher than in the other groups at the end of storage. The relative contents of hexanoic acid, 2-pentylfuran, 1-pentanol, and pentanal increased after 42 d of storage, while the relative contents of 1-octen-3-ol and octanal increased significantly after 70 d of storage. Thus, the turning point for flavor deterioration in the untreated highland barley flour was around 42 d. The relative contents of hexanal, 1-pentanol, and 1-octen-3-ol increased with storage time in the explosion puffed group, while fluctuations were noted in the relative contents of hexanoic acid and 2-pentylfuran. The relative content of pentanal increased rapidly after 56 d of storage, i.e., the turning point for flavor deterioration in the explosion puffed highland barley flour was around 56 d. The relative contents of the nine key volatile compounds were significantly higher in the extrusion puffed group than in the other groups. The relative contents of 2-pentylfuran and 2-butyl-2-octenal increased significantly after 42 and 70 d of storage, respectively, while the relative contents of the other compounds increased significantly after 14 d of storage. These results indicated that extrusion promoted flavor deterioration in highland barley flour, and the turning point was around 14 d. In the baked group, the relative contents of hexanal, hexanoic acid, and 2-pentylfuran increased slowly with storage time, indicating that the flavor deterioration in baked highland barley flour did not have a significant turning point. In the fried group, the relative contents of hexanoic acid and 1-pentanol increased with storage time, while that of hexanal increased and then decreased rapidly. The relative content of 2-pentylfuran began increasing after 42 d of storage, while the relative contents of pentanal and 1-octen-3-ol increased significantly after 28 d of storage, and that of octanal increased slightly at the end of storage. Thus, the turning point for flavor deterioration in fried highland barley flour was around 28 d. It could be seen that the upregulation times for some key volatile compounds differed, which might be due to the different precursors that form these compounds in the samples undergoing different changes during storage. Overall, extrusion accelerated flavor deterioration in highland barley flour, while baking effectively delayed flavor deterioration. We previously conducted research on the storage characteristics of highland barley flour after different processing methods. The results showed that the malondialdehyde levels remained consistently high during accelerated storage of the extrusion puffed and fried groups. The peroxide levels of the extrusion puffed and fried groups varied greatly during accelerated storage, while those of the explosion puffed and baked groups were relatively stable. The fatty acid levels of different treatment groups fluctuated within 42 d of storage and sharply increased after 84 d of storage. The explosion puffed group showed the longest storage time, followed by the baked and fried groups, while the extrusion puffed group had the shortest storage time, which could well support the flavor characterization results in this work [31]. The findings of this study differed from earlier results wherein extrusion processing was beneficial in improving the storage stability of millet and oats, probably due to differences in grain types and their chemical composition [14,23].

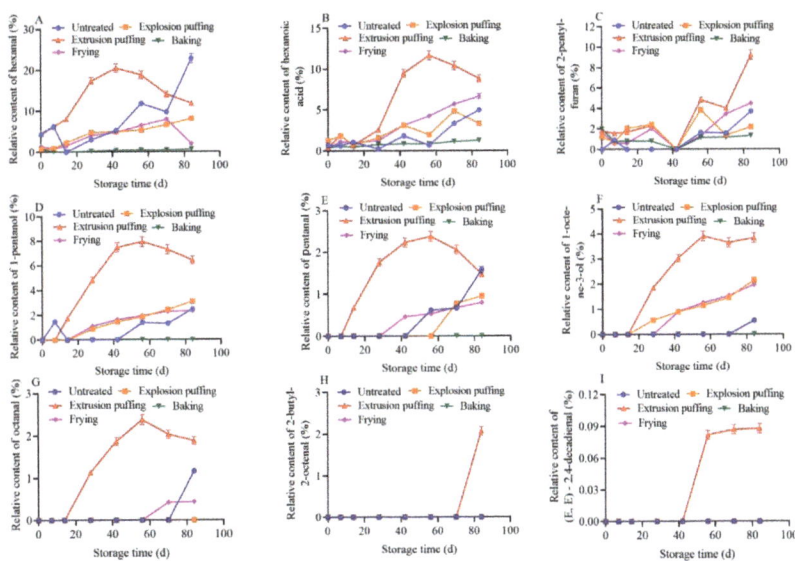

Figure 3. Changes in the relative contents of 9 key volatile compounds during storage of different highland barley flours. (**A–I**) represent the relative content changes in hexanal, hexanoic, 2-pentylfuran, 1-pentanol, pentanal, 1-octene-3-ol, octanal, 2-butyl-2-octenal, and (E,E)-2,4-decadienal during storage, respectively.

In addition, there was a positive correlation between the relative hexanal content and storage time in the untreated ($r = 0.811$, $p < 0.05$) and explosion puffed ($r = 0.971$, $p < 0.05$) groups, indicating that the changes in the relative hexanal content were more effective in reflecting the changes in the quality of the untreated and explosion puffed highland barley flours. A linear relationship between the relative content of hexanoic acid and storage time was observed in the fried ($r = 0.982$, $p < 0.01$) and baked ($r = 0.976$, $p < 0.01$) groups, indicating that the changes in the relative content of hexanoic acid could better reflect the changes in the quality of fried and roasted baked highland barley flours. In the extrusion puffed group, the linear relationships between the relative contents of 1-pentanol ($r = 0.961$, $p < 0.05$), pentanal ($r = 0.936$, $p < 0.05$), 1-octen-3-ol ($r = 0.986$, $p < 0.05$), and octanal ($r = 0.984$, $p < 0.05$) and the storage time were good in the range of 14–56 d, indicating that the changes in the relative contents of these four compounds could reflect the changes in the quality of the extrusion puffed highland barley flour over time. In summary, increases in the contents of the nine key volatile compounds during storage indicated increases in the degree of lipid oxidation and deterioration of the different highland barley flours, while the rate of flavor deterioration after different processing methods was in the order of the extrusion puffed group > fried group > untreated group > explosion puffed group > baked group. Some of the key compounds were found to be effective in reflecting the flavor deterioration in highland barley flour after different processing methods.

As can be seen from Figure 4A, H0, H14, and H84 were the most concentrated in the principal components analysis (PCA) plot, indicating a relatively higher flavor similarity between samples within the baked group during accelerated storage. The overall flavor change in the other groups started to increase after 14 d of storage and was most obvious in the extrusion puffed group, followed by the untreated group. This finding indicated a faster flavor change in the extrusion puffed group and a slower change in the baked group, which was consistent with the above analysis (Table 1 and Figure 3). However, PCA analysis using the overall volatile compounds provided poor separation of the samples. The key volatile compounds analyzed using the OPLS-DA model showed that the model reached significance (R2X = 0.952, R2Y = 0.993, Q2 = 0.890). The separation of highland barley flours at different storage times (Figure 4B) was significantly higher than that achieved with PCA analysis (Figure 4A), indicating that the key flavor compounds screened in this study could achieve the differentiation and prediction of highland barley flour samples at different stages of storage, and thus provide a theoretical basis for quality control and monitoring of highland barley products.

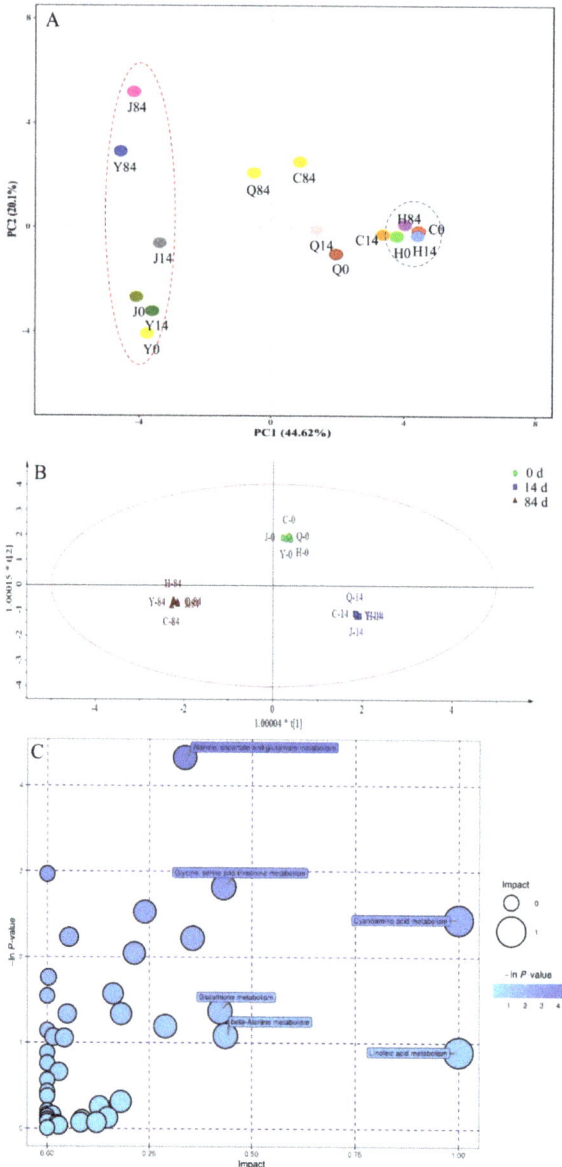

Figure 4. Principal component analysis (PCA) of overall volatile flavor compounds in different highland barley flours (**A**); OPLS-DA scores of key flavor compounds (**B**); Pathway enrichment results of differential metabolites (**C**).

3.5. Analysis of the Formation Mechanism of Key Volatile Compounds

As shown in Figure 4C, five amino acid metabolism pathways, including the metabolism of alanine, aspartate and glutamate; glycine, serine and threonine; cyanoamino acid; glutathione; and β-alanine, and one fatty acid metabolism pathway of linoleic acid had a significant effect on the formation of volatile flavor compounds at $p < 0.1$ and Impact > 0.4. These results indicated that amino acids and fatty acids are important for the formation of volatile compounds in different highland barley flours and their changes during storage. Highland barley is rich in amino acids and unsaturated fatty acids, and it may provide an important material basis for the metabolism of various volatile compounds in different samples [3]. The metabolic pathways of the nine key compounds identified in this study to be associated with flavor deterioration are shown in Figure 5. The conversion of amino acids to volatile compounds has been confirmed in a variety of plants and plant tissues [32]. In general, amino acids and their derivatives are subject to the removal of amino groups in the reaction catalyzed by branch chain aminotransferases, and the intermediates formed are further converted to aldehydes, alcohols, and esters by decarboxylases and esterification [33]. For example, benzyl alcohol, benzaldehyde, and phenylacetaldehyde may be derived from phenylalanine amino acids, and the chorismate synthase protein is a potential key protein for phenylalanine degradation [34]. Moreover, free amino acids can react with α-dicarbonyl to form Strecker aldehydes, such as benzaldehyde, phenylacetaldehyde, 2-methylbutyraldehyde, and 2-butyl-2-octenal [35].

Unsaturated fatty acids are less stable due to the presence of one or more double bonds and are susceptible to oxidation. Oxidation of oleic, linoleic, α-linolenic, and arachidonic acids with or without enzymatic catalysis (such as lipoxygenase) can form a variety of flavor compounds, including short-chain aldehydes and alcohols, a phenomenon that has been confirmed in grains, fruits, vegetables, vegetable oils, and others [36,37]. Fatty acid oxidation forms small low-threshold aldehydes with grassy, fatty, oily, and fruity flavors, which are common markers of lipid oxidative rancidity and food deterioration [25]. During storage, the lipids in highland barley flour are hydrolyzed by lipase to release unsaturated fatty acids, resulting in high levels of free fatty acids in highland barley flour during the middle and late stages of storage. These free fatty acids are subject to rearrangement, autoxidation, and enzymatic oxidation, usually to form the corresponding hydroperoxides (13-, 11-, or 10-ROOH), followed by the action of hydroperoxide lyase to produce aldehydes, which can be reduced to alcohols by alcohol dehydrogenase [38,39]. The metabolism of unsaturated fatty acids can ultimately produce undesirable flavor compounds, such as 1-octen-3-ol, 1-pentanol, hexanol, heptanol, nonanol, hexanal, nonanal, heptanal, octanal, and (E,E)-2,4-nonadienal, which was further confirmed by the results of this study [40].

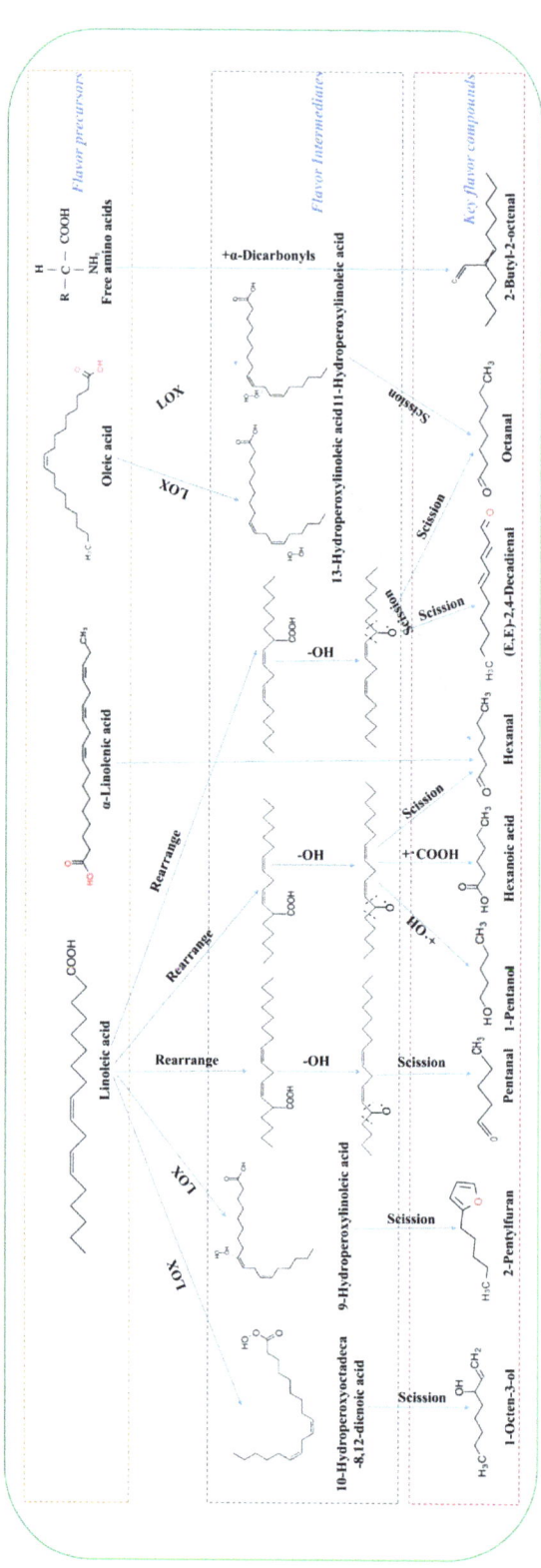

Figure 5. Formation mechanism of 9 key volatile compounds of stored highland barley flour after different processing methods. Note: Structures and metabolic processes of 9 kinds of the key compounds involved in the figure are mainly referred to reports [38,39,41,42].

4. Conclusions

The composition of volatile flavor compounds during storage differed significantly in highland barley flour treated with different processing methods. After heat processing, extrusion puffed highland barley flour formed more hydrocarbons and explosion puffed highland barley flour formed more ketones and heterocycles, while baked and fried highland barley flours formed a large number of heterocycles. With the extension in storage time, there was a dynamic change in the differential metabolic compounds in highland barley flour, flavor deterioration increased, and low-threshold compounds appeared at the end of storage. Extrusion and frying accelerated the flavor deterioration in highland barley flour and the effect was in the order of extrusion puffing > frying, while explosion puffing and baking slowed the flavor deterioration in highland barley flour, in the order of baking > explosion puffing. Five aldehydes (hexanal, pentanal, octanal, 2-butyl-2-octenal, and (E,E)-2,4-decadienal), two alcohols (1-pentanol and 1-octen-3-ol), one acid (hexanoic acid), and one heterocycle (2-pentylfuran) were the key volatile compounds observed to be involved in the flavor deterioration of differently treated highland barley flours. The formation of these compounds mainly involved amino acid metabolism and linoleic acid metabolism. The relative contents of these compounds could be used to reflect the flavor deterioration process in highland barley flour at different processing stages and to distinguish the quality of highland barley flour at different storage times. This study can provide a theoretical basis for understanding the storage quality and implementing flavor quality control of hot-processed highland barley products.

Author Contributions: Conceptualization, W.Z. and X.Y.; methodology, W.Z. and Y.L.; software, W.Z., B.D. and J.Z.; validation, W.Z., B.D. and X.Y.; investigation, W.Z. and J.Z.; data curation, W.Z. and J.Z.; writing-original draft preparation, W.Z.; writing-review and editing, B.D. and Y.L.; formal analysis, J.Z. and Y.L.; visualization, J.Z.; funding acquisition, X.Y.; project administration, B.D. All authors have read and agreed to the published version of the manuscript.

Funding: This study was supported by the Applied Basic Research of Qinghai Province (2021-ZJ-758).

Institutional Review Board Statement: Not applicable.

Informed Consent Statement: Not applicable.

Data Availability Statement: Data is within the article.

Acknowledgments: The authors would like to express their thanks to researcher Kunlun Wu at the Qinghai Academy of Agricultural and Forestry Sciences for providing the tested highland barley variety.

Conflicts of Interest: The authors declare no conflict of interest.

References

1. Dang, B.; Zhang, W.-G.; Zhang, J.; Yang, X.-J.; Xu, H.-D. Evaluation of nutritional components, phenolic composition, and antioxidant capacity of highland barley with different grain colors on the Qinghai Tibet Plateau. *Foods* **2022**, *11*, 2025. [CrossRef] [PubMed]
2. Zeng, X.; Guo, Y.; Xu, Q.; Mascher, M.; Guo, G.; Li, S.; Tashi, N. Origin and evolution of qingke barley in Tibet. *Nat. Commun.* **2018**, *9*, 5433. [CrossRef] [PubMed]
3. Obadi, M.; Sun, J.; Xu, B. Highland barley: Chemical composition, bioactive compounds, health effects, and applications. *Food Res. Int.* **2021**, *140*, 110065. [CrossRef] [PubMed]
4. Ikram, S.; Zhang, H.; Ming, H.; Wang, J. Recovery of major phenolic acids and antioxidant activity of highland barley brewer's spent grains extracts. *J. Food Process. Preserv.* **2019**, *44*, 14308. [CrossRef]
5. Tuersuntuoheti, T.; Wang, Z.; Zhang, M.; Pan, F.; Liang, S.; Sohail, A.; Wang, X. Different preparation methods affect the phenolic profiles and antioxidant properties of Qingke barley foods. *Cereal Chem.* **2021**, *98*, 729–739. [CrossRef]
6. Bi, S.; Wang, A.; Lao, F.; Shen, Q.; Liao, X.; Zhang, P.; Wu, J. Effects of frying, roasting and boiling on aroma profiles of adzuki beans (*Vigna angularis*) and potential of adzuki bean and millet flours to improve flavor and sensory characteristics of biscuits. *Food Chem.* **2021**, *339*, 127878. [CrossRef]
7. Diez-Simon, C.; Mumm, R.; Hall, R.D. Mass spectrometry-based metabolomics of volatiles as a new tool for understanding aroma and flavour chemistry in processed food products. *Metabolomics* **2019**, *15*, 41. [CrossRef]

8. Liu, K.; Li, Y.; Chen, F.; Yong, F. Lipid oxidation of brown rice stored at different temperatures. *Int. J. Food Sci. Technol.* **2017**, *52*, 188–195. [CrossRef]
9. Takemitsu, H.; Amako, M.; Sako, Y.; Kita, K.; Ozeki, T.; Inui, H.; Kitamura, S. Reducing the undesirable odor of barley by cooking with superheated steam. *J. Food Sci. Technol.* **2019**, *56*, 4732–4741. [CrossRef]
10. Tatsu, S.; Matsuo, Y.; Nakahara, K.; Hofmann, T.; Steinhaus, M. Key odorants in Japanese roasted barley tea (Mugi-Cha)-Differences between roasted barley tea prepared from naked barley and roasted barley tea prepared from hulled barley. *J. Agric. Food Chem.* **2020**, *68*, 2728–2737. [CrossRef]
11. Sun, X.; Li, W.; Hu, Y.; Zhou, X.; Ji, M.; Yu, D.; Luan, G. Comparison of pregelatinization methods on physicochemical, functional and structural properties of tartary buckwheat flour and noodle quality. *J. Cereal Sci.* **2018**, *80*, 63–71. [CrossRef]
12. Wang, L.; Wang, L.; Qiu, J.; Li, Z. Effects of superheated steam processing on common buckwheat grains: Lipase inactivation and its association with lipidomics profile during storage. *J. Cereal Sci.* **2020**, *95*, 103057. [CrossRef]
13. Wang, L.; Wang, L.; Wang, A.; Qiu, J.; Li, Z. Superheated steam processing improved the qualities of noodles by retarding the deterioration of buckwheat grains during storage. *LWT-Food Sci. Technol.* **2021**, *138*, 110746. [CrossRef]
14. Wang, R.; Chen, Y.; Ren, J.; Guo, S. Aroma stability of millet powder during storage and effects of cooking methods and antioxidant treatment. *Cereal Chem.* **2014**, *91*, 262–269. [CrossRef]
15. Jiao, S.; Zhu, D.; Deng, Y.; Zhao, Y. Effects of hot air-assisted radio frequency heating on quality and shelf-life of roasted peanuts. *Food Bioprocess Technol.* **2015**, *9*, 308–319. [CrossRef]
16. Kaneko, S.; Kodama, T.; Kohyama, N.; Watanabe, H.; Hayase, F. Aroma components characterizing the odor of cooked barley. *J. Jpn. Soc. Food Sci.* **2013**, *60*, 439–442. [CrossRef]
17. Adebowale, O.J.; Taylor, J.R.N.; de Kock, H.L. Stabilization of wholegrain sorghum flour and consequent potential improvement of food product sensory quality by microwave treatment of the kernels. *LWT-Food Sci. Technol.* **2020**, *132*, 109827. [CrossRef]
18. Zhang, G.; Zhang, C.; Feng, T.; Zhuang, H. Analysis of volatile flavor compounds of corn under different treatments by GC-MS and GC-IMS. *Front. Chem.* **2022**, *10*, 725208. [CrossRef]
19. Zheng, Q.; Wang, Z.; Xiong, F.; Zhang, G. Enzyme inactivation induced by thermal stabilization in highland barley and impact on lipid oxidation and aroma profiles. *Front. Nutr.* **2023**, *10*, 1097775. [CrossRef]
20. Chang, C.; Wu, G.; Zhang, H.; Jin, Q.; Wang, X. Deep-fried flavor: Characteristics, formation mechanisms, and influencing factors. *Crit. Rev. Food Sci. Nutr.* **2020**, *60*, 1496–1514. [CrossRef]
21. Franklin, L.M.; King, E.S.; Chapman, D.; Byrnes, N.; Huang, G.; Mitchell, A.E. Flavor and acceptance of roasted California almonds during accelerated storage. *J. Agric. Food Chem.* **2018**, *66*, 1222–1232. [CrossRef] [PubMed]
22. Warner, K.J.H.; Dimick, P.S.; Ziegler, G.R.; Mumma, R.O. 'Flavor-fade' and off flavors in ground roasted peanuts as related to selected pyrazines and aldehydes. *J. Food Sci. Technol.* **1996**, *61*, 469–472. [CrossRef]
23. Lampi, A.-M.; Damerau, A.; Li, J.; Moisio, T.; Partanen, R.; Forssell, P.; Piironen, V. Changes in lipids and volatile compounds of oat flours and extrudates during processing and storage. *J. Cereal Sci.* **2015**, *62*, 102–109. [CrossRef]
24. Bi, S.; Xu, X.; Luo, D.; Lao, F.; Pang, X.; Shen, Q.; Wu, J. Characterization of key aroma compounds in raw and roasted peas (*Pisum sativum* L.) by application of instrumental and sensory techniques. *J. Agric. Food Chem.* **2020**, *68*, 2718–2727. [CrossRef] [PubMed]
25. Wang, Y.-H.; Yang, Y.-Y.; Zhang, J.-Y.; Zhang, Q.-D.; Xu, F.; Li, Z.-J. Characterization of volatiles and aroma in Chinese steamed bread during elaboration. *J. Cereal Sci.* **2021**, *101*, 103310. [CrossRef]
26. Franklin, L.M.; Chapman, D.M.; King, E.S.; Mau, M.; Huang, G.; Mitchell, A.E. Chemical and sensory characterization of oxidative changes in roasted almonds undergoing accelerated shelf life. *J. Agric. Food Chem.* **2017**, *65*, 2549–2563. [CrossRef]
27. Wilkins, C.K.; Scholl, S. Volatile metabolites of some barley storage molds. *Int. J. Food Microbiol.* **1989**, *8*, 11–17. [CrossRef]
28. Bi, S.; Wang, A.; Wang, Y.; Xu, X.; Luo, D.; Shen, Q.; Wu, J. Effect of cooking on aroma profiles of Chinese foxtail millet (*Setaria italica*) and correlation with sensory quality. *Food Chem.* **2019**, *289*, 680–692. [CrossRef]
29. McGorrin, R.J. Key aroma compounds in oats and oat cereals. *J. Agric. Food Chem.* **2019**, *67*, 13778–13789. [CrossRef]
30. Li, Y.; Li, Q.; Zhang, B.; Shen, C.; Xu, Y.; Tang, K. Identification, quantitation and sensorial contribution of lactones in brandies between China and France. *Food Chem.* **2021**, *357*, 129761. [CrossRef]
31. Dang, B.; Zhang, W.-G.; Zhang, J.; Yang, X.-J.; Xu, H.-D. Effect of thermal treatment on the internal structure, physicochemical properties and storage stability of whole grain highland barley flour. *Foods* **2022**, *11*, 2021. [CrossRef]
32. Maoz, I.; Lewinsohn, E.; Gonda, I. Amino acids metabolism as a source for aroma volatiles biosynthesis. *Curr. Opin. Plant Biol.* **2022**, *67*, 102221. [CrossRef]
33. Xi, Y.; Li, Q.; Yan, J.; Baldwin, E.; Plotto, A.; Rosskopf, E.; Li, J. Effects of harvest maturity, refrigeration and blanching treatments on the volatile profiles of ripe "Tasti-Lee" tomatoes. *Foods* **2021**, *10*, 1727. [CrossRef]
34. Li, Q.; Yang, S.; Zhang, R.; Liu, S.; Zhang, C.; Li, Y.; Li, J. Characterization of honey peach (*Prunus persica* (L.) Batsch) aroma variation and unraveling the potential aroma metabolism mechanism through proteomics analysis under abiotic stress. *Food Chem.* **2022**, *386*, 132720. [CrossRef]
35. Rizzi, G.P. The strecker degradation of amino acids: Newer avenues for flavor formation. *Food Rev. Int.* **2008**, *24*, 416–435. [CrossRef]
36. Suzuki, T.; Honda, Y.; Mukasa, Y.; Kim, S.J. Effects of lipase, lipoxygenase, peroxidase, and rutin on quality deteriorations in buckwheat flour. *J. Agric. Food Chem.* **2005**, *53*, 8400–8405. [CrossRef]

37. Contreras, C.; Tjellström, H.; Beaudry, R.M. Relationships between free and esterified fatty acids and LOX-derived volatiles during ripening in apple. *Postharvest Biol. Technol.* **2016**, *112*, 105–113. [CrossRef]
38. Ao, X.; Mu, Y.; Xie, S.; Meng, D.; Zheng, Y.; Meng, X.; Lv, Z. Impact of UHT processing on volatile components and chemical composition of sea buckthorn (*Hippophae rhamnoides*) pulp: A prediction of the biochemical pathway underlying aroma compound formation. *Food Chem.* **2022**, *390*, 133142. [CrossRef]
39. Suh, J.H.; Madden, R.T.; Sung, J.; Chambers, A.H.; Crane, J.; Wang, Y. Pathway-based metabolomics analysis reveals biosynthesis of key flavor compounds in mango. *J. Agric. Food Chem.* **2022**, *70*, 10389–10399. [CrossRef]
40. Wu, S.; Yang, J.; Dong, H.; Liu, Q.; Li, X.; Zeng, X.; Bai, W. Key aroma compounds of Chinese dry-cured Spanish mackerel (*Scomberomorus niphonius*) and their potential metabolic mechanisms. *Food Chem.* **2021**, *342*, 128381. [CrossRef]
41. Zhang, C.; Hua, Y.; Li, X.; Kong, X.; Chen, Y. Key volatile off-flavor compounds in peas (*Pisum sativum* L.) and their relations with the endogenous precursors and enzymes using soybean (*Glycine max*) as a reference. *Food Chem.* **2020**, *333*, 127469. [CrossRef] [PubMed]
42. Teshima, T.; Funai, R.; Nakazawa, T.; Ito, J.; Utsumi, T.; Kakumyan, P.; Matsui, K. Coprinopsis cinerea dioxygenase is an oxygenase forming 10(S)-hydroperoxide of linoleic acid, essential for mushroom alcohol, 1-octen-3-ol, synthesis. *J. Biol. Chem.* **2022**, *298*, 102507. [CrossRef] [PubMed]

Disclaimer/Publisher's Note: The statements, opinions and data contained in all publications are solely those of the individual author(s) and contributor(s) and not of MDPI and/or the editor(s). MDPI and/or the editor(s) disclaim responsibility for any injury to people or property resulting from any ideas, methods, instructions or products referred to in the content.

Article

Characterisation of the Aroma Profile and Dynamic Changes in the Flavour of Stinky Tofu during Storage

Huaixiang Tian [1], Ling Zou [1], Li Li [2], Chen Chen [1], Haiyan Yu [1], Xinxin Ma [2], Juan Huang [1], Xinman Lou [1] and Haibin Yuan [1,*]

[1] Department of Food Science and Technology, Shanghai Institute of Technology, Haiquan Road 100, Shanghai 201418, China
[2] Shanghai Tramy Green Food Co., Ltd., No. 201, Xuanchun Road, Sanzao Industrial Park, Xuanqiao Town, Pudong New Area, Shanghai 201314, China
* Correspondence: yuanhb@sit.edu.cn

Abstract: Stinky tofu is a traditional Chinese food with wide consumption in China. Nevertheless, the dynamic changes in the flavour of stinky tofu during storage have yet to be investigated. In this study, the flavour changes of stinky tofu over six different storage periods were comprehensively analysed through sensory, electronic nose and gas chromatography-mass spectrometry (GC-MS) analyses. The results of the sensory and electronic nose analyses confirmed the changes in the flavour of stinky tofu across different storage periods. In the GC-MS analysis, 60 volatile compounds were detected during storage, and the odour activity values indicated that 29 of these 60 compounds significantly contributed to the aroma profile. During storage, the alcohol concentration of the stinky tofu gradually decreased while the acid and ester concentrations increased. According to a partial least squares analysis, 2-phenylethyl acetate, 2-phenylethyl propanoate, p-cresol, and phenylethyl alcohol, which were detected after 10 days of storage, promoting the release of an overripe apple-like odour from the stinky tofu. Findings regarding the flavour changes and characteristics of stinky tofu during different storage periods can provide a potential reference for recognising the quality of these products.

Keywords: stinky tofu; food flavour; volatile compounds; GC-MS

1. Introduction

As a type of traditional food, stinky tofu has been a popular snack since the Wei Dynasty in 220 A.D. in China [1]. Stinky tofu is prepared from soybeans, which contain a large amount of protein that decomposes into amino acids after fermentation. Moreover, stinky tofu contains vitamin B12, which can help prevent Alzheimer's disease [2]. Stinky tofu also has a high content of s-equol, which has been proven to promote the health of menopausal women [3,4]. At present, stinky tofu has become a household snack in China and is increasingly popular.

Stinky tofu is prepared by soaking tofu in a specially fermented brine for several hours to several days. The fermented brine leads to mild fermentation of the tofu and endows it with a unique flavour [5]. This brine is typically prepared by mixing tempeh, shiitake mushrooms, amaranth, bamboo shoots, and other edible plants with water; this mixture is naturally fermented for several months to 3 years. During this period, microorganisms in the mixture grow and produce various enzymes, such as proteases and lipases, whose reaction products will contribute to the special odour profile of the brine [6]. In general, the core microbes that contributed to brine fermentation were *Enterococcus*, *Lactobacillus*, *Lactococcus* and *Leuconostoc* [5]. The characteristic aroma of stinky tofu is generated during the process of soaking tofu in fermented brine. Considering the significant influence of fermented brine on the flavour of stinky tofu, many researchers have focused on extracting and analysing the volatile compounds in different types of stinky tofu and fermented brine, for instance, by examining the characteristic flavour compounds in stinky tofu samples [7], dynamic changes in the volatile compounds in brine during fermentation [8], differences in the volatile compounds of stinky tofu sold under different brand names [9], and the flavours of fermented brine prepared by laboratory lactic fermentation [10].

The shelf life of stinky tofu is typically short (approximately 5–13 days). This short lifespan is attributable to the high water and protein content of tofu and the fermented brine remaining in the packaging, which leads to the continuous fermentation of products during transportation and storage. The post-fermentation process quickly and considerably changes the flavour of stinky tofu. However, none of the existing studies have investigated the nature of these changes.

In this study, volatile compounds were extracted from stinky tofu over a 13-day storage period through headspace solid-phase microextraction (HS-SPME). Moreover, gas chromatography-mass spectrometry (GC-MS) and odour active value (OAV) analyses were performed to clarify the compositions and intensities of the volatile compounds. A quantitative descriptive analysis (QDA) and an electronic nose (e-nose) analysis were performed to examine the aroma profiles of the stinky tofu during storage. We speculate that the types and concentrations of volatile flavour substances in Stinky tofu will change regularly during storage. The findings can allow researchers to compare the dynamic changes in stinky tofu flavour during storage and provide a reference for recognizing the quality of products.

2. Materials and Methods

2.1. Materials and Reagents

Alkane standards (C6–C30) were purchased from Sigma–Aldrich (St. Louis, MO, USA), and 2-octanol (internal standard, IS) was purchased from Dr. Ehrenstorfer GmbH, Augsburg, Germany. All chemicals were of analytical reagent grade with a purity higher than 98%. Stinky tofu samples were produced according to the traditional production process by soaking tofu in brine and provided by a local tofu manufacturer in Shanghai. The general production process of stinky tofu: take the fresh tofu cut into uniform blocks, completely soaked in the brine for 10–30 min. The brine was mainly fermented with amaranth and other various spices for nearly 1 year. Then the soaked tofu was immediately packaged and stored at 4 °C for sale. The storage of samples is calculated from the date of production, and the dynamic changes in aroma were examined after 1, 3, 5, 7, 10 and 13 days of storage at 4 °C.

2.2. Quantitative Descriptive Analysis

The sensory descriptors were determined based on international standards (ISO 8589-2007.14) after administering a check-all-that-apply survey and training the sensory evaluators, who included 19 panellists from the Shanghai Institute of Technology (10 women and 9 men aged 22–28 years), with reference to our previous study [11]. The sensory panellists were selected by assessing the sensory discrimination abilities of 32 candidates with experience in sensory testing who participated in the training process. The candidates were not informed of the purpose of the study and were trained for 15 days (60 min/day) to describe and identify aromas. Considering the descriptors recorded in previous studies and the results of preliminary experiments, descriptors agreed upon by more than 50% of the panellists were retained [12,13]. Nine sensory descriptors (overripe apple-like, rotten egg-like, mellow, winey, beany, rotten plant-like, musty, rancid, and sweaty) were reserved in the formal test. The following odorants were used to compare the nine odour descriptors: overripe apple-like, overripe apples; rotten egg-like, 0.05% hydrogen sulfide in water; mellow, cotton balls soaked with 5 mL of alcohol in a sealed container for 24 h; winey, 5 mL of whisky in a brown glass bottle; beany, fresh beans soaked in water for 12 h; rotten plant-like, stems soaked in water for 2–3 d; musty, enoki mushroom root; rancid, 0.5% butyric acid solution in propylene glycol; and sweaty, 0.01% butyric acid solution in water.

The aroma profile was analysed in an individual sensory testing laboratory at a constant temperature of 20 °C, according to the ISO 4121-2003 standard. The intensity of each aroma attribute was rated on a scale of 0–5 (0 = not perceivable, 3 = medium perceivable, 5 = strongly perceivable), with the final score of each aroma attribute corresponding to the average for all panellists.

2.3. E-Nose Detection

In order to verify the flavour changes of stinky tofu in different storage periods, a HERACLES e-nose from Alpha-MOS (Toulouse, France) was used to analyse the volatile compounds in the stinky tofu. The instrument was equipped with 18 metal oxide sensors and a headspace autosampler (HS100) that could perform data processing. The GC function and e-nose olfactory fingerprint software were installed in the e-nose. Stinky tofu samples (1 g) were placed in 10-mL glass vials with Teflon rubber caps. Each vial was incubated at 40 °C for 10 min with stirring (500 rpm). The headspace (5000 µL) was carried by air (150 mL/min) and injected into the nose. The sensor resistance was measured within 100 s with an acquisition frequency of once per second. The performance characteristics of electronic nose sensors are listed in Table 1.

Table 1. Performance characteristics of electronic nose sensors.

No.	Sensor Name	Type of Sensitive Substance
1	LY2/LG	Chlorine, Fluorine, Sulfide
2	LY2/G	Ammonia, Amine compounds
3	LY2/AA	Ethanol, Ammonia
4	LY2/Gh	Ammonia, Amine compounds
5	LY2/gCTI	Sulfide
6	LY2/gCT	Propane, Butane
7	T30/1	Propanol, Hydrogen chloride
8	P10/1	Hydrocarbons, n-octane
9	P10/2	Methane, n-heptane
10	P40/1	Fluorine, Chlorine, Methyl furfural
11	T70/2	Xylene, Toluene
12	PA/2	Acetaldehyde, Amine compounds
13	P30/1	Ammonia, Ethanol
14	P40/2	Chlorine, Methyl mercaptan
15	P30/2	Hydrogen sulfide, Copper
16	T40/2	Chlorine
17	T40/1	Fluorine
18	TA/2	Alcohol

2.4. Aroma Extraction through HS-SPME

The volatile compounds in the stinky tofu were extracted through HS-SPME with a divinylbenzene/carboxen/polydimethylsiloxane (DVB/CAR/PDMS) fused silica (75 µm)-coated fibre (1-cm-long; Supelco, Inc., Bellefonte, PA, USA) [14]. Prior to extraction, the stinky tofu was evenly crushed into a paste in a mortar. Subsequently, 3 g of the crushed stinky tofu and 7 µg of 2-octanol solution (220 µg/kg) were placed in a 15-mL headspace vial that was later sealed with a Teflon cover. The extraction fibre types were optimised according to the type and quantity of volatile compounds in the preliminary experiment. The number of volatile compounds extracted by 3 different SPME fibres was 34 (PDMS/DVB), 60 (DVB/CAR/PDMS), and 28 (CAR/PDMS), respectively. The final optimised extraction conditions were as follows: the headspace vial containing the stinky tofu sample was equilibrated in a water bath at 60 °C for 15 min. Subsequently, the SPME fibre was inserted in the headspace of the vial for 50 min to extract the volatile compounds. The SPME fibre was directly introduced into the injection port of the GC-MS after extraction.

2.5. GC-MS Analysis

The GC-MS analysis was performed using an Agilent 7890 gas chromatograph (5973C, Agilent Technologies, Santa Clara, CA, USA) equipped with a mass spectrometer. Separations were performed using HP-Innowax analytical fused silica capillary columns (60 m × 0.25 mm × 0.25 μm) from Agilent Technologies. A splitless mode was used for the injection for 4 min at 40 °C. The operating conditions for the GC-MS were as follows: the flow rate of helium was 1 mL/min, the mass spectrum was recorded in the electron impact mode (70 eV) in a scan range of 35–350 m/z with the ion source maintained at 250 °C. The initial oven temperature for the HP-Innowax column was set at 40 °C and maintained for 4 min. Subsequently, the temperature was increased to 130 °C at a rate of 4 °C/min, maintained at this value for 5 min, increased to 200 °C at a rate of 4 °C/min, and maintained at this value for 5 min [14,15]. Kovats' retention indices (RIs) were calculated using mixtures of n-alkanes(C6–C30) for both stationary phases [16]. All volatile compounds were quantified using 2-octanol as an internal standard [17]. Compound identification was based on mass spectra matching with the standard NIST 17 MS library and on the comparison of RI sourced from the NIST Standard Reference Database. The volatile compounds were quantified by comparison of peak areas in the ion extraction chromatogram (IEC), which was obtained by selecting target ions for each compound to that of internal standard (2-octanol). These ions corresponded to base ion (m/z 100% intensity), molecular ion (M^+) or another characteristic ion for each molecule. Hence, some peaks that could be co-eluted in scan mode can be integrated with a value of resolution greater than 1. The minimum detection limit for this method was 3 μg/kg, relative standard deviation (RSD) < 6.0. The concentration of volatile compounds was calculated as the following formula:

$$Wi = ((Cs \times Vs)/m) \times (APi/APs) \tag{1}$$

where Wi is the concentration of the volatile component to be measured, μg/L; Cs is the concentration of the internal standard substance, μg/L; Vs is the volume of internal standard added, μL; m is the weight of the sample, g; APi is the peak area of the volatile component; APs is the peak area of the internal standard.

2.6. OAV Analysis

OAV is the ratio of the concentration of an aromatic substance to its threshold value in water [18], which indicates the minimum concentration of volatile compounds that can be smelled. A compound with an OAV greater than 1 is considered to influence the aroma profile. The magnitude of this value reflects the contribution of the compound to the aroma profile. The OAV can be calculated as the following formula.

$$OAV = C_i/OT_i \tag{2}$$

where C_i is the concentration of the volatile compounds, and OT_i is the threshold of the compound in water.

2.7. Statistical Analysis

All data were evaluated through analysis of variance (ANOVA) and Duncan's multiple range tests, implemented using SPSS software (v. 19.0, SPSS Inc., Chicago, IL, USA). Alpha Soft (Alpha-MOS proprietary software) was used to process the e-nose data, and chemometric methods executed automatically by the software were used to interpret the e-nose measurement results. Simca-p soft (v14.1, MKS Umetrics AB) was used to derive the partial least squares (PLS) plots. Origin 2018 software (OriginLab, Northampton, MA, USA) was used to obtain radar charts of the sensory attribute scores and histograms for the GC-MS results. Each test was performed in triplicate.

3. Results and Discussion

3.1. Sensory Analysis

QDA was performed to explore the aroma profiles of the stinky tofu during the tested storage periods and differences in the sensory characteristics. Nine descriptors, namely overripe apple-like, rotten egg-like, mellow, winey, beany, rotten plant-like, musty, rancid, and sweaty, were used to analyse the aroma characteristics of the samples. As shown in Figure 1, all aroma attributes except sweaty changed significantly across the storage periods ($p < 0.05$). In the descriptive analysis, the score for beany was significantly higher ($p < 0.05$) than the scores of the other attributes in the Day 5 samples. The scores for rotten egg-like, rotten plant-like, and overripe apple-like were significantly higher, and the scores for mellow and winey were lower in the Day 13 samples than in the other samples ($p < 0.01$). Samples stored for less than 7 days exhibited similar aroma profiles. The beany attribute decreased significantly in the samples after Day 10, and the flavour profile changed significantly on Day 13. Increases in unpleasant flavour attributes, such as overripe apple-like, musty, rotten egg-like, and rotten plant-like, deteriorated the flavour of the stinky tofu.

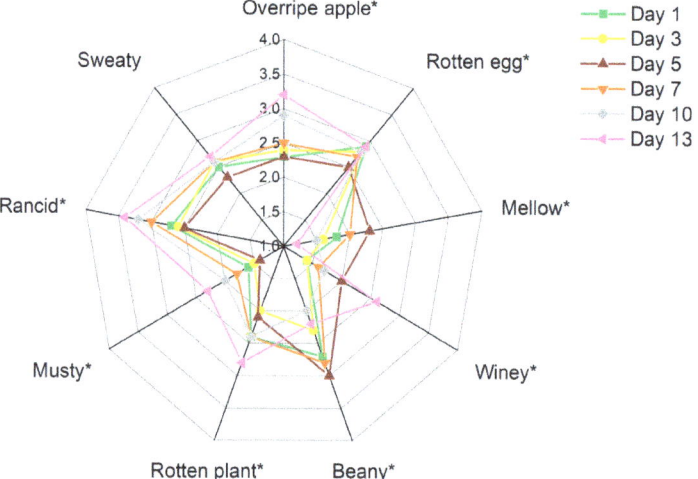

Figure 1. Sensory aroma profiles of the stinky tofu samples in different storage periods. '*' indicates the significance ($p < 0.05$) according to ANOVA and Duncan's multiple comparison tests.

3.2. E-Nose Analysis

To verify the difference in the flavour of stinky tofu across different storage periods, a dynamic factor analysis (DFA) was performed using an e-nose to identify correlations between the individual composition variables of the stinky tofu over time [19]. The two-dimensional DFA plot is shown in Figure 2. The sum of the first two discriminant factor values was 91.936% (DF1: 54.942% and DF2: 36.994%). The stinky tofu samples stored for less than 7 days are distributed on the positive semi-axis of DF1, whereas the samples stored for less than 5 days are concentrated in the first quadrant and are easily distinguishable from the samples stored for more than 10 days. The flavour characteristics of stinky tofu samples stored for 1 to 7 days were similar, whereas significant differences were observed after 10 days of storage. Both of the stinky tofu samples stored for 10 and 13 days could be distinct from earlier time points. Overall, the results of the e-nose analysis verified the flavour of the stinky tofu changed significantly with increasing storage time, and significant flavour differences were observed between the samples in the early and late stages of storage.

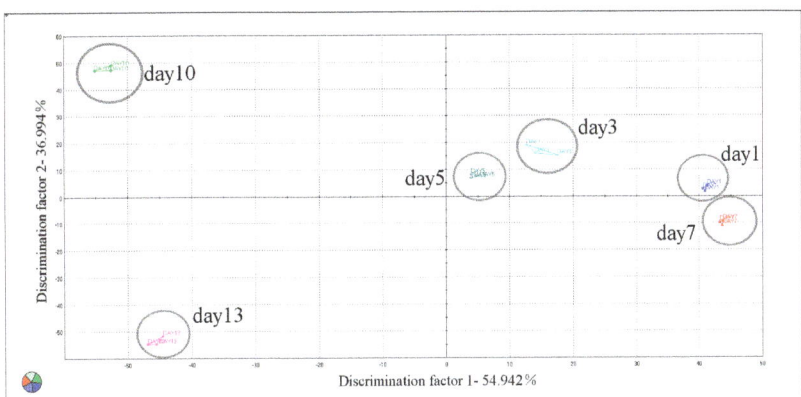

Figure 2. Results of e–nose of DFA plot of the stinky tofu samples in different storage periods.

3.3. Volatile Compounds in the Stinky Tofu Samples

The volatile compounds in the stinky tofu were analysed by HS-SPME-GC-MS in the 13-day storage period and are listed in Table 2. A total of 60 volatile compounds were detected, including 10 acids, 17 alcohols, 11 aldehydes, 13 esters, three phenols, three ketones and three compounds of other types. Among them, 29 volatile compounds with OAVs greater than 1 were detected (Table 3), including three acids, six alcohols, 10 aldehydes, six esters, two phenols, one ketone and one furan substance. In particular, 1-octen-3-ol, (E,E)-2,4-decadienal, (E,E)-2,4-nonadienal, nonanal, and p-cresol contributed significantly to the aroma profile of the stinky tofu (OAV > 100). Among these compounds with OAV greater than 100, 1-octen-3-ol typically has a mushroom-like smell, (E,E)-2,4-decadienal typically has a chicken and fatty smell, whereas the other two aldehydes have a green and fatty smell. All of these compounds except p-cresol were considered the main sources of the beany odour.

Table 2. The volatile compounds detected by GC-MS in stinky tofu during storage.

No.	RI ** Calc	RI ** Ref	Compound	Concentration (mg/kg) * Day 1	Day 3	Day 5	Day 7	Day 10	Day 13
1	1455	1461	Acetic acid	1.014 ± 0.822 a	1.723 ± 0.132 a	2.035 ± 0.242 a	2.754 ± 0.003 c	3.299 ± 0.597 b	3.471 ± 0.320 c
2	1541	1564	Propanoic acid	0.498 ± 0.310 a	0.652 ± 0.319 bc	0.906 ± 0.161 ab	1.212 ± 0.010 c	1.588 ± 0.467 c	2.825 ± 0.258 d
3	1571	1544	2-methyl-propanoic acid	0.052 ± 0.019 b	0.052 ± 0.013 b	0.082 ± 0.010 ab	0.088 ± 0.008 b	0.036 ± 0.003 a	0.027 ± 0.006 b
4	1630	1639	Butanoic acid	1.367 ± 0.555 a	2.495 ± 1.201 c	3.471 ± 0.666 b	4.173 ± 0.124 b	5.692 ± 0.418 c	7.501 ± 0.360 d
5	1672	1674	2-methyl-butanoic acid	0.182 ± 0.048 a	0.374 ± 0.012 b	0.263 ± 0.031 b	0.118 ± 0.018 b	—	—
6	1673	1679	3-methyl-butanoic acid	—	0.331 ± 0.060	—	—	—	—
7	1738	1734	Pentanoic acid	0.066 ± 0.015 a	0.287 ± 0.051 c	0.273 ± 0.007 a	0.164 ± 0.007 a	0.067 ± 0.018 a	0.019 ± 0.012 b
8	1801	1800	4-methyl-pentanoic acid	—	0.021 ± 0.001 ab	0.099 ± 0.003 c	0.167 ± 0.013 bc	0.149 ± 0.020 c	0.132 ± 0.012 abc
9	1843	1831	Hexanoic acid	0.125 ± 0.055 a	0.266 ± 0.071 bc	0.17 ± 0.047 ab	0.162 ± 0.006 ab	0.105 ± 0.039 a	0.066 ± 0.004 cd
10	2055	2039	Octanoic acid	—	0.038 ± 0.014 ab	0.018 ± 0.002 a	—	—	—
11	928	939	Ethanol	0.686 ± 0.087 bc	0.585 ± 0.090 ab	0.532 ± 0.088 ab	0.553 ± 0.006 ab	0.541 ± 0.490 c	0.309 ± 0.035 a
12	1033	1037	1-propanol	—	0.011 ± 0.002	—	0.009 ± 0.001	—	—
13	1142	1150	1-butanol	—	0.014 ± 0.001	—	0.012 ± 0.001	—	—
14	1205	1206	3-methyl-1-butanol	0.041 ± 0.028 c	0.029 ± 0.002 b	0.011 ± 0.001 a	—	—	—
15	1246	1254	1-pentanol	0.016 ± 0.000 a	0.023 ± 0.014 b	0.022 ± 0.017	0.022 ± 0.003 c	—	—
16	1349	1340	1-hexanol	0.704 ± 0.316 cd	0.765 ± 0.013 bc	0.827 ± 0.022 d	0.955 ± 0.107 d	0.234 ± 0.097 a	0.157 ± 0.003 ab
17	1389	1394	3-octanol	0.035 ± 0.004 ab	0.029 ± 0.020 a	0.092 ± 0.002 d	0.046 ± 0.010 bc	0.034 ± 0.002 ab	0.011 ± 0.001 c
18	1447	1447	1-octen-3-ol	0.186 ± 0.042 ab	0.262 ± 0.070 d	0.259 ± 0.021 d	0.211 ± 0.005 c	0.056 ± 0.001 a	—
19	1451	1459	1-heptanol	0.021 ± 0.048 a	0.022 ± 0.030 a	0.034 ± 0.008 a	0.051 ± 0.022 d	0.223 ± 0.015 c	0.044 ± 0.007 b
20	1508	1504	(E)-2-hepten-1-ol	0.037 ± 0.001 a	0.029 ± 0.011 a	0.015 ± 0.009 b	—	—	—
21	1554	1554	1-octanol	0.024 ± 0.009	0.017 ± 0.005	—	—	—	—
22	1609	1609	(E)-2-octen-1-ol	0.062 ± 0.001 a	0.042 ± 0.003 b	0.011 ± 0.001 c	—	—	—
23	1656	1666	1-nonanol	—	0.122 ± 0.001 a	0.134 ± 0.007 a	0.056 ± 0.001 b	0.028 ± 0.001 c	0.304 ± 0.068 b
24	1912	1935	Phenylethyl alcohol	0.018 ± 0.009 a	0.044 ± 0.006 a	0.042 ± 0.001 a	0.064 ± 0.040 a	0.093 ± 0.035 a	0.412 ± 0.012
25	1957	1935	1-dodecanol	—	—	—	—	0.128 ± 0.001	—
26	1878	1889	Benzyl alcohol	0.032 ± 0.021 a	0.021 ± 0.001 a	0.025 ± 0.001 a	0.027 ± 0.001 a	0.031 ± 0.001 a	0.049 ± 0.004 b
27	1957	1935	1-Dodecanol	—	—	—	0.138 ± 0.001 a	0.228 ± 0.024 b	0.412 ± 0.012 c
28	1528	1529	Benzaldehyde	0.208 ± 0.197 b	0.165 ± 0.006 a	0.128 ± 0.001 ab	0.055 ± 0.438 a	0.027 ± 0.001 a	—
29	1712	1730	4-ethyl-benzaldehyde	0.056 ± 0.018 a	0.042 ± 0.008 a	0.032 ± 0.008 b	0.013 ± 0.002 c	—	—
30	1217	1220	(E)-2-hexenal	0.014 ± 0.000	0.002 ± 0.001	—	—	—	—

Table 2. Cont.

No.	RI ** Calc	Ref	Compound	Concentration (mg/kg) * Day 1	Day 3	Day 5	Day 7	Day 10	Day 13
31	1322	1332	(E)-2-heptenal	0.052 ± 0.023 a	0.059 ± 0.027 a	0.064 ± 0.017 b	0.054 ± 0.001 a	—	—
32	1390	1396	Nonanal	0.049 ± 0.011 a	0.062 ± 0.001 b	0.189 ± 0.020 c	0.049 ± 0.032 ab	0.014 ± 0.001 b	—
33	1429	1434	(E)-2-octenal	0.056 ± 0.000 a	0.071 ± 0.001 a	0.097 ± 0.007 b	0.052 ± 0.001 a	—	—
34	1496	1500	Decanal	—	—	0.035 ± 0.008	—	—	—
35	1535	1542	(E)-2-nonenal	—	—	0.048 ± 0.007	—	—	—
36	1642	1643	(E)-2-decenal	0.022 ± 0.001 a	0.037 ± 0.008 a	0.055 ± 0.438 b	0.085 ± 0.001 c	0.074 ± 0.011 c	0.051 ± 0.001 b
37	1701	1706	(E,E)-2,4-nonadienal	0.007 ± 0.001 a	0.014 ± 0.002 a	0.018 ± 0.002 b	—	—	—
38	1809	1827	(E,E)-2,4-decadienal	0.043 ± 0.033 a	0.065 ± 0.002 a	0.113 ± 0.019 ab	0.298 ± 0.044 c	0.314 ± 0.001 c	0.241 ± 0.021 bc
39	889	891	Ethyl acetate	—	—	—	—	0.011 ± 0.001	0.329 ± 0.001
40	951	964	Propanoic acid ethyl ester	—	—	—	—	—	0.094 ± 0.011
41	1069	964	Acetic acid butyl ester	—	—	—	—	—	0.014 ± 0.002
42	1171	1176	Acetic acid pentyl ester	—	—	—	—	—	0.074 ± 0.002
43	1268	1269	Acetic acid hexyl ester	—	0.011 ± 0.001 a	0.014 ± 0.003 a	0.019 ± 0.001 a	0.095 ± 0.055 b	1.862 ± 0.032 c
44	1336	1342	Propanoic acid, hexyl ester	—	—	—	—	—	0.327 ± 0.012
45	1349	1336	Formic acid, hexyl ester	—	—	—	—	—	0.197 ± 0.004
46	1371	1392	Acetic acid heptyl ester	—	—	—	—	—	0.018 ± 0.001
47	1554	1560	Formic acid octyl ester	—	—	0.057 ± 0.001	0.025 ± 0.003	—	—
48	1580	1579	Bornyl acetate	—	0.005 ± 0.001 a	0.018 ± 0.009 a	0.002 ± 0.001 a	—	—
49	1638	1635	Butyrolactone	0.032 ± 0.002 a	0.069 ± 0.009 b	0.031 ± 0.001 a	—	—	—
50	1818	1825	2-phenylethyl acetate	—	—	—	—	0.734 ± 0.169	12.172 ± 0.684
51	1884	/	2-phenyl ethyl propanoate	—	—	—	—	0.078 ± 0.022	3.515 ± 0.699
52	1864	1889	2-methoxy phenol	—	0.017 ± 0.001 a	0.041 ± 0.007 b	0.018 ± 0.004 a	—	—
53	2009	2004	Phenol	0.119 ± 0.050 a	0.125 ± 0.022 a	0.178 ± 0.001 ab	0.105 ± 0.514 ab	0.077 ± 0.021 a	0.061 ± 0.010 c
54	2086	2094	P-cresol	0.367 ± 0.189 a	1.014 ± 0.141 b	1.316 ± 0.255 b	1.716 ± 0.102 b	1.807 ± 0.503 b	2.031 ± 0.080 c
55	1281	1278	2-octanone	0.041 ± 0.011 a	0.022 ± 0.002 b	0.028 ± 0.014 b	0.023 ± 0.514 b	0.022 ± 0.002 b	0.017 ± 0.004 a
56	1656	1647	Acetophenone	0.124 ± 0.012 a	0.082 ± 0.008 b	0.039 ± 0.009 b	—	—	—
57	1968	1988	Maltol	0.019 ± 0.013 b	0.026 ± 0.014 b	0.022 ± 0.010 b	0.045 ± 0.027 bc	0.003 ± 0.002 a	—
58	1224	1244	2-pentyl-furan	0.088 ± 0.032 cd	0.084 ± 0.002 a	0.102 ± 0.016 d	0.082 ± 0.001 cd	0.062 ± 0.008 bc	—
59	1257	1257	Styrene	0.051 ± 0.016 a	0.011 ± 0.008 b	0.005 ± 0.001 c	—	—	0.046 ± 0.029 b
60	1618	1619	2-(2-ethoxyethoxy)-ethanol	0.046 ± 0.002 a	0.077 ± 0.031 a	0.031 ± 0.001 a	—	—	—

Note: *: Volatile compounds identification based on the NIST17 mass spectral database. **: The retention index of volatile compounds on HP-Innowax columns. —: Not detected in the sample. /: Not found in references. Values with different letters (a to d) in a row are significantly different ($p \leq 0.05$).

Table 3. Thresholds and OAVs (≥1) of the volatile compounds in the stinky tofu during storage.

NO.	Compound [A]	Threshold (mg/kg) [B]	OAV Day 1	Day 3	Day 5	Day 7	Day 10	Day 13	Description
1	3-methyl-1-butanol	0.02	2	1	1	—	3	—	Apple brandy-like, spicy
2	1-hexanol	0.07	10	7	7	6	3	2	Resin-like, flowery, green
3	1-octen-3-ol	0.0015	57	74	93	120	137	182	Mushroom-like
4	1-heptanol	0.003	7	7	—	12	—	15	Green, chemical-like
5	(E)-2-octen-1-ol	0.02	3	2	1	1	—	—	Mellow
6	1-dodecanol	0.016	—	—	—	—	16	26	Mellow
7	Ethyl acetate	0.005	—	—	—	11	21	66	Mellow, fruity
8	Propanoic acid ethyl ester	0.01	—	—	—	—	—	9	Mellow, fruity
9	Acetic acid pentyl ester	0.043	—	6	8	—	48	2	Fruity, banana- and pear-like
10	Acetic acid hexyl ester	0.002	—	—	—	10	21	931	Fruity
11	Propanoic acid, hexyl ester	0.008	—	—	—	7	13	41	Sweet, fruity
12	2-phenylethyl acetate	0.24959	4	2	2	—	—	24	Sweet, slightly green leafy
13	4-ethyl-benzaldehyde	0.013	10	17	44	26	11	2	Bitter almond-like, slightly sweet
14	Hexanal	0.005	19	9	—	—	—	—	Green, citrusy and fatty
15	(E)-2-octenal	0.003	45	57	172	41	17	—	Green, citrusy, and fatty
16	Nonanal	0.0011	5	5	5	4	—	7	Green, citrusy, and fatty
17	(E)-2-heptenal	0.013	—	—	10	1	—	—	Irritating, green grass-like
18	Decanal	0.003	97	162	251	180	97	12	Green and fatty
19	(E)-2-nonenal	0.00019	73	108	167	284	220	169	Green
20	(E)-2-decenal	0.0003	70	123	180	110	42	—	Green and fatty
21	(E,E)-2,4-nonadienal	0.0001	1593	1928	4169	11,032	4235	2924	Green and fatty
22	(E,E)-2,4-decadienal	0.000027	7	12	17	22	28	37	Spicy, geranium-like
23	Butanoic acid	0.204	2	4	3	3	—	—	Strongly sour, cheese-like
24	2-methyl-butanoic acid	0.1	—	7	12	4	—	—	Sour, cheese-like
25	3-methyl-butanoic acid	0.046	2	—	—	—	—	—	Milk-like, sour, slightly sweet
26	Acetophenone	0.065	3	11	18	8	—	—	Hawthorn-like
27	2-methoxyphenol	0.0016	37	101	132	122	111	103	Sweet, woody, slightly smoky
28	P-cresol	0.01	5	14	18	14	11	8	Phenolic, sour
29	2-pentyl-furan	0.0058	5	14	18	14	11	8	Oily, soy, green

[A] The aroma compounds identified on the HP-Innowax column; [B] The threshold of volatile compounds in water referred to in the literature [19]. —: Not detected in the sample.

The stinky tofu over 6 different storage periods exhibited high concentrations of acids, including acetic acid, propanoic acid, and butanoic acid. The maximum concentration and OAV corresponded to butyric acid, which has a strongly acidic and sweat-like smell and likely contributes significantly to the overall aroma profile of stinky tofu. In addition, 2-methylbutanoic acid and 3-methylbutanoic acid, which have pungent, sour, and rancid odours, had high OAVs (OAV $\geq$ 1). The branched-chain fatty acids in stinky tofu are typically produced by the fermentation of amino acids by anaerobic microorganisms [19]. The obtained results are similar to those obtained by Wang et al. in a study of fermented brine [20].

The alcohol compounds, which are mainly derived from the fermentation of raw materials in the brine, endowed the stinky tofu with a mellow aroma and functioned as the main components of the volatile compounds in the stinky tofu. Among these alcohols, hexanol and 1-octen-3-ol may promote the beany smell of stinky tofu through synergy or masking effects [20]. Phenylethyl alcohol, which has a sweet rose-like fragrance, is a typical volatile compound in food and is likely derived from benzenesulfonic acid through a series of reactions [21,22].

Aldehydes present in the sample are widely recognized as off-flavour substances produced during food storage [23,24]. Aldehyde compounds usually exhibit green, herbal and fatty odours [20], in which benzaldehyde is a typical volatile compound in food and has a bitter almond-like nutty aroma, and hexanal is considered to be a typical compound to enhance the beany odour of the stinky tofu [25].

Esters, which typically exhibit floral and fruity scents, contributed to the 'fermented' and 'over-ripe apple' odour in the aroma profile of the stinky tofu. Although the pleasant floral and fruity aromas of esters have been noted to enhance food flavours [26], the excessive ester concentration in the Day 13 samples did not result in a pleasant odour, likely because of the dynamic trends of volatile compounds and masking of the ester aromas by acids.

Phenol and p-cresol, which exhibit unpleasant odours, were found to contribute to the characteristic odour of stinky tofu. These conclusions were consistent with those of existing studies [14,26].

There are many kinds of volatile compounds in stinky tofu, which was very similar to some existing research results. However, some compounds considered the core flavour substance of stinky tofu was not detected, such as indole and skatole [9,14]. The volatile compounds in stinky tofu may relate to the processing technology, raw materials and geographic area [5]. The stinky tofu used in this research was sourced from the southeast coastal regions of China, and the brine was fermented using an amaranth-based mixture of plants. This mixture may impart a different aroma compound than that found in the central and western regions of China.

3.4. Changes in the Aroma Profile during Storage

The dynamic changes in the contents and ratios of volatile compounds in the stinky tofu during 6 different storage periods were determined by comparing and analysing the volatile compounds. As shown in Figure 3, as the storage period increased, the number of volatile compounds in the stinky tofu first increased and then decreased. Furthermore, the compositions of volatile compounds in the stinky tofu stored for less than 7 days were similar, and the composition of the compounds changed significantly after 10 days. The number of volatile compounds was maximised in the Day 5 samples (40 compounds), likely because of the incomplete fermentation of the freshly prepared stinky tofu and, consequently, the incomplete production of the volatile compounds. Alcohols, esters, and phenols exhibited significant fluctuations in numbers and concentrations ($p \leq 0.05$). The numbers and concentrations of esters increased significantly after Day 10, whereas those of alcohols decreased. In terms of concentration, the significant increases in ester concentrations in post-fermentation samples likely occurred because esters are generated by bacterial esterification, which may have been promoted by the high-water activity of the stinky tofu during storage [27]. In general, stinky tofu samples stored for different time periods had distinct flavours and a regular dynamic trend in these flavours was observed with changes in the storage time. During storage, the stinky tofu exhibited high concentrations of acids, alcohols, and aldehydes. The acid concentration in nearly all samples was high and contributed significantly to the aroma profile of the stinky tofu owing to their low threshold. The acid concentrations peaked (72.8%) in the Day 3 sample. The ester concentrations increased significantly after Day 7 ($p < 0.05$), and extremely high concentrations were observed in the Day 13 sample. These high concentrations likely occurred because the acid produced in the later stage of fermentation provided a sufficient substrate for the formation of esters [28].

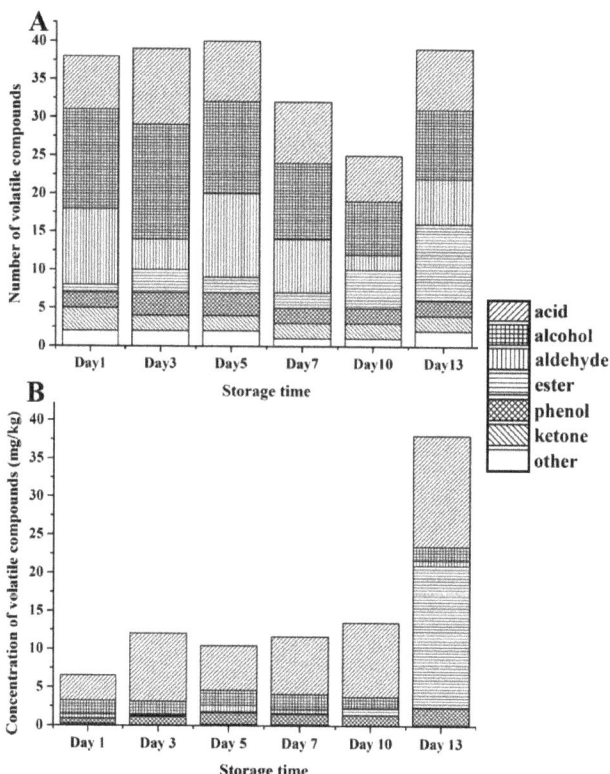

Figure 3. (A) Numbers and (B) concentrations of volatile compounds in the stinky tofu samples during storage.

As shown in Figure 3, esters were mainly detected in the post-fermentation samples (storage period of 10–13 days), and the numbers of esters increased as the storage period was prolonged, likely because of the delay in esterification between acids and alcohols [22]. High concentrations of the esters 2-phenethyl acetate and 2-phenethyl propionate were detected on Day 7 and Day 10. Notably, these two volatile compounds, which represent the final products of benzenesulfonic acid conversion, only appeared after Day 10 and may indicate over-fermentation of the stored stinky tofu [21]. The p-cresol was observed in all storage periods, and its concentrations increased with the storage time, with the maximum values observed in the Day 13 sample. The p-cresol exhibits a phenolic odor, and it has been reported that it can contribute to the unique aroma of stinky tofu [13].

3.5. PLS Analysis

PLS analyses were performed to examine the correlations between the sensory attributes and the volatile compounds of the stinky tofu. The concentrations of 29 volatile compounds with OAV greater than 1 were represented as the X-matrix, and the nine sensory descriptors were represented as the Y-matrix. Among these 29 compounds, 15 compounds exhibited VIP values greater than 1, indicating the significant contribution of these compounds to the aroma of the stinky tofu. Volatile compounds and sensory analysis data were used to establish a model to identify the relationships between the volatile compounds and sensory attributes (Figure 4). The results indicated that most of the sensory attributes were correlated with the volatile compounds, and nearly all X and Y variables lay within the ellipse (R^2 = 100%; R^2 represents the degree of interpretation).

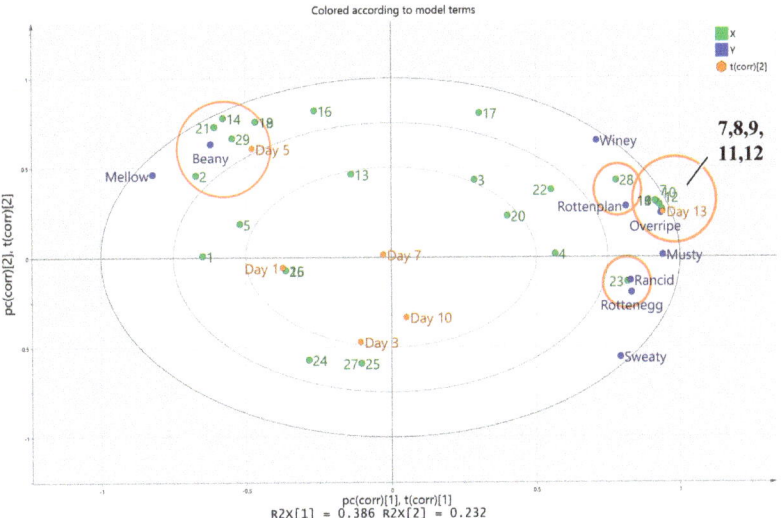

Figure 4. Partial least squares correlation plot from the analysis of the stinky tofu samples. The model was derived from 29 volatile compounds with OAV greater than 1 as the X–matrix and the result of sensory analysis as the Y–matrix. Codes 1–29 refer to the volatile compounds listed in Table 3.

The concentrations of volatile compounds that contributed to beany flavour were higher in the early periods of storage, such as 1-hexanol (2), 2-pentyl-furan (29), hexanal (14), decanal (18), (E)-2-nonenal (19) and (E,E)-2,4-nonadienal (21), which consistent with previous results [21]. In addition, ethyl acetate (7), propanoic acid ethyl ester (8), acetic acid pentyl ester (9), acetic acid hexyl ester (10), propanoic acid hexyl ester (11) and 2-phenylethyl acetate (12), were considerably influenced the score for the overripe apple-like attribute. It is worth mentioning that all of the substances related to overripe apple-like flavor were esters, with rapidly increased concentrations in the late fermentation stage. Generally, the sweet, fruity flavour of esters had a positive effect on food flavour [29]; however, the sensory of esters was perceived as an unpleasant "overripe" odour in the samples, likely due to the inadequate coordination of esters with other volatile compounds in stinky tofu. This finding is also consistent with the results of sensory analysis for stinky tofu samples. Butyric acid (23) was significantly correlated with the descriptors of rotten eggs and rancidity in the samples, and the point representing p-cresol (28) was close to the descriptors of "Rottenplant." These two compounds were demonstrated to be strongly associated with the characteristic odor of stinky tofu, a finding that is similar to those reported in other studies [21]. Therefore, the PLSR analysis results and the key aroma compounds identified by OAV indicated that the different types and concentrations of volatile compounds have a significant impact on the aroma attributes of stinky tofu.

4. Conclusions

The aroma profile and dynamic changes in the flavour of the stinky tofu during storage were investigated based on QDA and SPME-GC-MS. Sixty volatile compounds were isolated and identified in stinky tofu samples stored for six periods by SPME-GC-MS, and the contributions of these compounds to the aroma were investigated by calculating the OAVs. According to the results of sensory evaluation, the scores of unpleasant flavour attributes such as sweaty, rancid, rotten plant-like and rotten egg-like tended to increase at the late stage of storage. Based on the PLS analysis, the esters in the stinky tofu generated an overripe apple-like aroma, and stinky tofu stored for more than 7 days exhibited significant increases in the ester concentration. The ester concentration thus can be considered an indicator of the product storage time. The concentrations of 1-hexanol, 2-pentyl-furan, hexanal, decanal, (E)-2-nonenal and (E,E)-2,4-nonadienal were positively correlated with beany characteristics and were higher in the early stages of storage than in later stages. The results regarding the flavour changes and characteristics of stinky tofu during different storage periods can provide a reference for recognizing the quality of products. In addition, the established correlation between aroma compounds and sensory descriptors of stinky tofu can be leveraged for further research, such as targeted regulation of certain aroma characteristics in stinky tofu based on the concentrations of volatile compounds associated with descriptors.

Author Contributions: Conceptualization, H.Y. (Haibin Yuan) and H.T.; methodology, L.Z.; software, J.H. and X.L.; validation, L.L. and X.M.; formal analysis, C.C. and H.Y. (Haiyan Yu); investigation, L.Z.; resources, H.T.; data curation, L.Z.; writing—original draft preparation, L.Z. and H.Y. (Haibin Yuan); writing—review and editing, H.Y. (Haibin Yuan); visualization, H.Y. (Haibin Yuan); supervision, H.T.; project administration, H.T.; funding acquisition, H.T. All authors have read and agreed to the published version of the manuscript.

Funding: This research was funded by a Horizontal Scientific Research Project Grant [J2021-26].

Data Availability Statement: The data presented in this study are available on request from the corresponding author.

Acknowledgments: We thank Shanghai Tramy Green Food (Group) Co., Ltd. for their help with and support of this research.

Conflicts of Interest: The authors declare no conflict of interest.

References

1. Gu, J.; Liu, T.; Sadiq, F.A.; Yang, H.; Yuan, L.; Zhang, G.; He, G. Biogenic amines content and assessment of bacterial and fungal diversity in stinky tofu—A traditional fermented soy curd. *LWT* **2018**, *88*, 26–34. [CrossRef]
2. Mo, H.; Kariluoto, S.; Piironen, V.; Yang, Z.; Sanders, M.G.; Vincken, J.P. Effect of soybean processing on content and bioaccessibility of folate, vitamin B12 and isoflavones in tofu and tempe. *Food Chem.* **2013**, *141*, 2418–2425. [CrossRef]
3. Abiru, Y.; Kumemura, M.; Ueno, T.; Uchiyama, S.; Masaki, K. Discovery of an S-equol rich food stinky tofu, a traditional fermented soy product in Taiwan. *Int. J. Food Sci. Nutr.* **2012**, *63*, 964–970. [CrossRef]
4. Xu, L.; Du, B.; Xu, B. A systematic, comparative study on the beneficial health components and antioxidant activities of commercially fermented soy products marketed in China. *Food Chem.* **2015**, *174*, 202–213. [CrossRef] [PubMed]
5. Chao, S.H.; Tomii, Y.; Watanabe, K.; Tsai, Y.C. Diversity of lactic acid bacteria in fermented brines used to make stinky tofu. *Int. J. Food Micribiol.* **2008**, *123*, 134–141. [CrossRef]
6. Chang, H.O.; Wang, S.W.; Chen, J.C.; Hsu, L.F.; Hwang, S.M. Mutagenic analysis of fermenting strains and fermented brine for stinky tofu. *J. Food Drug. Anal.* **2020**, *9*, 45–49. [CrossRef]
7. Lucas, S.; Omata, Y.; Hofmann, J.; Bttcher, M.; Iljazovic, A.; Sarter, K.; Albrecht, O.; Schulz, O.; Krishnacoumar, B.; Krnke, G. Short-chain fatty acids regulate systemic bone mass and protect from pathological bone loss. *Nat. Commum.* **2018**, *9*, 55. [CrossRef] [PubMed]
8. Wang, Y.; Gao, Y.; Liang, W.; Liu, Y.; Gao, H. Identification and analysis of the flavor characteristics of unfermented stinky tofu brine during fermentation using SPME-GC–MS, e-nose, and sensory evaluation. *J. Food Meas. Charact.* **2020**, *14*, 597–612. [CrossRef]
9. Li, P.; Xie, J.; Tang, H.; Shi, C.; Xie, Y.; He, J.; Zeng, Y.; Zhou, H.; Xia, B.; Zhang, C.; et al. Fingerprints of volatile flavor compounds from southern stinky tofu brine with headspace solid-phase microextraction/gas chromatography-mass spectrometry and chemometric methods. *Food Sci. Nutr.* **2019**, *7*, 890–896. [CrossRef]

10. Deng, S.; Du, L.; Wei, D. Study of stinky tofu brine fermented with lactic acid bacteria and the analysis of its aroma components. *Food Ferment. Ind.* **2020**, *46*, 107–113+118.
11. Tian, H.X.; Xu, X.L.; Chen, C.; Yu, H.Y. Flavoromics approach to identifying the key aroma compounds in traditional Chinese milk fan. *J. Dairy Sci.* **2019**, *102*, 9639–9650. [CrossRef]
12. Poveda, J.M.; Palomo, E.S.; Sánchez-Palomo, M.S.; Cabezas, L. Volatile composition, olfactometry profile and sensory evaluation of semi-hard Spanish goat cheeses. *Dairy Sci. Technol.* **2008**, *88*, 355–367. [CrossRef]
13. Pelin, S.; Gokce, G.; Perihan, K.; Deniz, B.; Zafer, E. Composition, proteolysis, lipolysis, volatile compound profile and sensory characteristics of ripened white cheeses manufactured in different geographical regions of Turkey. *Int. Dairy J.* **2018**, *87*, 26–36.
14. Tang, H.; Ma, J.K.; Chen, L.; Jiang, L.W.; Xie, J.; Li, P.; He, J. GC-MS characterization of volatile flavor compounds in stinky tofu brine by optimization of headspace solid-phase microextraction conditions. *Molecules* **2018**, *23*, 3155. [CrossRef]
15. Chen, C.; Huang, K.; Yu, H.Y.; Tang, H.X. The diversity of microbial communities in Chinese milk fan and their effects on volatile organic compound profiles. *J. Dairy Sci.* **2021**, *104*, 2581–2593. [CrossRef]
16. Linstrom, P.J.; Mallard, W.G. *NIST Standard Reference Database Number 69*; National Institute of Standards & Technology: Gaithersburg, MD, USA, 2001.
17. Tian, H.X.; Shen, Y.; Yu, H.Y.; He, Y.; Chen, C. Effects of 4 probiotic strains in coculture with traditional starters on the flavor profile of yogurt. *J. Food Sci.* **2017**, *82*, 1693–1701. [CrossRef]
18. Gemert, L.J.V. *Compilations of Odour Threshold Values in Air, Water and Other Media*; Oliemans Punter & Partners BV: Houten, The Netherlands, 2003.
19. Sun, J.; Yang, K.; Yanmin, L.I.; Chen, Y.; Liu, Y.; Zhang, Y. Analysis of characteristic aroma compounds in wangzhihe stinky tofu by simultaneous distillation-extraction and gas chromatography-mass spectrometry (SDE-GC-MS). *Food Sci.* **2015**, *16*, 127–131.
20. Wang, Y.; Yang, X.; Li, L. A new style of fermented tofu by *Lactobacillus casei* combined with salt coagulant. *3 Biotech* **2020**, *10*, 81. [CrossRef]
21. Liu, P.; Xiang, Q.; Gao, L.; Wang, X.; Li, J.; Cui, X.; Lin, J.; Che, Z. Effects of different fermentation strains on the flavor characteristics of fermented soybean curd: Effects of strains on flavor quality. *J. Food Sci.* **2019**, *84*, 154–164. [CrossRef] [PubMed]
22. Wei, G.; Yang, Z.; Regenstein, J.M.; Liu, X.; Liu, D. Characterizing aroma profiles of fermented soybean curd with ageing solutions during fermentation. *Food Biosci.* **2020**, *33*, 100508. [CrossRef]
23. Roh, H.S.; Park, J.Y.; Park, S.Y.; Chun, B.S. Isolation of off-flavors and odors from tuna fish oil using supercritical carbon dioxide. *Biotechnol. Bioproc. E* **2006**, *11*, 496–502. [CrossRef]
24. Lin, S.; Yang, R.; Cheng, S.; Wang, K.; Qin, L. Decreased quality and off-flavour compound accumulation of 3–10 kDa fraction of pine nut (*Pinus koraiensis*) peptide during storage. *LWT* **2017**, *84*, 23–33. [CrossRef]
25. Shi, Y.G.; Yang, Y.; Piekoszewski, W.; Zeng, J.H.; Guan, H.N.; Wang, B.; Liu, L.L.; Zhu, X.Q.; Chen, F.L.; Zhang, N. Influence of four different coagulants on the physicochemical properties, textural characteristics and flavour of tofu. *Int. J. Food Sci. Tech.* **2020**, *55*, 1218–1229. [CrossRef]
26. Rêgo, E.; Rosa, C.A.; Freire, A.L.; Machado, A.; Padilha, F.F. Cashew wine and volatile compounds produced during fermentation by non-Saccharomyces and Saccharomyces yeast. *LWT* **2020**, *126*, 109291. [CrossRef]
27. Montel, M.C.; Masson, F.; Talon, R. Bacterial role in flavour development. *Meat. Sci.* **1998**, *S49*, S111–S123. [CrossRef]
28. Yin, H.; Liu, L.P.; Yang, M.; Ding, X.T.; Jia, S.; Dong, J.J.; Zhong, C. Enhancing medium chain fatty acid ethyl ester production during beer fermentation through EEB1 and/or ETR1 overexpression in *saccharomyces pastorianus*. *J. Agri. Food Chem.* **2019**, *7*, 231. [CrossRef] [PubMed]
29. Xu, Y.; Zhao, J.; Liu, X.; Zhang, C.; Sun, B. Flavor mystery of Chinese traditional fermented baijiu: The great contribution of ester compounds. *Food Chem.* **2021**, *7*, 130920. [CrossRef]

Disclaimer/Publisher's Note: The statements, opinions and data contained in all publications are solely those of the individual author(s) and contributor(s) and not of MDPI and/or the editor(s). MDPI and/or the editor(s) disclaim responsibility for any injury to people or property resulting from any ideas, methods, instructions or products referred to in the content.

Article

Determination of the Antifungal, Antibacterial Activity and Volatile Compound Composition of *Citrus bergamia* Peel Essential Oil

Nur Cebi [1,*] and Azime Erarslan [2]

[1] Food Engineering Department, Chemical-Metallurgical Faculty, Yıldız Technical University, Istanbul 34210, Turkey
[2] Bioengineering Department, Chemical-Metallurgical Faculty, Yıldız Technical University, Istanbul 34210, Turkey
* Correspondence: nurcebi@yildiz.edu.tr; Tel.: +90-543-467-6691

Abstract: Safe and health-beneficial citrus oils can be employed as natural preservatives, flavorings, antioxidants, and as antibacterial and antifungal agents in a wide variety of food products. In this research, using GC–MS methodology, the major volatile composition of *Citrus bergamia* EO, obtained by hydro-distillation, was determined to consist of limonen (17.06%), linalool (46.34%) and linalyl acetate (17.69%). The molecular fingerprint was obtained using FTIR spectroscopy. The antibacterial effect of *C. bergamia* EO at different concentrations (0.5, 1, 2.5 and 5 µg/mL) was tested against different pathogen species (*Salmonella typhimurium*, *Bacillus cereus*, *Staphylococcus aureus*, *Escherichia coli*, *Listeria monocytogenes*), based on disc diffusion assay. The in vitro antifungal activity of *C. bergamia* EO oil against *Aspergillus niger* and *Penicillium expansum* was evaluated using agar disc diffusion assay. Clear inhibition zones were formed by *C. bergamia* EO against selected species of pathogens. Almost all of the concentrations were revealed to have antifungal activity against selected fungal pathogens. The highest inhibition rate of *A. niger* at 6 incubation days was 67.25 ± 0.35 mm with a 20 µL dose, while the growth in the control was 90.00 ± 0.00 mm. In addition, the highest inhibition rate of *P. expansum* was 26.16 ± 0.76 mm with a 20 µL dose, while the growth was 45.50 ± 2.12 mm in the control fungus. A higher antifungal effect of *C. bergamia* EO against *P. expansum* was obtained. It was observed that the growth of fungi was weakened with increasing concentrations (5, 10, 15 and 20 µL dose) of *C. bergamia* EO. Statistically significant ($p < 0.05$) results were obtained for the antibacterial and antifungal effects of *C. bergamia* EO. The findings from the research may shed light on the further use of *C. bergamia* EO obtained from peels in innovative food engineering applications in order to maintain food quality, food safety, and food sustainability.

Keywords: GC–MS; volatile composition; *C. bergamia*; antimicrobial; essential oil; post-harvest fungi

1. Introduction

Essential oils obtained from plants are natural valuable products that have economic and commercial importance in industrial applications. With their unique volatile compounds, essential oils have pioneering importance in foods, drinks, perfumeries, pharmaceuticals and cosmetics [1]. Several studies have documented the fact that essential oils can be employed as natural preservatives, flavorings, antioxidants, and antibacterial and antifungal agents in a wide variety of food products. In particular, citrus oils have been found to have tremendous applications in the area of food production because of their safe and health-beneficial properties [2].

To date, essential oils have attracted attention in both the academic field and industrial applications, due to their properties, including volatility and safety. The main active EO components are phenols, terpenes, aldehydes and ketones, whose actions are directed against the cytoplasmic membranes of target microorganism cells [3]. Factors found to influence essential oil composition have been explored in several studies, and a number of researchers have demonstrated that genetic variation, geographical location of the plants, seasonal variations, and climate change are some of the major factors that determine the uniqueness and chemistry of essential oils [1]. The chemical composition of essential oils determines their antifungal, antibacterial, antioxidant and insecticidal capabilities. Consequently, various studies have been dedicated to exploring the volatile composition of essential oils [4].

Today, there is a growing interest, in both academic institutions and in industry, in determining the safe and green capabilities of using essential oils to combat and control pests and diseases in agriculture [5]. Economic losses can result from the growth of fungal microorganisms in edible agricultural products, causing plant diseases and food spoilage. In agriculture, fruit and vegetables are often exposed to microbial contamination from pathogenic fungi during post-harvest storage [6,7]. Due to the widespread application of chemical fungicides, most pathogenic strains have developed a resistance to them. In addition, there are reports of effects, such as carcinogenesis, residual toxicity and environmental pollution, caused by the continued use of such fungicides [8]. In order to minimize the undesirable side effects of synthetic fungicide applications, researchers have focused on evaluating alternative natural biofungicides. With a view to ensuring consumer safety when consuming fruit and vegetables protected by using synthetic fungicides, natural extracts (EOs) have been investigated as healthy and non-toxic alternatives for the past few decades [9].

Recently, in many countries around the world, there has been an apparent increase in demand for organic products that have not been treated with agrochemicals, especially after harvest. Therefore, essential oils can be evaluated as safe and reliable antimicrobial and antifungal agents to effectively manage major post-harvest diseases. Related to the demand for natural and reliable control agents, essential oils have become prominent as a means of safe alternatives to synthetic fungicides [10].

C. bergamia, also known as "Bergamot", is a plant belonging to the Rutaceae family, and takes the form of a hybrid of bitter orange and lemon. The main producer countries are Italy, countries in East Africa, Ivory Coast, Argentina, Brazil and Turkey. The essential Bergamot oil (*C. bergamia* EO) has commercial use in the perfumery and essence industries due to its favorable aroma and fragrance properties. Other usage areas can be listed as the pharmaceutical industry and the food industry. As is recognized as safe, *C. bergamia* EO can be used in the latter as a flavoring agent in a wide range of foodstuffs, such as liqueurs, tea, coffee, ice cream, confectionery and drinks [11].

The essential oil quality varies depending on the geographical origin, climate change, seasonal factors and soil properties. Consequently, the volatile composition, and the bioactive, antifungal and antibacterial properties exhibit changes based on these factors. The above-mentioned functional properties of essential oils determine the industrial quality and their use in high-value commercial products. Additionally, at the present time, waste management and valorization are important problems. In this context, the determination of the chemical composition of the waste parts of fruits, such as peels, is gaining growing importance. To the best of our knowledge, this study is the first attempt to comprehensively evaluate the antimicrobial, antifungal, molecular and volatile properties of *C. bergamia* peel essential oil, obtained by the hydro-distillation of fresh bergamot peels (Turkey, Hatay). There is a need for thorough studies in which the specific properties of *C. bergamia* peel essential oil is investigated in detail using robust analytical techniques. The aim of this study was to first evaluate the antifungal activity of *C. bergamia* peel essential oil from Turkey by in vitro methods, against *A. niger* and *P. expansum* fungal pathogens and *E. coli*, *L. monocytogenes*, *B. cereus* and *S. aureus* pathogenic bacteria. The second aim of the study was to determine the volatile compound composition of *C. bergamia* essential oil from Turkey (Hatay) using the robust GC–MS technique.

2. Materials and Methods

2.1. Essential Oil and Chemicals

C. bergamia organic fruits originating from Hatay (Turkey) were used in this study. Potato dextrose agar (PDA) and other chemicals were procured from Merck (Darmstadt, Germany). *E. coli* ATCC 8739, *L. monocytogenes* ATCC 13,932, *B. cereus* ATCC 11,778.

The *S.* Typhimurium ATCC 14,028 and *S. aureus* ATCC 6538 and *A. niger* and *P. expansum* were obtained from Yildiz Technical University, Turkey. Samples were diluted with diethyl ether (Merck-Schuchardt, FRG, GC > 98%) prior to the GC–MS analysis.

2.2. Isolation and Identification of Pathogenic Fungi

Fungi were previously isolated from infected apples at the correct stage of maturity to isolate *A. niger* and *P. expansum*, and stored at room temperature until spoilage. These fungi were identified using the available literature, which describes their colony and hyphae morphology, conidial structure and characteristic features [6,12]. Pure cultures were maintained on Potato Dextrose Agar (PDA, Merck, Darmstadt, Germany) at 27 °C with 50 mg/L streptomycin (Merck, Darmstadt, Germany) for 7 days. Spores were collected by filling the media surface with sterile distilled water and gently shaking the plate to remove and separate the spores. A conidial suspension was prepared in sterile Ringer's solution (Merck, Darmstadt, Germany). Spores were counted and the final inoculum concentration was adjusted to a concentration of 1×10^5 spores/mL per pathogen. Suspensions obtained prior to inoculation were shaken using a vortex mixer for 30 s [13].

2.3. In Vitro Antifungal Assay

The disk evaporation method was used to determine the antifungal activity of *C. bergamia* EO against *A. niger* and *P. expansum*. Fungal plugs from a 7-day old, actively growing, culture, 6 mm in diameter, were inoculated into petri dishes containing 20 mL of fresh Potato Dextrose Agar (PDA) medium. Diluted essential oils, ranging from 5 to 20 µL/petri, were then adsorbed onto blank antimicrobial disks. The disks were placed on the petri dishes by inverting the petri dishes. Sterile distilled water was used as a control. The petri dishes were covered with parafilm immediately after the addition of the essential oil to allow effective exposure of the vapors to the fungal mycelia, followed by incubation at 27 °C for 7 days. This assay per test pathogen was triplicated to arrive at a statistically sound conclusion [7,9,14].

2.4. Linear Polynomial Contrast

The linear relationships between the applied doses of *C. bergamia* essential oil and the growth of the fungal pathogens *(A. niger* and *P. expansum)* were established using a simple linear regression analysis, performed using Origin 6.0 software.

2.5. Determination of Antibacterial Activity

The antibacterial activity of *C. bergamia* EO at different concentrations from 0.5 to 5 µg/mL was evaluated by the agar disc diffusion method. Nutrient Agar (NA) medium was used to cultivate the bacteria, and each bacterial suspension was diluted and adjusted to the equivalent of the 0.5 McFarland standard (10^8 CFU/mL). A sterile 6 mm paper disc, impregnated with 20 µL of each tested *C. bergamia* EO was placed on the surface of the inoculated plates. A disk impregnated with 20 µL of distilled water was used as a negative control. The plates were incubated at 37 °C for 24 h. Microbial inhibition was assessed visually in terms of the diameter of the zones of inhibition surrounding the discs, including the intervertebral discs, and recorded in millimeters according to NCCLS (2015) [15].

2.6. Statistical Analysis

The size of the mycelial growth levels (mm) was expressed as the mean of three recordings with their standard deviation. The significance of the mean differences was compared statistically using Student's t test at $p < 0.05$. Experimental data were subjected to one-way analysis of variance (ANOVA) using the JMP (release 6.0.0, SAS) software package.

2.7. FTIR Data Acquisition

FTIR measurements were performed by using the ATR accessory of the equipment. Samples were kept in amber vials until analysis at 4 °C. The spectral parameters of resolution and accumulation were selected as 4 cm^{-1} and 16 scans, respectively. Data were acquired and processed by using the OPUS program Version 7.2 (Bruker Gmbh, Bremen-Germany). A quantity of 100 mL of each sample was dripped on the crystal, after which the diamond crystal of the ATR accessory was cleaned with pure ethyl alcohol. An air spectrum was used as a background spectrum prior to each acquisition.

2.8. GC–MS Analysis

2.8.1. C. bergamia EO Extraction from Fresh Peels

The essential oil extraction from *C. bergamia* fruits was performed using both a hydro-distillation Clevenger apparatus system and a microwave extraction system (Milestone, Italy). In the traditional hydro-distillation, Clevenger apparatus system was used, *C. bergamia* peel was subjected to hydro-distillation for 2 h with a peel:weight ratio of 1:1.

The microwave assisted extraction of *C. bergamia* fruit peels was performed in the following procedure. The system included a cooling system outside the microwave oven, with the EO being obtained from a Clevenger-type equipment connected to the oven. The operation temperature was 100 °C, with 600 W power being applied for 35 min. The operation was stopped when no more essential oil was obtained. The water and *C. bergamia* peel weight ratio was selected as 1:1 and 4 mL of essential oil was obtained from 400 g of fresh *C. bergamia* peel. The *C. bergamia* essential oil was stored at 4 °C until the GC–MS analysis.

2.8.2. GC–MS Data Acquisition of C. bergamia EO

In the sample preparation, diethyl ether was used as a diluent (1:20). An Rtx-5MS capillary column (30 m × 0.25 mm × 0.25 µm) was used in measurements. The oven temperature increased gradually from 40 °C for 3 min at the beginning of the temperature program, then increased to 100 °C at an 8 °C/min rate, and was then raised to 200 °C at a rate of 5 °C/min, and, finally, to 250 °C at a rate of 10 °C/min. The temperature of the injection block and the flow rate of the carrier gas were 250 °C and 1 mL/min, respectively. The samples were scanned at a mass range of 35 and 650 (m/z). The composition of the essential oil was determined by comparing the obtained GC–MS total ion chromatogram with those included in the NIST (National Institute of Standard and Technology) and Wiley Library of GCMS-QP2010 equipment (Shimadzu, Milan, Italy). The quantitative amount of each compound was calculated on the basis of the percentage area of each identified compound chromatogram compared to the areas of total peaks (100%). Samples were scanned three times, and all the peaks detected in at least two of the total ion chromatograms (TIC) were used for the calculation of relative abundance.

3. Results

3.1. Antibacterial Activity Assay

In evaluating the antibacterial effect, *C. bergamia* EO was prepared in various concentrations, such as 0.5 µg/mL, 1 µg/mL, 2.5 µg/mL and 5 µg/mL, against *E. coli*, *L. monocytogenes*, *B. cereus*, *S. typhimurium* and *S. aureus* by the disc diffusion method. The results shown in Table 1 indicate that the bacterial strains expressed a varied range of susceptibilities to the actions of *C. bergamia* EOs (Figure 1).

At a concentration of 0.5 µg/mL, the lowest inhibition rate of 8.50 ± 0.70 mm was observed for *B. cereus*. The lowest inhibition rate for *S. aureus* and the highest inhibition rate for *L. monocytogenes* was achieved at concentrations of 1 and 2.5 µg/mL. At a concentration of 5 µg/mL, the highest inhibition zone of 13.50 ± 0.70 mm was observed for

L. monocytogenes. As a result of this study, *L. monocytogenes* was identified as the most sensitive bacterium.

Table 1. Zone of inhibition (mm) of *C. bergamia* EO against bacterial strains.

Species	0.5 µg/mL	1 µg/mL	2.5 µg/mL	5 µg/mL
S. typhimurium	9.25 ± 0.35 [a]	10.87 ± 1.23 [b]	11.50 ± 0.70 [c]	11.50 ± 0.35 [c]
B. cereus	8.50 ± 0.70 [a]	12.00 ± 0.00 [b]	12.50 ± 0.70 [c]	12.50 ± 0.35 [c]
S. aureus	10.00 ± 0.00 [a]	10.50 ± 0.70 [b]	10.50 ± 2.12 [b]	11.50 ± 0.00 [c]
E. coli	10.00 ± 0.00 [a]	11.50 ± 0.70 [b]	12.50 ± 0.70 [c]	12.50 ± 0.00 [c]
L. monocytogenes	10.00 ± 0.00 [a]	12.75 ± 0.35 [b]	13.00 ± 0.70 [b]	13.50 ± 0.70 [c]

All data are presented as mean ± standard deviation. [a-c]: in each row, lowercase superscript letters indicate differences between doses of *C. bergamia* EO against each bacteria species. $p < 0.05$ was considered statistically significant.

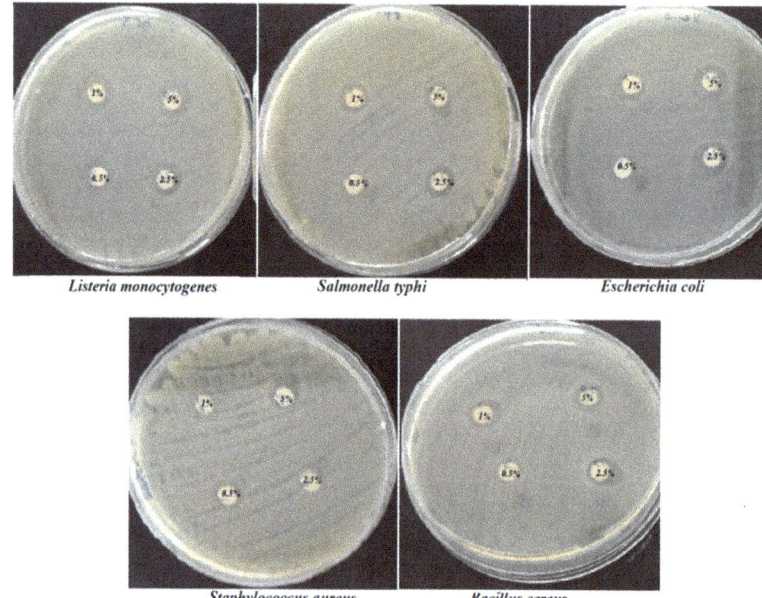

Figure 1. Clear inhibition zones formed by *C. bergamia* EO against different species of pathogens.

3.2. In Vitro Antifungal Assay

From Tables 2 and 3, it is apparent that the *C. bergamia* essential oil showed antifungal properties with regard to both the fungal pathogens. Raising the concentration of essential oil resulted in a higher growth inhibition ratio (%)As a result, almost all of the concentrations were revealed to have antifungal activity against the selected fungal pathogens. Increasing the dose from 5 to 20 µL/petri of *C. bergamia* resulted in a weaker growth of the fungi. The vapors of the essential oil of *C. bergamia* could inhibit the pathogen growth even more at a concentration of 20 µL/petri.

In the 6 days of incubation of *A. niger*, inhibition rates of 78.00 ± 0.00 mm at a 5 µL dose, 77.00 ± 2.82 mm at a 10 µL dose, 75.00 ± 0.00 mm at a 15 µL dose and 67.25 ± 0.35 mm at a 20 µL dose and 90.00 ± 0.00 mm growth of control were observed. In the 6 incubation days of *P. expansum*, inhibition rates of 32.00 ± 2.82 mm at 5 µL dose, 31.50 ± 0.00 mm at 10 µL dose, 27.00 ± 0.00 mm at 15 µL dose and 26.16 ±0.76 mm at 20 µL dose and 45.50 ± 2.12 mm of control fungi were observed. A higher antifungal effect on the part of *C. bergamia* EO was obtained against *P. expansum*. The experimental results suggested that increasing the doses (5, 10, 15, 20 µL/petri) of *C. bergamia* EO resulted in weaker growth of both *A. niger* and *P. expansum* (Figures 2 and 3).

Table 2. Inhibitory effect of *C. bergamia* EO against in vitro mycelial growth (mm) of *A. niger* at different incubation periods ($n = 3$).

Incubation Days	Control	5 µL	10 µL	15 µL	20 µL
3	41.66 ± 1.25 [a]	31.00 ± 3.04 [b]	27.83 ± 1.60 [c]	22.75 ± 0.35 [d]	19.83 ± 1.04 [e]
4	58.66 ± 1.52 [a]	46.5 ± 2.12 [b]	43.25 ± 1.76 [c]	40.00 ± 0.00 [d]	34.50 ± 0.70 [e]
5	77.5 ± 3.53 [a]	64.00 ± 2.82 [b]	59.5 ± 2.12 [c]	57.75 ± 0.35 [c]	50.00 ± 1.41 [d]
6	90.00 ± 0.00 [a]	78.00 ± 0.00 [b]	77.00 ± 2.82 [b]	75.00 ± 0.00 [c]	67.25 ± 0.35 [d]

All data are presented as mean ± standard deviation. [a–e]: in each row, lowercase superscript letters indicate differences between doses at each incubation period. $p < 0.05$ was considered statistically significant.

Table 3. Inhibitory effect of *C. bergamia* EO against in vitro mycelial growth (mm) of *P. expansum* at different incubation periods ($n = 3$).

Incubation Days	Control	5 µL	10 µL	15 µL	20 µL
3	19.00 ± 0.00 [a]	19.50 ± 1.50 [b]	17.50 ± 0.00 [c]	14.66 ± 0.76 [d]	13.83 ± 0.76 [e]
4	25.50 ± 0.35 [a]	23.83 ± 2.56 [b]	22.50 ± 0.00 [c]	18.33 ± 0.76	17.66 ± 0.57 [e]
5	30.75 ± 2.47 [a]	28.00 ± 2.00 [b]	26.50 ± 0.00 [c]	23.83 ± 1.60 [c]	21.83 ± 0.76 [e]
6	45.50 ± 2.12 [a]	32.00 ± 2.82 [b]	31.50 ± 0.00 [b]	27.00 ± 0.00 [c]	26.16 ± 0.76 [d]

All data are presented as mean ± standard deviation. [a–e]: in each row, lowercase superscript letters indicate differences between doses at each incubation period. $p < 0.05$ was considered statistically significant.

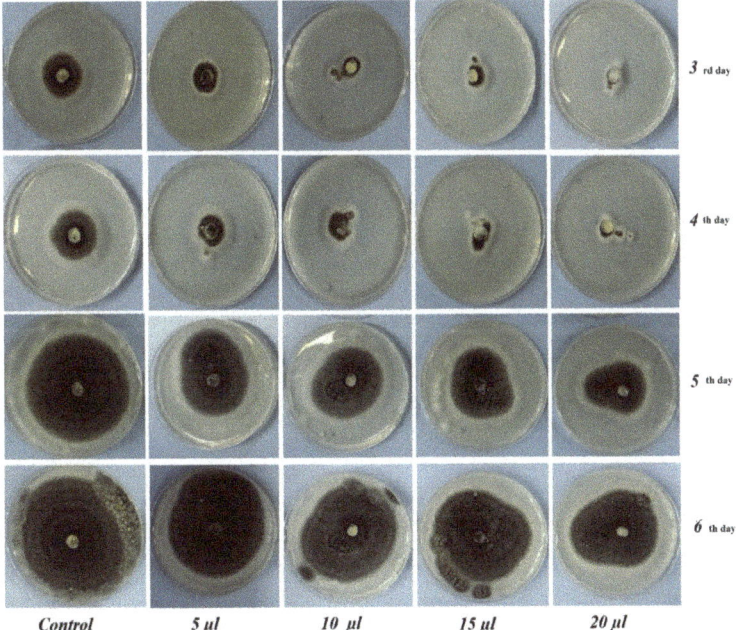

Figure 2. The antifungal effects of *C. bergamia* EO at different doses against *A. niger*.

In Figure 4, the results of the linear regression analysis are displayed as plots of the dose of *C. bergamia* EO versus in vitro mycelial growth of *A. niger* and *P. expansum*. It was determined that the relationship between dose application and fungal growth was linear and negative, based on the linear regression analysis, which was well explained by the high coefficients of determination (R^2), which ranged from 0.80 and 0.99. The values of the linear determination coefficients obtained as a response to linear regression between fungal growth and dose application, suggested that the essential oil of C. bergamia possessed strong antifungal properties, which remarkably inhibited the growth of the fungal species considered.

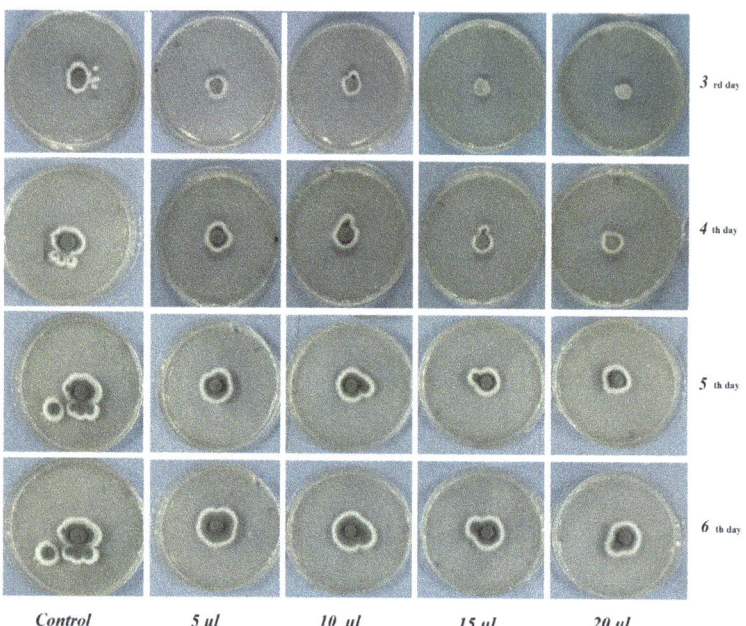

Figure 3. The antifungal effects of C. bergamia EO at different doses against P. expansum.

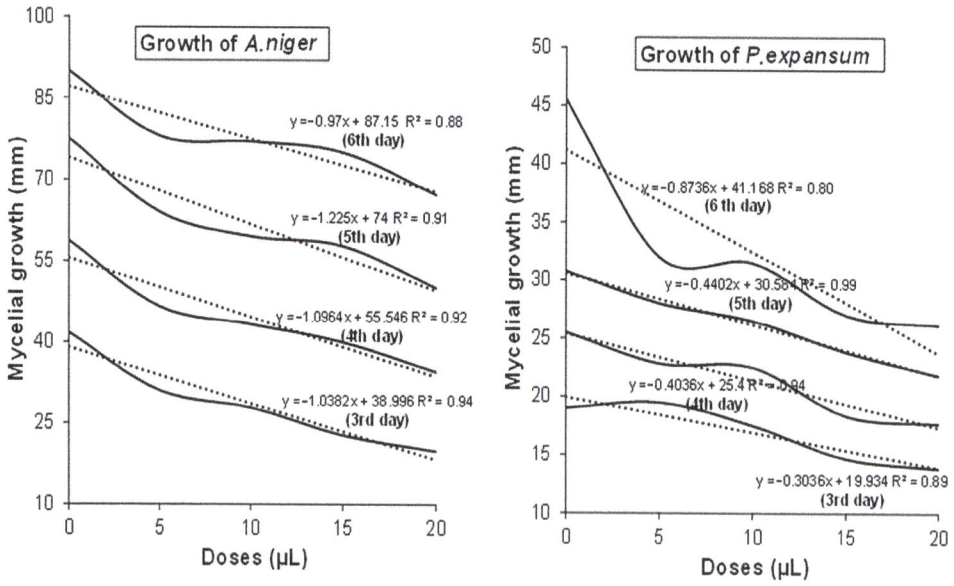

Figure 4. Plots of the dose (μL) of C. bergamia EO versus in vitro mycelial growth (mm) of A. niger and P. expansum, respectively (obtained by linear regression), showing a negative linear polynomial relationship between mycelial growth and doses, along with linear equations and R^2 values.

3.3. FTIR Characterization of C. bergamia EO

The ATR-FTIR spectrum of *C. bergamia* EO essential oil at the spectral range of 4000–400 cm^{-1} is presented in Figure 5. The major bands were observed at 2967, 2918, 1724, 1643, 1450, 1373, 1240, 1112, 994, 918, 887, 834 and 800 cm^{-1}. In the previous studies, the FTIR band assignments, of a citrus species lemon essential oil, were presented [2]. Common FTIR bands were observed in the FTIR spectrum of *C. bergamia* EO. The band, with a peak point at 2967 cm^{-1}, corresponded to –CH$_3$ asymmetric and symmetric stretching vibrations [2]. The bands at 2918, 1724, 1643 cm^{-1} were due to the C–H stretching vibrations, the C=O stretching vibrations of ester groups, and ν(–C=C–, cis-) and δ(–OH) vibrations, respectively [16]. The band at 1450 cm^{-1} might be related to the bending vibrations of CH$_2$ and CH$_3$ of aliphatic groups [17]. The significant peaks at 1376 cm^{-1} and 887 cm^{-1} arose from C-H bending vibrations and (–HC = CH–, trans-) bending vibrations, respectively [16,18]. The significant bands at 994 and 918 cm^{-1} were probably due to δ(–HC=CH–, trans-) bending vibrations [16]. Lastly, the bands at 834 and 800 cm^{-1} corresponded to C–H stretching vibrations and C=C bending vibrations, respectively [2,19].

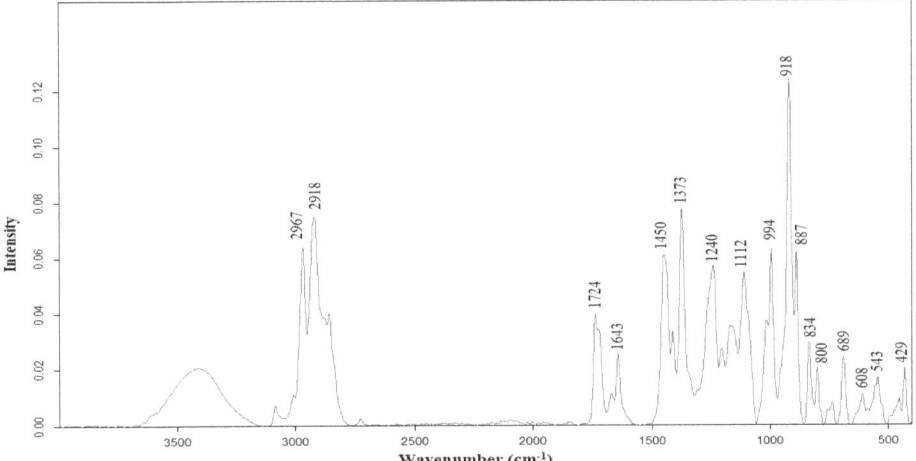

Figure 5. ATR-FTIR spectrum of *C. bergamia* EO essential oil at the spectral range of 4000–400 cm^{-1}.

3.4. GC–MS Characterization of C. bergamia EO

The volatile compounds of *C. bergamia* essential oil obtained by hydro-distillation is presented in Table 4. These compounds were assigned by a comparison of GC–MS data with the commercial libraries (NIST27 and WILEY7) of the equipment. The volatile compounds presented in Table 4 explained 99.85% of all volatile ingredients in the composition of *C. bergamia* essential oil. The major volatile compounds were determined to be limonene (17.06%), linalool (46.34%) and linalyl acetate (17.69%).

Table 4. Composition (%), retention time (R.T.) and volatile compounds in the *C. bergamia* peel EO.

Content (%)	Chemical Name	RT [a]	RI [b]
0.14 ± 0.02	α-Thujene	8.541	924
0.57 ± 0.02	α-Pinene	8.701	932
0.45 ± 0.05	Sabinene	9.615	969
2.51 ± 0.03	β-Pinene	9.698	974
1.18 ± 0.07	Myrcene	9.983	988
0.04 ± 0.01	Octanal	10.234	998
0.16 ± 0.09	α-Terpinene	10.567	1014
17.06 ± 0.06	Limonene	10.947	1024
0.2 ± 0.04	Z-β-Ocimene	11.021	1032

Table 4. Cont.

Content (%)	Chemical Name	RT [a]	RI [b]
0.54 ± 0.05	E-β-Ocimene	11.277	1044
4.11 ± 0.06	γ-Terpinene	11.604	1054
0.39 ± 0.06	Terpinolene	12.389	1086
46.34 ± 0.02	Linalool	13.046	1095
0.28 ± 0.03	Trans-sabinene hydrate	15.103	1098
2.41 ± 0.02	Exo-fenchol	15.522	1118
0.05 ± 0.09	Decanal	15.819	1201
0.08 ± 0.05	Acetic acid	15.969	1211
1.23 ± 0.01	Nerol	16.569	1227
0.56 ± 0.03	Z-citral	16.903	1235
17.69 ± 0.06	Linalyl acetate	17.414	1254
0.80 ± 0.05	Dimethoxy-(E)-citral	17.757	1338
0.95 ± 0.05	Neryl acetate	20.306	1359
1.13 ± 0.04	Geranyl acetate	20.822	1379
0.30 ± 0.05	Z-Caryophyllene	21.954	1408
0.27 ± 0.06	Cis-α-Bergamotene	22.290	1411
0.05 ± 0.07	Z-β-Farnesene	22.755	1440
0.36 ± 0.08	β-Bisabolene	24.126	1505
Total: 99.85	Total		
27.35	Monoterpene hydrocarbons		
71.52	Oxygenated monoterpenes		
0.98	Sesquiterpene hydrocarbons		
0	Oxygenated sesquiterpenes		

RT [a] Retention time of the compounds. RI [b] Reported retention indices in the literature [20].

4. Discussion

The antimicrobial effects of essential oils suggest that their activity stems from their lipophilic properties. Interactions between antimicrobial compounds and cell membranes affect both lipid assembly and bilayer stability. Their mode of action occurs in the phospholipid bilayer, which is related to disruption of the cell membrane. This is caused by biochemical mechanisms catalyzed by the cell's phospholipid bilayer. These processes include the inhibition of electron transport, protein translocation, phosphorylation steps, and other enzyme-dependent reactions [21].

Marotta et al. [15] evaluated the in vitro antimicrobial activity of various *C. bergamia* EOs against several strains of *L. monocytogenes*. Fisher and Phillips [22] demonstrated the antimicrobial activity of *C. bergamia* EO against *C. jejuni, E. coli* 0157, *L. monocyto-genes, B. cereus*, and *S. aureus*, in both direct oil and vapor forms. They stated that gram-positive bacteria were more susceptible in vitro than gram-negative bacteria. This may be due to the relatively impermeable outer membrane that surrounds gram-negative bacteria and restricts the diffusion of hydrophobic compounds [23,24].

Mycelium growth inhibition data, recorded at 6 days after inoculation at 25 ± 2 °C treated with *C. bergamia* EO at a 20 μL/petri concentration, showed the strongest mycelium growth inhibition of *P. expansum*. At 20 μL/petri concentrations, the most significant micelle inhibitory effect was observed for both fungi.

C. bergamia oil at a 0.8% v/v concentration inhibited the mycelium growth of pathogenic fungi. However, a powerful control could not be observed for *P. grisea* and *R. solani* [21].

Kulkarni et al. [9] investigated the antifungal activity of wild *C. bergamia* essential oil against the postharvest fungal pathogens, *Colletotrichum musae* and *Lasiodiplodia theobromae*, of banana fruit at concentrations of 1 mL to 10 mL. Their results exhibited 100% growth inhibition for both fungal pathogens at 4 mL.

Table 4 displays the chemical names, contents (%), and retention times (RTa) for the Rtx-5MS GC column, and the reported retention indices noted in the literature [20]. Monoterpene hydrocarbons, oxygenated monoterpenes and sequiterpene hydrocarbons constituted the volatile composition at percentages of 27.35%, 71.52 and 0.9%, respectively.

This also accorded with some earlier observations, which showed that hydro-distillated *C. bergamia* essential oil involves limonene (32.29%), linalool (33.64%) and linalyl acetate (9.22%) [25]. However significant differences were observed between the limonene contents of *C. bergamia*. In accordance with the present results, previous studies showed that the major volatile compounds of *C. bergamia* essential oil were limonene, linalool and linalyl acetate [26].

The literature highlighted the importance of terpenes and terpenoids with respect to other EO components. Phenol ring structure-rich molecules were found to have considerable antibacterial activity, but enrichment with OH groups might increase their antibacterial properties [3]. The studies presented thus far provide evidence that the antifungal activities of essential oils are clearly related to their chemistry, including their chemical compounds, the percentage structure of each component, as well as their structure [9,27]. Kulkarni et al. [9] indicated that the chemical composition of wild bergamot (*Monarda fistulosa*) essential oil was determined by the GC–MS technique, and revealed that the main compounds of the essential oil, thymol, carvacrol and cinnamyl carbanilate, showed the best antifungal activities. Additionally, the antibacterial effect of terpenes is well known. They destroy the multi-layered structure of polysaccharides, fatty acids and phospholipids, penetrate the cell wall, and make the cytoplasmic membrane permeable. In bacteria, these events are associated with ion loss and decreased membrane potential, leading to breakdown of the proton pump and depletion of the ATP pool and lysis [15,28].

5. Conclusions

Volatile compound composition was determined using the GC–MS technique. The most abundant volatile constituents were determined to be limonene (17.06%), linalool (46.34%) and linalyl acetate (17.69%). The molecular fingerprint of the *C. bergamia* EO obtained by hydro-distillation was determined on the basis of FTIR spectroscopy. The essential oils from the *C. bergamia* peel provided anti-mycelial and anti-bacterial growth, and disrupted spore germination activity in the case of pathogenic fungi [29]. According to the results obtained, *C. bergamia* EO provides a good basis for the formulation of products, with potential effectiveness in terms of combatting bacteria and fungi. In addition, using smaller amounts of oil can have significant economic implications. These findings increase the opportunities and possibilities of exploiting essential oils as promising candidates for use in crop production systems, as alternative safe natural antifungal agents. In other words, the results suggested that bergamot peel essential oil could be used as a natural alternative for food preservation The findings from the current research may shed light with regard to the further use of *C. bergamia* peel EO in innovative food engineering applications as a means of maintaining food quality, ensuring food safety and improving food sustainability.

Author Contributions: Conceptualization, N.C. and A.E.; methodology, N.C. and A.E.; resources, N.C. and A.E.; software, N.C. and A.E.; validation, N.C. and A.E.; writing—review and editing, N.C. and A.E.; supervision, N.C. All authors have read and agreed to the published version of the manuscript.

Funding: This research received no external funding.

Data Availability Statement: No new data were created or analyzed in this study. Data sharing is not applicable to this article.

Acknowledgments: The authors would like to thank the Anamed and Analitik Group for their help and technical support.

Conflicts of Interest: The authors declare no conflict of interest.

References

1. Raut, J.S.; Karuppayil, S.M. A status review on the medicinal properties of essential oils. *Ind. Crops Prod.* **2014**, *62*, 250–264. [CrossRef]
2. Cebi, N.; Taylan, O.; Abusurrah, M.; Sagdic, O. Detection of orange essential oil, isopropyl myristate, and benzyl alcohol in lemon essential oil by ftir spectroscopy combined with chemometrics. *Foods* **2021**, *10*, 27. [CrossRef] [PubMed]
3. Hyldgaard, M.; Mygind, T.; Meyer, R.L. Essential oils in food preservation: Mode of action, synergies, and interactions with food matrix components. *Front. Microbiol.* **2012**, *3*, 12. [CrossRef]
4. El Ouadi, Y.; Manssouri, M.; Bouyanzer, A.; Majidi, L.; Bendaif, H.; Elmsellem, H.; Shariati, M.A.; Melhaoui, A.; Hammouti, B. Essential oil composition and antifungal activity of Melissa officinalis originating from north-Est Morocco, against postharvest phytopathogenic fungi in apples. *Microb. Pathog.* **2017**, *107*, 321–326. [CrossRef] [PubMed]
5. Kordali, S.; Cakir, A.; Mavi, A.; Kilic, H.; Yildirim, A. Screening of chemical composition and antifungal and antioxidant activities of the essential oils from three Turkish Artemisia species. *J. Agric. Food Chem.* **2005**, *53*, 1408–1416. [CrossRef]
6. Yilmaz, A.; Bozkurt, F.; Cicek, P.K.; Dertli, E.; Durak, M.Z.; Yilmaz, M.T. A novel antifungal surface-coating application to limit postharvest decay on coated apples: Molecular, thermal and morphological properties of electrospun zein–nanofiber mats loaded with curcumin. *Innov. Food Sci. Emerg. Technol.* **2016**, *37*, 74–83. [CrossRef]
7. Vehapi, M.; Koçer, A.T.; Yılmaz, A.; Özçimen, D. Investigation of the antifungal effects of algal extracts on apple-infecting fungi. *Arch. Microbiol.* **2020**, *202*, 455–471. [CrossRef]
8. Singh, S.; Singh, N.; Kumar, V.; Datta, S.; Wani, A.B.; Singh, D.; Singh, K.; Singh, J. Toxicity, monitoring and biodegradation of the fungicide carbendazim. *Environ. Chem. Lett.* **2016**, *14*, 317–329. [CrossRef]
9. Kulkarni, S.A.; Sellamuthu, P.S.; Nagarajan, S.K.; Madhavan, T.; Sadiku, E.R. Antifungal activity of wild bergamot (*Monarda fistulosa*) essential oil against postharvest fungal pathogens of banana fruits. *S. Afr. J. Bot.* **2022**, *144*, 166–174. [CrossRef]
10. Lee, J.E.; Seo, S.M.; Huh, M.J.; Lee, S.C.; Park, I.K. Reactive oxygen species mediated-antifungal activity of cinnamon bark (*Cinnamomum verum*) and lemongrass (*Cymbopogon citratus*) essential oils and their constituents against two phytopathogenic fungi. *Pestic. Biochem. Physiol.* **2020**, *168*, 104644. [CrossRef]
11. Marzocchi, S.; Baldi, E.; Crucitti, M.C.; Toselli, M.; Caboni, M.F. Effect of harvesting time on volatile compounds composition of bergamot (*Citrus* × *bergamia*) essential oil. *Flavour Fragr. J.* **2019**, *34*, 426–435. [CrossRef]
12. Yilmaz, A.; Ermis, E.; Boyraz, N. Investigation of in vitro and in vivo antifungal activities of different plant essential oils against postharvest apple rot diseases-Colletotrichum gloeosporioides, Botrytis cinerea and Penicillium expansum. *J. Food Saf. Food Qual.* **2016**, *67*, 122–131.
13. Vehapi, M.; Yilmaz, A.; Özçimen, D. Antifungal Activities of *Chlorella vulgaris* and *Chlorella minutissima* microalgae cultivated in bold basal medium, wastewater and tree extract water against *Aspergillus niger* and antifungal activities of *Chlorella vulgaris* and *Chlorella minutissima* microalgae cultivated in bold basal medium, wastewater and tree extract water against *Aspergillus niger* and *Fusarium oxysporum*. *Rom. Biotechnol. Lett.* **2018**, *1*, 1–8. [CrossRef]
14. Zhang, J.; Ma, S.; Du, S.; Chen, S.; Sun, H. Antifungal activity of thymol and carvacrol against postharvest pathogens *Botrytis cinerea*. *J. Food Sci. Technol.* **2019**, *56*, 2611–2620. [CrossRef] [PubMed]
15. Marotta, S.M.; Giarratana, F.; Parco, A.; Neri, D.; Ziino, G.; Giuffrida, A.; Panebianco, A. Evaluation of the antibacterial activity of bergamot essential oils on different *Listeria monocytogenes* strains. *Ital. J. Food Saf.* **2016**, *5*, 3. [CrossRef] [PubMed]
16. Taylan, O.; Cebi, N.; Sagdic, O. Rapid screening of *Mentha spicata* essential oil and l-menthol in *Mentha piperita* essential oil by atr-ftir spectroscopy coupled with multivariate analyses. *Foods* **2021**, *10*, 202. [CrossRef]
17. Bounaas, K.; Bouzidi, N.; Daghbouche, Y.; Garrigues, S.; de la Guardia, M.; El Hattab, M. Essential oil counterfeit identification through middle infrared spectroscopy. *Microchem. J.* **2018**, *139*, 347–356. [CrossRef]
18. Berechet, M.D.; Calinescu, I.; Stelescu, M.D.; Manaila, E.; Craciun, G.; Purcareanu, B.; Mihaiescu, D.E.; Rosca, S.; Fudulu, A.; Niculescu-Aron, I.G.; et al. Composition of the essential oil of Rosa damascena Mill. cultivated in Romania. *Rev. Chim.* **2015**, *66*, 1986–1991.
19. Benoudjit, F.; Maameri, L.; Ouared, K.; History, A. Evaluation of the quality and composition of lemon (*Citrus limon*) peel essential oil from an Algerian fruit juice industry. *Alger. J. Environ. Sci. Technol.* **2020**, *6*, 1575–1581.
20. Adams, R.P. *Identification of Essential oil Components by Gas Chromatography/Mass Spectrometry*, 4th ed.; Allured Publishing Corporation: Carol Stream, IL, USA, 2017.
21. Thobunluepop, P.; Udumsilp, J.; Piyo, A.; Khaengkhan, P. Screening for The antifungal activity of essential oils from bergamot oil (*Citrus hystrix* DC.) and tea tree oil (*Melaleuca alternifolia*) against economically rice pathogenic fungi: A driving force of organic rice cv. KDML 105 Production. *Asian J. Food Agro-Ind.* **2009**, *2*, 374–380.
22. Fisher, K.; Phillips, C.A. The effect of lemon, orange and bergamot essential oils and their components on the survival of *Campylobacter jejuni, Escherichia coli* O157, *Listeria monocytogenes, Bacillus cereus* and *Staphylococcus aureus* in vitro and in food systems. *J. Appl. Microbiol.* **2006**, *101*, 1232–1240. [CrossRef] [PubMed]
23. Cirmi, S.; Bisignano, C.; Mandalari, G.; Navarra, M. Anti-infective potential of *C. bergamia* Risso et Poiteau (bergamot) derivatives: A systematic review. *Phyther. Res.* **2016**, *1411*, 1404–1411. [CrossRef] [PubMed]
24. Oussalah, M.; Caillet, S.; Lacroix, M. Mechanism of action of Spanish oregano, Chinese cinnamon, and savory essential oils against cell membranes and walls of *Escherichia coli* $O_{157}:H_7$ and *Listeria monocytogenes*. *J. Food Prot.* **2006**, *69*, 1046–1055. [CrossRef]

25. Caputo, L.; Cornara, L.; Bazzicalupo, M.; De Francesco, C.; De Feo, V.; Trombetta, D.; Smeriglio, A. Chemical composition and biological activities of essential oils from peels of three Citrus species. *Molecules* **2020**, *25*, 1890. [CrossRef] [PubMed]
26. Nabiha, B.; Kachouri, F.; Herve, C. Chemical Composition of Bergamot (*C. bergamia* Risso) Essential oil chemical composition of bergamot *(C. bergamia* Risso) essential oil obtained by hydrodistillation. *J. Chem. Chem. Eng.* **2010**, *4*, 60–62.
27. Radulovic, N.S.; Blagojevic, P.D.; Stojanovic-Radic, Z.Z.; Stojanovic, N.M. Antimicrobial plant metabolites: Structural diversity and mechanism of action. *Curr. Med. Chem.* **2013**, *20*, 932–952.
28. Oussalah, M.; Caillet, S.; Salmiéri, S.; Saucier, L.; Lacroix, M. Antimicrobial effects of alginate-based films containing essential oils on *Listeria monocytogenes* and *Salmonella typhimurium* present in bologna and ham. *J. Food Prot.* **2007**, *70*, 901–908. [CrossRef]
29. Cebi, N.; Erarslan, A. Determination of the Chemical Fingerprint, Volatile Composition, Antibacterial and Antifungal Effects of Bergamot *Essential oil*. In Proceedings of the 6th International Academic Studies Congress, Online, 26–28 July 2021.

Disclaimer/Publisher's Note: The statements, opinions and data contained in all publications are solely those of the individual author(s) and contributor(s) and not of MDPI and/or the editor(s). MDPI and/or the editor(s) disclaim responsibility for any injury to people or property resulting from any ideas, methods, instructions or products referred to in the content.

Article

The Effect of Cow Breed and Wild Garlic Leaves (*Allium ursinum* L.) on the Sensory Quality, Volatile Compounds, and Physical Properties of Unripened Soft Rennet-Curd Cheese

Agnieszka Pluta-Kubica [1,*], Dorota Najgebauer-Lejko [1], Jacek Domagała [1], Jana Štefániková [2] and Jozef Golian [3]

[1] Department of Animal Product Processing, Faculty of Food Technology, University of Agriculture in Krakow, Balicka 122, 30-149 Krakow, Poland
[2] AgroBioTech Research Centre, Slovak University of Agriculture in Nitra, Tr. A. Hlinku 2, 949 76 Nitra, Slovakia
[3] Department of Food Hygiene and Safety, Faculty of Biotechnology and Food Sciences, Slovak University of Agriculture in Nitra, Tr. A. Hlinku 2, 949 76 Nitra, Slovakia
* Correspondence: agnieszka.pluta-kubica@urk.edu.pl; Tel.: +48-126624805

Abstract: The aim of this study was to investigate the effects of cow breed and the addition of wild garlic on the sensory quality, volatile compounds, and physical properties of soft rennet-curd cheese. Cheese was produced from the milk of the Polish Holstein-Friesian breed Black-and-White type and the Polish Red breed, with or without the addition of wild garlic leaves. The samples were analyzed for their sensory quality, volatile compounds (using an electronic nose and GC/MS), color, and texture. The intensity of taste and smell characteristics depended only on the addition of wild garlic. PCA showed that the differences in volatile profiles resulted both from the milk cow breed and the use of wild garlic. Breed influenced almost all color parameters, while the addition of wild garlic affected all of them. The milk source, wild garlic addition, and storage duration influenced the majority of the textural parameters of the cheeses. The research conducted indicates that the addition of wild garlic leaves results in the enrichment of the volatile compound profile of cheese, making its taste and smell less milky and sour ($p \leq 0.001$), while modifying its color and some textural properties ($p \leq 0.001$); while, at the same time, not adversely affecting the sensory assessment of the color, appearance, texture, smell, or taste of the cheese ($p > 0.05$).

Keywords: soft cheese; herbs; wild garlic; chemical composition; volatile organic compounds; flavor; color; texture

Citation: Pluta-Kubica, A.; Najgebauer-Lejko, D.; Domagała, J.; Štefániková, J.; Golian, J. The Effect of Cow Breed and Wild Garlic Leaves (*Allium ursinum* L.) on the Sensory Quality, Volatile Compounds, and Physical Properties of Unripened Soft Rennet-Curd Cheese. *Foods* 2022, 11, 3948. https://doi.org/10.3390/foods11243948

Academic Editor: Dippong Thomas

Received: 2 November 2022
Accepted: 3 December 2022
Published: 7 December 2022

Publisher's Note: MDPI stays neutral with regard to jurisdictional claims in published maps and institutional affiliations.

Copyright: © 2022 by the authors. Licensee MDPI, Basel, Switzerland. This article is an open access article distributed under the terms and conditions of the Creative Commons Attribution (CC BY) license (https://creativecommons.org/licenses/by/4.0/).

1. Introduction

Unripened soft rennet-curd cheese can be manufactured from the raw or pasteurized milk of various animal species: cows, sheep, and goats. Its production involves the utilization of mesophilic lactic acid bacteria (LAB) as a starter culture and rennet. Unlike mold- or smear-ripened soft rennet-curd cheese, it is consumed fresh and no maturation is required [1]. Herbs are often added to soft cheese, in order to enrich its flavor and increase variety. This is sometimes related to the traditions of the production region; e.g., traditional "Otlu cheese", in the eastern part of Turkey, is produced with *Allium* sp., *Ferula* sp., *Tymus* sp., *Prangos* sp., *Antriscus nemorosa*, *Chaerophyllum macropodum*, *Silene vulgaris*, and *Mentha* sp. [2]. Moreover, wild garlic is added to camembert cheese for barbecuing in the Czech Republic [3]. Fresh leaves of wild garlic can also be added as a spice to flavor cottage cheese [4]. Herbs contain many aroma compounds. Moreover, their presence in cheese increases the hydrolysis of fat, which causes the release of higher levels of free fatty acids [2]. Therefore, the addition of herbs changes the profile of volatile compounds, both directly and indirectly.

Wild garlic (*Allium ursinum* L.), also known as bear garlic, ramson, or broad-leaved garlic, is used in local cuisine in Eastern Europe, as well as in Poland, Germany, the Czech Republic, and Turkey. The leaves, bulbs, and flowers of wild garlic are edible; however, its leaves have the greatest consumer use. For example, in Turkey and Poland, wild garlic leaves are used in the manufacturing of local rennet-curd cheeses [3]. Fresh soft rennet-curd cow milk cheese with wild garlic has not previously been investigated. On the other hand, researching the influence of wild garlic leaves on the properties of herby-pickled (Otlu) cheese revealed their effect on the color, determined using the CIELAB system, as well as on sensory-assessed body and texture. Nevertheless, no influence on acceptability was found [2]. Moreover, wild garlic contains many sulfur compounds. Their hydrolysis gives rise to various volatile compounds, e.g., (poly)sulfides and thiosulfinates, responsible for the specific odor and taste of this herb [4].

Cheese properties can be affected by several factors: genetic, environmental, and/or technological. The main factor among genetic aspects is breed, which indirectly affects cheese quality through its influence on milk characteristics [5]. Moreover, environmental factors are connected with feeding systems. Grazed multifloral pastures, hay, and silage fed to animals influence the milk composition in different ways. Feeding affects the basic chemical composition and sensory properties of milk, as well as its volatile profile [6]. Milk origin was found to influence the sensory and textural properties of fromage frais type cheeses. The cheese, made from Polish Red breed (RP) milk, was characterized by a more pleasant smell and lower values of hardness and chewiness than that made from Polish Holstein-Friesian breed Black-and-White type (HF) milk [7]. On the other hand, breed did not affect the color indices or firmness of camembert cheese made from Holstein and Normande cows [8].

We hypothesized that the addition of wild garlic leaves could have a positive influence on the quality of soft cow milk rennet-curd cheese. Therefore, the aim of this study was to investigate the effects of cow milk source and the addition of wild garlic leaves on the sensory quality, volatile compound profile, and physical properties of soft rennet-curd cheese.

2. Materials and Methods

2.1. Materials

Milk for production of the soft rennet-curd cheese came from two sources. Cow milk from HF was obtained directly from a farm located in Dziekanowice near Krakow (Poland), while from RP it came from the Dairy Cooperative in Bochnia (Poland). HF cows were grazed on pastures located in Dziekanowice near Krakow (Lesser Poland Voivodeship, Poland), as well as fed on freshly mown vegetation from the pasture, GMO free fodders, and hay. RP cows were grazed on fresh mountain meadows near Bochnia in the mountain and sub-mountain areas on the border of the Beskid Wyspowy mountain range and Pogórze Wiśnickie (Lesser Poland Voivodeship, Poland), as well as fed GMO free fodders and hay. The raw bovine milk was obtained in one season, the summer (from July to September).

Cheese was produced under laboratory conditions and in two independent series at the Faculty of Food Technology, University of Agriculture in Krakow, Poland. The production process was the same as previously described by Pluta-Kubica et al. [1], with minor modifications. Briefly, the milk was standardized to 2.9% fat content; pasteurized at 72 °C for 15 s; cooled down to 32 °C; enriched with anhydrous calcium chloride (0.2 g/kg of the vat milk); inoculated with a mesophilic mixed strain starter culture containing *Lactococcus lactis* subsp. *cremoris*, *Lactococcus lactis* subsp. *lactis*, *Leuconostoc mesenteroides* subsp. *cremoris*, and *Lactococcus lactis* subsp. *diacetylactis* (CHN-19, Chr. Hansen, Hørsholm, Denmark); fermented at 32 °C for 1 h; and coagulated (32 °C, 2.5 h) using a microbial rennet with activity of 2200 IMCU/g (Fromase 2200TL, Specialties, Heerlen, Denmark). Afterwards, the curd was subjected to cutting and stirring (32 °C, 0.5 h). The curd was divided into two groups before molding. Half of it was utilized for natural cheese production (N), while the remaining part was gently mixed with wild garlic leaves (5 g/kg of the curd), to obtain herbal cheese (H). Cylindrical molds with a diameter of 8 cm were used. The cheeses were drained (28 °C, 1 h; 20 °C, 18 h), brined (16% NaCl, 16–18 °C, pH 5.1–5.2, 20 min), dripped (4–6 °C, 22 h), individually packed in plastic bags, and then stored for 2 weeks at 4 °C. An average of 18.4 L and 19.3 L of milk was used to produce the HF and RP cheese, respectively. On average, 1005.0 g and 1160.8 g of HF N and H cheese was obtained, respectively. Regarding RP cheese, 1479.4 g (N) and 1539.3 g (H) of cheese were produced. The weight of cheese was determined after dripping.

The wild garlic (*Allium ursinum* L.) used in this research consisted of whole leaves (the above-ground part). Fresh material (20 kg) was collected in a privately owned forest located in the village Ropa, Lesser Poland Voivodeship, in Poland, at an altitude of 600 m above sea level. The village is located in the West Beskid Mountains, a part of the West Outer Carpathians. The leaves were harvested before the plant flowered, in early April. After picking, the fresh material was packed into a cooled box and transferred to the laboratory within 2 h. The raw material was processed on the same day. As part of preparing the raw material, the leaves were washed in cold tap water. Surface water was removed by gentle centrifugation in a leafy-vegetable centrifuge and with ambient air blowing from a fan. Any leaves that were mechanically damaged or diseased during these actions were removed. Convection drying was carried out in a ProfiLine-type chamber dryer with airflow parallel to the sieves (Hendi, The Netherlands). The charge of the material was 2 kg per 1 m^2 of the screen. The drying temperature was 40 ± 1 °C, and the time was about 40 h, until the humidity reached 10%. Dried leaves were stored in a dry and cool place, away from light, in glass jars, and chopped using a sharp knife just before the production of cheese. The chopped leaves were 5 to 10 mm in length and width.

All reagents were of analytical grade and utilized as received, without further purification. Chemicals were produced by POCH S.A. (Gliwice, Poland), unless otherwise stated.

2.2. Chemical Composition, Acidity, and Water Activity Evaluation

The content of water was determined according to ISO 5534:2004 [9]. The total protein, sodium chloride, and ash contents in the cheese were analyzed according to AOAC [10]. The amount of fat was determined according to ISO 3433:2008 [11]. The acidity (pH) was electrometrically assessed using a pH-meter (CP-411, Elmetron, Poland). Water activity was measured according to ISO 18787:2017 [12] using a LabMaster-aw (Novasina AG, Lachen, Switzerland). The chemical composition of the cheese was evaluated on the day following brining. Additionally, the water content, pH, and water activity were estimated at the end of the storage period.

2.3. Sensory Quality Assessment

Sensory quality assessment was only performed on fresh samples (on the day following production). All cheese groups were evaluated by eight trained panelists (of age from 21 to 58) in two series (n = 64; two sources of milk × two kinds of cheese × two series of production × eight panelists). The participants were tested for their taste and smell detection thresholds, as well as ageusia and anosmia. They were familiar with the descriptive terms used and instructed about the process of evaluating the different sensory attributes. The analysis was conducted in a sensory laboratory equipped with six individual boxes. It took about 10 min per panelist. Potable water was available to the evaluators during the analysis.

The sensory evaluation of cheese samples was performed using two methods. First, a 5 point scale (1—"bad quality", 2—"insufficient quality", 3—"satisfactory quality", 4—"good quality", and 5—"very good quality") [13] was applied to evaluate the color, external and cross-sectional appearance, texture, taste, and smell. Definitions for five quality levels for each selected trait were established (Table S1). The following indices of significance: 0.15, 0.20, 0.15, 0.25, and 0.25, respectively, were adopted. Afterwards, the overall quality was calculated (the sum of the individual scores for the properties multiplied by the corresponding indices) [13].

Second, an assessment regarding the intensity (with boundary terms from 0—'imperceptible', to 5—'very intense') of taste and smell discriminants was performed using the profiling method of quantitative descriptive analysis [14]. The following discriminants of taste were taken into account: milky, sour, herbal, bitter, piquant, salty, and pleasant; and of smell: milky, sour, herbal, and pleasant. These qualitative features were established during a special session.

2.4. Volatile Compound Analysis

Analysis of volatile compounds was performed using an electronic nose (e-nose) and gas chromatography-mass spectrometry (GC/MS). The samples were frozen prior to analysis, stored at −20 °C and later thawed.

The electronic nose (Heracles II, Alpha M.O.S., Toulouse, France) method employed in this study has previously been described [15,16]. Briefly, 2.5 g of the cheese sample was dynamically incubated (250 rpm) in a 20 mL vial (in a thermostat block) at 50 °C for 15 min (Autosampler, Alpha M.O.S.). Afterwards, a volume totaling of 5 mL of the headspace gaseous compounds was withdrawn using a headspace autosampler syringe and dispensed into the e-nose injector for each analysis. The selected compounds with a discriminant >0.950 were identified by matching the measured peaks using Kovats retention indices with the NIST (National Institute of Standards and Technology) library (>50%), implementing Alpha Soft V14 software (Alpha M.O.S.). The analysis was repeated three times for each sample (n = 48; two sources of milk × two kinds of cheese × two storage periods × two series of production × threefold analysis).

The head-space solid-phase microextraction (HS–SPME) method, previously described [17], was used with some modifications. Briefly, 2.5 g of sample was dynamically incubated (250 rpm) with an SPME fiber (1 cm; 50/30 μm DVB/CAR/PDMS) (Supelco, Bellefonte, PA, USA) in a 20 mL vial in a thermostat block at 50 °C for 30 min (CombiPal automated sample injector 120, CTC Analytics AG, Zwingen, Switzerland). The initial conditioning of the fiber was performed by heating the sample to 270 °C for 1 h in a SPME Fiber Cleaning and Conditioning Station (placed in the CombiPal). SPME extracts were desorbed in a GC injector at 250 °C for 1 min. Post-desorption, the fiber was cleaned at 230 °C for 10 min using the SPME Fiber Cleaning and Conditioning Station. The relative volatile profile of cheese samples was determined by GC-MS (GC 7890B–MS 5977A; Agilent Technologies Inc., Santa Clara, CA, USA), equipped with a DB–WAXms column (30 m × 0.32 mm × 0.25 μm; Agilent Technologies Inc., Santa Clara, CA, USA) and operating with a previously reported temperature program and MS conditions [17]. The identification of compounds was carried out by comparing the mass spectra (over 80% match) with a commercial database NIST®2017 and the Wiley library. The relative content of the determined compounds was calculated by dividing the individual peak area by the total area of all peaks. Peaks under 1% were not counted. The analysis was repeated three times for each sample (n = 48; two sources of milk × two kinds of cheese × two storage periods × two series of production × threefold analysis).

2.5. Analysis of Physical Properties

The cheese color was determined as previously described [18], using a Konica Minolta CM-3500d spectrophotometer (Konica Minolta Sensing Inc., Osaka, Japan) in reflectance mode under the illuminant D65/10°. The following parameters in the CIE L* a* b* system were determined: L*—lightness (from 0—"black" to 100—"white"), a* coordinate—from greenness (negative values) to redness (positive values), and b* coordinate—from blueness (negative values) to yellowness (positive values). Additionally, h°—hue angle and C*—chroma (saturation) were calculated. All cheese samples were cut into four cubes with a side length of 2 cm and the color was determined twice (n = 128; two sources of milk × two kinds of cheese × two storage periods × two series of production × four cubes × two repetitions). Each sample's temperature was adjusted to 20 °C before conducting measurements.

Additionally, the total color difference value (ΔE) was calculated using the following equation (Equation (1)):

$$\Delta E = \sqrt{(\Delta L*)^2 + (\Delta a*)^2 + (\Delta b*)^2} \qquad (1)$$

where $\Delta L*$, $\Delta a*$, and $\Delta b*$ are the differences between the color value parameters of the compared cheeses. The interpretation of ΔE was as follows: $\Delta E < 1$ means that the color differences could not be perceived by the human eye, values of ΔE within the range of 1–3 mean that minor color differences could be detected by the human eye, while $\Delta E > 3$ means that color differences could be easily noticed by the human eye [19].

Instrumental texture profile analysis (TPA) was performed, as previously described [20], using a Universal Texture Analyser TA-XTPlus (Stable Micro Systems, Surrey, UK), controlled by a computer. Each cheese sample was cut into four cubes with a side length of 2 cm (n = 64; two sources of milk × two kinds of cheese × two storage periods × two series of production × four cubes). Their temperature was adjusted to approximately 20 °C. Compression at 60% deformation of the baseline sample height was carried out at a test speed of 1 mm/s. The test was conducted using a cylindrical compression plate 10 cm in diameter and 1 cm in height (P/100). All samples were compressed in two consecutive compression cycles. The obtained diagrams of the force dependence on time were analyzed using Texture Expert for Windows v. 1.05 (Stable Micro Systems, Surrey, UK). The Fracture TPA algorithm was applied, which allowed assignment of measures for the hardness, adhesiveness, springiness, cohesiveness, and chewiness of the cheeses.

2.6. Statistical Analysis

Analyses were performed in duplicate, unless otherwise stated. The obtained results, except for volatile compound determination, were statistically analyzed using Statistica version 13.3 (TIBCO Software Inc., Palo Alto, CA, USA). Means and standard deviations were calculated. Two-way ANOVA was applied for the results of the basic chemical composition and sensory quality assessment. The independent variables were milk source and wild garlic addition. Moreover, the results for water content, pH, water activity, color, and textural evaluation were subjected to three-way ANOVA. The third independent variable was storage duration. The null hypothesis was discarded for $p \leq 0.05$ in all statistical analyses.

The results obtained during analysis of volatiles, using an e-nose (compounds with a discriminant > 0.900 were selected), were subjected to PCA (principal component analysis, Alpha M.O.S.) using Alpha Soft V14 software (Alpha M.O.S.).

3. Results and Discussion

3.1. Chemical Composition, Acidity, and Water Activity of Cheese

The basic chemical composition of the fresh cheeses is given in Table 1. A significant interaction ($p \leq 0.05$) between the effects was only found for fat content. Milk source had a significant effect on fat ($p \leq 0.001$) and NaCl ($p \leq 0.05$) contents, while wild garlic addition affected the amount of protein ($p \leq 0.001$) and ash ($p \leq 0.05$). HF cheeses contained more fat and NaCl than RP ones. Nevertheless, the differences, although statistically significant, were somewhat small. Cheeses with wild garlic demonstrated higher protein and ash contents than cheeses without herbs. Likewise, a higher ash content due to wild garlic addition was previously reported in herby pickled cheese, and it was concluded that this herb is an important source of mineral matter [2].

Table 1. Basic chemical composition, acidity, and water activity of cheeses (mean ± SD).

Feature	Storage Time (Weeks)	Milk Source				p-Value		
		HF		RP				
		N	H	N	H	S	A	T
Water content (%)	0	53.5 ± 1.3	47.2 ± 1.9	58.2 ± 2.3	51.5 ± 3.0	NS	***	NS
	2	56.3 ± 1.3	47.1 ± 10.1	57.4 ± 1.8	48.5 ± 3.7			
pH	0	4.5 ± 0.2	4.6 ± 0.2	4.7 ± 0.1	4.5 ± 0.0	NS	NS	NS
	2	4.8 ± 0.6	4.4 ± 0.2	4.5 ± 0.1	4.6 ± 0.1			
Water activity	0	0.95 ± 0.03	0.94 ± 0.02	0.92 ± 0.00	0.93 ± 0.02	NS	NS	NS
	2	0.94 ± 0.02	0.94 ± 0.02	0.95 ± 0.03	0.95 ± 0.03			
Fat content (%)	0	24.8 ± 2.1	26.4 ± 2.5	24.1 ± 1.8	21.1 ± 0.9	***	NS	ne
Protein content (%)	0	18.8 ± 1.6	23.4 ± 1.9	18.5 ± 2.3	21.2 ± 1.5	NS	***	ne
Ash content (%)	0	2.32 ± 0.27	2.63 ± 0.31	2.23 ± 0.05	2.57 ± 0.23	NS	*	ne
NaCl content (%)	0	0.6 ± 0.2	0.7 ± 0.1	0.5 ± 0.1	0.4 ± 0.2	*	NS	ne

Abbreviations: HF—cheese from Polish Holstein-Friesian breed Black-and-White type milk; RP—cheese from Polish Red breed milk; 0—fresh samples; 2—samples stored for 2 weeks at 4 °C; N—cheese with no wild garlic addition; H—cheese with wild garlic; S—breed effect; A—wild garlic addition effect; T—storage duration effect; ne—not examined. * $p \leq 0.05$; *** $p \leq 0.001$; NS: $p > 0.05$. Significant interaction ($p \leq 0.05$): S×A for fat content. Values are expressed as the mean of four determinations (two series of production × two repetitions) ± standard deviation (SD).

The water content, acidity, and water activity of cheeses during storage is presented in Table 1. Neither milk source nor storage duration had any significant effects on the water content, pH, or water activity. The addition of wild garlic only had an influence on water content ($p \leq 0.001$), causing its decrease. No significant interactions were found between the effects.

The decrease in water content of the cheese with wild garlic was expected. In general, dried leaves of herbs are characterized by high dry matter contents. Thus, the incorporation of wild garlic caused an increase in the dry matter content of the cheeses examined in this study. Likewise, Gliguem et al. [21] reported the higher dry matter content of double cream cheese supplemented with *Allium roseum*, in comparison to the control. Moreover, mixing the cheese curd with wild garlic leaves, although gently performed, could have promoted whey release. Nevertheless, there was no whey release during refrigerated storage.

3.2. Sensory Quality of Cheese

In Table 2, the results of the sensory quality assessment are presented for the fresh cheeses on a five-point scale. No significant interactions were found between effects. Milk source only had significant effects on the smell and overall quality ($p \leq 0.05$). The smell was more desirable in the HF than in RP cheeses. Moreover, the HF cheese was characterized by a better overall quality than that made from RP milk. This was most likely a consequence of the significantly higher grades for smell appointed to the HF cheese by the panelists. The addition of wild garlic had no significant influence on the sensory features assessed using the five-point scale. Photos of the obtained cheeses are presented in Figure S1.

It is advantageous that the addition of wild garlic did not compromise the sensory quality of the cheese. Furthermore, cheese containing herbs can exhibit a rancid, too sour, too pungent, and/or unidentified bitter taste, causing low overall acceptability [22].

The discriminant intensity of the fresh cheese's taste and smell is demonstrated in Table 2. No significant interactions were found between effects. Milk source did not affect any of the features ($p > 0.05$). On the other hand the addition of wild garlic influenced a lot of the flavor characteristics. Obviously, a herbal taste and smell were imperceptible in the natural cheeses and intense in those with wild garlic ($p \leq 0.001$). Moreover, milky and sour tastes, as well as smell, were significantly more pronounced in the natural cheeses. Furthermore, cheeses containing wild garlic were characterized by a more intense piquant taste ($p \leq 0.001$).

Table 2. Sensory quality of cheeses (mean ± SD).

Feature	Milk Source				p-Value	
	HF		RP			
	N	H	N	H	S	A
Sensory quality of cheeses on a five-point scale						
Color	4.91 ± 0.20	4.81 ± 0.36	4.63 ± 0.62	4.66 ± 0.60	NS	NS
Appearance	4.56 ± 0.54	4.69 ± 0.48	4.41 ± 0.61	4.34 ± 0.57	NS	NS
Texture	4.53 ± 0.62	4.63 ± 0.43	4.38 ± 0.56	4.31 ± 0.44	NS	NS
Taste	4.59 ± 0.46	4.66 ± 0.44	4.53 ± 0.62	4.66 ± 0.44	NS	NS
Smell	4.75 ± 0.37	4.84 ± 0.35	4.41 ± 0.61	4.53 ± 0.59	*	NS
Overall quality	4.66 ± 0.27	4.73 ± 0.23	4.47 ± 0.43	4.51 ± 0.33	*	NS
Discriminant intensity of cheese taste and smell						
Milky taste	3.25 ± 1.39	2.47 ± 1.77	3.19 ± 1.11	2.06 ± 1.06	NS	***
Sour taste	2.81 ± 1.11	2.09 ± 1.10	3.25 ± 1.00	2.31 ± 0.95	NS	***
Herbal taste	0.00 ± 0.00	4.19 ± 0.68	0.00 ± 0.00	4.06 ± 0.93	NS	***
Bitter taste	0.47 ± 0.88	0.75 ± 0.68	0.63 ± 1.02	1.00 ± 1.03	NS	NS
Piquant taste	0.31 ± 0.60	2.22 ± 1.48	0.50 ± 0.89	1.94 ± 1.53	NS	***
Salty taste	2.94 ± 0.98	2.69 ± 0.95	2.56 ± 0.96	2.56 ± 0.89	NS	NS
Pleasant taste	4.44 ± 0.73	4.47 ± 0.56	4.25 ± 0.77	4.19 ± 0.66	NS	NS
Milky smell	3.25 ± 1.34	1.94 ± 1.69	3.31 ± 1.01	1.94 ± 1.18	NS	***
Sour smell	2.50 ± 1.21	2.16 ± 1.21	2.81 ± 1.38	1.81 ± 1.11	NS	*
Herbal smell	0.00 ± 0.00	4.41 ± 0.71	0.00 ± 0.00	4.25 ± 1.18	NS	***
Pleasant smell	4.53 ± 0.62	4.44 ± 0.81	4.19 ± 0.83	4.44 ± 0.73	NS	NS

Abbreviations: HF—cheese from Polish Holstein-Friesian breed Black-and-White type milk; RP—cheese from Polish Red breed milk; N—cheese with no wild garlic addition; H—cheese with wild garlic; S—breed effect; A—wild garlic addition effect. * $p \leq 0.05$; *** $p \leq 0.001$; NS: $p > 0.05$. No significant interactions were found between effects ($p > 0.05$). Values are expressed as the mean of sixteen determinations (two series of production × eight panelists) ± standard deviation (SD).

A milky and sour flavor is typical for fresh soft rennet-curd cheese. A lower intensity of these taste and smell discriminants in the H compared to N cheese could have been caused by the intense herbal flavor. Likewise a spicy hint was more pronounced in the double cream cheeses with *Allium roseum* in comparison to the plain one [21].

3.3. Volatile Compounds in Cheese

Fresh samples of RP cheese without the addition of wild garlic (RP0N) were not significantly different (positive PC1 and PC2 axis scores) compared to the stored RP2N samples (Figure 1). On the contrary, fresh samples made from RP milk with the addition of wild garlic (RP0H) (positive PC1 and PC2 axis scores) had significantly different aroma profiles than the samples stored at 4 °C for 2 weeks. In this case, RP cheese with wild garlic made in two production series had significantly different aroma profiles following storage. The samples from the first series had an aroma profile within the positive PC1 axis score and negative PC2 axis score. In contrast, the samples from the second series had an aroma profile within the negative PC1 and positive PC2 axis scores.

Fresh samples of HF cheese without the addition of wild garlic (positive PC1 and PC2 axis scores) were significantly different from the stored ones (negative PC1 and positive PC2 axis scores). Samples of HF cheeses containing wild garlic (HF0H) had similar aromatic profiles (positive PC1 axis score 68.2%), which were significantly different from the HF2H samples of stored HF cheeses (negative PC1 and PC2 axis scores).

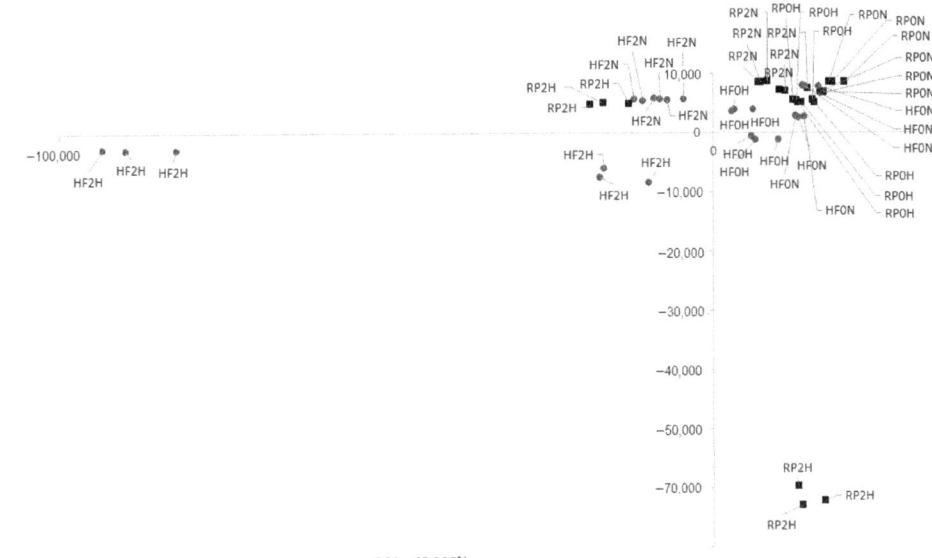

Figure 1. PCA of the cheese aromatic profile. Abbreviations: HF—cheese from Polish Holstein-Friesian breed Black-and-White type milk (gray); RP—cheese from Polish Red breed milk (black); 0—fresh samples; 2—samples stored for 2 weeks at 4 °C; N—cheese with no wild garlic addition; H—cheese with wild garlic.

The volatile compounds in fresh cheese samples determined using an electronic nose and HS–SPME GC/MS are listed with their sensory descriptors in Table 3. The sensory descriptors are from the AroChemBase database (Alpha M.O.S., Toulouse, France) or The Good Scents Company Information System [23]. The obtained results enabled the indication of volatile compounds, the presence of which was related to the addition of wild garlic. These were allyl (E)-1-propenyl disulfide ((E)-1-propenyl 2-propenyl disulfide), diallyl disulfide (di-2-propenyl disulfide), and allyl methyl disulfide (methyl 2-propenyl disulfide). These sulfur compounds were previously determined in essential oils isolated from *Allium ursinum* L. leaves [4]. Moreover, other compounds, such as 3-heptanol, 3-methyl butanal, isoamyl acetate, 3-methyl butanoic acid, propanoic acid, heptanoic acid, α-pinene, and 2,3-dimethyl pyrazine were only determined in cheese samples with the addition of wild garlic. However, these volatiles are common in many types of cheeses not containing any herbs [24–28].

Table 3. Volatile organic compounds in fresh cheese samples determined using an electronic nose with D > 0.9500 and via HS–SPME GC/MS.

	Compounds	Determination Method	Sensory Descriptors [1]	HF N	HF H	RP N	RP H
Alcohols	2-propanol	e-nose	alcoholic, ethereal	+	+	+	+
	2-methyl-propanol	e-nose	alcoholic, bitter, chemical, glue, leek, licorice, solvent	+	+	+	+
	2,3-butanediol	HS–SPME GC/MS	fruity, onion			+	
	3-heptanol	e-nose	green, herbaceous		+		+

Table 3. Cont.

	Compounds	Determination Method	Sensory Descriptors [1]	HF N	HF H	RP N	RP H
Aldehydes	acetaldehyde	e-nose	ethereal, fresh, fruity, pungent	+	+	+	+
	propanal	e-nose	ethereal, plastic, pungent, solvent	+	+	+	+
	2-methyl propanal	e-nose	burnt, fruity, green, malty, pungent, spicy, toasted	+	+	+	+
	3-methyl butanal	e-nose	almond, fruity, green, herbaceous, malty, toasted		+		
	benzaldehyde	HS–SPME GC/MS	almond, burnt sugar, fruity, woody	+		+	
Ketones	2,3-butanedione	e-nose and HS–SPME GC/MS	butter, caramelized, creamy, fruity, pineapple, spirit	+/+	+/+	+/+	+/+
	3-hydroxy-2-butanone	HS–SPME GC/MS	sweet buttery creamy, dairy, milky, fatty	+	+	+	+
Esters	ethyl acetate	e-nose	acidic, caramelized, fruity, pineapple, solvent, butter, ethereal, orange, pungent, sweet	+	+		+
	ethyl acrylate	e-nose	fruity	+	+	+	+
	ethyl isobutyrate	e-nose	fruity, rubber, strawberry, sweet		+	+	+
	ethyl propanoate	e-nose	acetone, fruity, solvent	+	+	+	+
	isoamyl acetate	e-nose	banana, fresh, fruity, pear, sweet		+		+
	α-terpineol acetate	HS–SPME GC/MS	-	+		+	
Free fatty acids	2-methyl propanoic acid	e-nose	acidic, butter, cheese, fatty, phenolic, rancid, sweaty	+	+	+	+
	3-methyl butanoic acid	e-nose	acidic, cheese, rancid, sweaty		+		+
	acetic acid	e-nose and HS–SPME GC/MS	acidic, pungent, sour, vinegar	+	+/+	+/+	+/+
	propanoic acid	e-nose	acidic, pungent, rancid, soy				+
	butanoic acid	e-nose and HS–SPME GC/MS	butter, cheese, rancid, sweaty	+/+	+/+	+/+	+/+
	hexanoic acid	HS–SPME GC/MS	cheese, fatty, goat, pungent, rancid, sweaty	+	+	+	−
	heptanoic acid	HS–SPME GC/MS	cheese, fatty, rancid, sour-sweat				+
	octanoic acid	HS–SPME GC/MS	cheese, fatty, fatty acid, fresh, mossy, sweaty	+	+	+	+
Sulfur compounds	allyl (E)-1-propenyl disulfide	HS–SPME GC/MS	sulfurous, alliaceous		+		+
	diallyl disulfide	HS–SPME GC/MS	alliaceous, onion, garlic, metallic		+		+
	allyl methyl disulfide	HS–SPME GC/MS	alliaceous, onion, garlic, green onion		+		+
Terpenes	D-limonene	HS–SPME GC/MS	citrus, fruity, minty, orange, peely	+		+	+
	α-pinene	e-nose	pine, terpenic		+		+

Table 3. Cont.

	Compounds	Determination Method	Sensory Descriptors [1]	HF N	HF H	RP N	RP H
Furans	dihydro-2,2-dimethyl-5-phenyl-3(2H)-furanone	HS–SPME GC/MS	-	+	+		+
Pyrazines	2,3-dimethyl pyrazine	e-nose	baked, cocoa, coffee, nutty, caramelized, meaty, peanut, butter		+		+
Hydrocarbons	6-methyl-octadecane	HS–SPME GC/MS	-	+		+	
Oxime	methoxy-phenyl-oxime	HS–SPME GC/MS	-	+	+	+	+

[1] Sensory descriptors are from the AroChemBase database (Alpha M.O.S., Toulouse, France) or The Good Scents Company Information System [23]. Abbreviations: HF—cheese from Polish Holstein-Friesian breed Black-and-White type milk; RP—cheese from Polish Red breed milk; N—cheese with no wild garlic addition; H—cheese with wild garlic. "+" means that the compound was detected.

Significant differences (PCA) in aromatic profiles were confirmed between the fresh N cheeses made from cow milk of the RP breed (PC1 axis positive score) and the HF breed (PC1 axis negative score) (Figure 2). These profiles were different regarding the presence of four compounds: 2,3-butanediol and ethyl isobutyrate were detected in RP but absent in the HF cheese; while ethyl acetate and dihydro-2,2-dimethyl-5-phenyl-3(2H)-furanone were identified in HF but were not present in the RP cheese (Table 3).

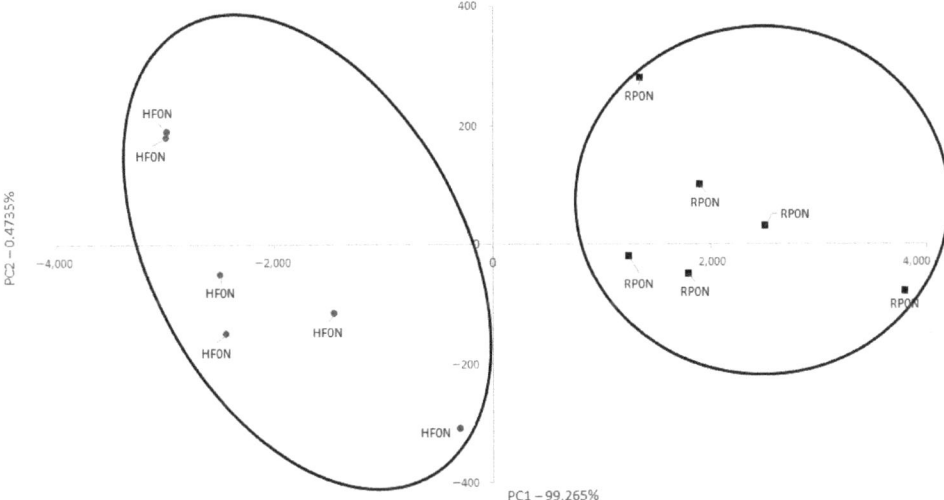

Figure 2. Aromatic profile PCA of fresh natural cheeses from Polish Holstein-Friesian breed Black-and-White type milk and from Polish Red breed milk. Abbreviations: HF—cheese from Polish Holstein-Friesian breed Black-and-White type milk (gray); RP—cheese from Polish Red breed milk (black); 0—fresh samples; N—cheeses with no wild garlic addition.

Both the source of cow milk and storage duration had a significant influence on the volatile compound profiles in N and H cheeses. The volatile pattern of milk used for cheese manufacturing could have been one of the possible reasons for the influence of the milk source. This depends on the animal breed, age, stage of lactation, and diet (which was different for the HF and RP cows, although the research was conducted during the same season) [29,30]. Moreover, storage duration typically influences the volatile profile in dairy products, such as e.g., fresh goat cheese and yoghurt [31].

3.4. Physical Properties of Cheese

The color parameters of the fresh and stored cheeses are presented in Table 4. Milk source influenced almost all of them ($p \leq 0.001$), with the exception of L*. The addition of wild garlic affected all parameters ($p \leq 0.001$). L* and a* values were stable during storage; however, the b* and C* values increased ($p \leq 0.001$), while the hue angle decreased ($p \leq 0.01$), after two weeks of refrigeration.

Table 4. Color and textural parameters of the cheeses (mean ± SD).

Feature	Storage Duration (Weeks)	Milk Source				p-Value		
		HF		RP		S	A	T
		N	H	N	H			
Color parameters								
L*	0	85.43 ± 3.04	74.10 ± 5.23	83.46 ± 2.59	73.92 ± 6.32	NS	***	NS
	2	83.68 ± 1.79	73.78 ± 7.53	84.38 ± 2.05	73.82 ± 5.17			
a*	0	−0.26 ± 0.45	−2.19 ± 0.67	0.63 ± 0.20	−1.06 ± 0.63	***	***	NS
	2	−0.09 ± 0.40	−1.79 ± 0.77	0.67 ± 0.15	−1.00 ± 0.44			
b*	0	11.78 ± 2.44	13.23 ± 2.38	14.71 ± 1.61	15.34 ± 1.66	***	***	***
	2	12.91 ± 1.59	14.59 ± 2.02	15.06 ± 1.53	15.99 ± 1.25			
h	0	91.72 ± 2.45	100.07 ± 4.35	87.64 ± 0.64	94.07 ± 2.53	***	***	**
	2	90.58 ± 1.86	97.38 ± 3.85	87.55 ± 0.54	93.62 ± 1.57			
C*	0	11.79 ± 2.42	13.44 ± 2.25	14.71 ± 1.61	15.39 ± 1.63	***	***	***
	2	12.91 ± 1.59	14.73 ± 1.94	15.07 ± 1.53	16.03 ± 1.23			
Textural parameters								
Hardness (kG)	0	2.18 ± 0.21	3.51 ± 0.50	3.06 ± 0.71	4.40 ± 0.58	***	***	NS
	2	2.14 ± 0.37	2.96 ± 0.47	3.07 ± 0.44	4.08 ± 0.39			
Adhesiveness (|kG × s|)	0	0.12 ± 0.05	0.07 ± 0.05	0.13 ± 0.08	0.12 ± 0.08	***	***	NS
	2	0.12 ± 0.03	0.11 ± 0.06	0.21 ± 0.10	0.13 ± 0.06			
Springiness (-)	0	0.79 ± 0.06	0.61 ± 0.04	0.65 ± 0.16	0.59 ± 0.12	NS	***	*
	2	0.64 ± 0.08	0.55 ± 0.05	0.71 ± 0.07	0.53 ± 0.05			
Cohesiveness (-)	0	0.23 ± 0.03	0.21 ± 0.02	0.24 ± 0.12	0.25 ± 0.12	NS	NS	***
	2	0.19 ± 0.02	0.17 ± 0.01	0.18 ± 0.01	0.16 ± 0.04			
Chewiness (kG)	0	0.40 ± 0.05	0.44 ± 0.09	0.44 ± 0.14	0.60 ± 0.13	***	*	***
	2	0.25 ± 0.07	0.28 ± 0.06	0.38 ± 0.07	0.34 ± 0.06			

Abbreviations: HF—cheese from Polish Holstein-Friesian breed Black-and-White type milk; RP—cheese from Polish Red breed milk; 0—fresh samples; 2—samples stored for 2 weeks at 4 °C; N—cheese with no wild garlic addition; H—cheese with wild garlic addition; S—breed effect; A—wild garlic addition effect; T—storage duration effect. * $p \leq 0.05$; ** $p \leq 0.01$; *** $p \leq 0.001$; NS: $p > 0.05$. Significant interactions ($p \leq 0.05$): S×T and S×A×T for springiness, A×T for chewiness. Values of color parameters are expressed as the mean of sixteen determinations (two series of production × four cubes × two repetitions) ± standard deviation (SD). Values of textural parameters are expressed as the mean of eight determinations (two series of production × four cubes) ± standard deviation (SD).

HF cheeses were more greenish and less yellowish than the RP ones. The red color coordinate prevailed over the green one in RP cheeses without wild garlic. Moreover, HF cheeses were characterized by a higher hue angle of approximately 90.58–100.07°, which is between the range of yellow (90°) and green (180°). Wild garlic addition triggered a lowering of L* values and made the cheeses more greenish and yellowish. Likewise, a decrease of L* and increase in the green color saturation were observed in a pickled cheese, due to herb addition [2].

RP cheeses had a higher color saturation intensity than HF ones, and the addition of wild garlic was associated with an increase in the value of this parameter. The chroma value also increased during the storage of cheeses. These differences were statistically significant; however, taking the general C* range from 0 to 60 into, they were somewhat small.

The calculation of ΔE demonstrated that the fresh N HF cheeses showed a clear color difference when compared to the fresh N RP samples ($\Delta E = 3.64$). Notwithstanding, the total color difference between the cheeses after storage was within the range of 1–3 ($\Delta E = 2.39$), meaning that only minor color differences could be detected by the human eye.

Similarly, the comparison between the fresh and stored HF cheeses (separately for N and H) revealed that the ΔE was between 1 and 3 (2.09 and 1.46, respectively). In comparison, storage did not trigger a noticeable color change in N and H RP cheeses (ΔE equaled 0.98 and 0.66, respectively).

The textural parameters of the fresh and stored cheeses are presented in Table 4. Milk source and wild garlic addition affected the hardness and adhesiveness ($p \leq 0.001$), as well as the chewiness ($p \leq 0.001$ and $p \leq 0.05$, respectively). Furthermore, the addition of herbs influenced the springiness of cheeses ($p \leq 0.001$). The hardness and adhesiveness did not change during the storage period; while the springiness, cohesiveness, and chewiness experienced a decrease ($p \leq 0.05$, $p \leq 0.001$ and $p \leq 0.001$, respectively).

HF cheeses were characterized by a lower hardness, adhesiveness, and chewiness than the RP ones. The addition of wild garlic resulted in an increase in hardness and chewiness. It also triggered a decrease in the adhesiveness and springiness.

A higher hardness noted in RP cheese, compared to HF, was probably caused by their lower fat content (Table 1). In general, fat reduction is related to an increase in hardness [32]. Moreover, the lower hardness of the N compared to H cheeses was most likely influenced by the water content, which was higher in the N cheeses (Table 1), making them softer. Likewise, the lower springiness value of the H cheeses in comparison to those from the N group was probably caused by the lower water content. H cheese samples crumbled during TPA. Differences in the chewiness values were the result of multiplying the hardness by springiness and by the cohesiveness [32].

4. Conclusions

To the best of our knowledge, fresh soft rennet-curd cow milk cheese with wild garlic leaves has not previously been investigated. The research conducted allowed demonstrating that the source of cow milk (cow breed), wild garlic addition, and storage duration significantly influenced various cheese characteristics. The addition of wild garlic leaves has a positive influence on the quality of soft cow milk rennet-curd cheese. It results in the enrichment of the volatile compound profile of cheese, making its taste and smell less milky and sour, while modifying its color and textural properties, which, at the same time, does not adversely affect the sensory assessment of the color, appearance, texture, smell, or taste of cheese with herbs. Therefore, wild garlic leaves can be recommended as an additive in the production of soft cow milk rennet-curd cheese, regardless of the cow milk source (HF or RP).

Supplementary Materials: The following are available online at https://www.mdpi.com/article/10.3390/foods11243948/s1, Figure S1. The photos of natural and herbal cheese during molding (A) and during dripping after brining (B), Table S1. The definitions of five quality levels for each selected trait in sensory analysis using a 5-point scale.

Author Contributions: Conceptualization, A.P.-K. and D.N.-L.; methodology, A.P.-K., D.N.-L. and J.Š.; validation, A.P.-K and D.N.-L.; formal analysis, A.P.-K.; investigation, A.P.-K., D.N.-L. and J.Š.; resources, A.P.-K. and D.N.-L.; writing—original draft preparation, A.P.-K.; writing—review and editing, D.N.-L., J.D., J.Š. and J.G.; visualization, A.P.-K.; supervision, A.P.-K.; funding acquisition, J.D. and J.G. All authors have read and agreed to the published version of the manuscript.

Funding: This work was financially supported by the Ministry of Science and Higher Education of the Republic of Poland and co-financed as a part of the project "Cultural heritage of small homelands" No. PPI/APM/2018/1/00010/U/001 funded by the Polish National Agency for Academic Exchange as a part of the International Academic Partnerships. This publication was supported by the Operational Program Integrated Infrastructure within the project: Demand-driven research for the sustainable and innovative food, Drive4SIFood 313011V336, co-financed by the European Regional Development Fund.

Data Availability Statement: The data used to support the findings of this study can be made available by the corresponding author upon request.

Acknowledgments: The authors would like to thank the Dairy Cooperative in Bochnia for supplying milk from Polish Red cows. Additionally, the authors would like to thank Emilia Bernaś for collecting and preparing the wild garlic utilized in this study. The authors would also like to extend their gratitude to Barbara Rusek and Joanna Sapeta for their assistance in the sensory analysis of the cheese.

Conflicts of Interest: The authors declare no conflict of interest. The funders had no role in the design of the study; in the collection, analysis, or interpretation of data; in the writing of the manuscript, or in the decision to publish the results.

References

1. Pluta-Kubica, A.; Jamróz, E.; Kawecka, A.; Juszczak, L.; Krzyściak, P. Active Edible Furcellaran/Whey Protein Films with Yerba Mate and White Tea Extracts: Preparation, Characterization and Its Application to Fresh Soft Rennet-Curd Cheese. *Int. J. Biol. Macromol.* **2020**, *155*, 1307–1316. [CrossRef] [PubMed]
2. Tarakci, Z.; Temiz, H.; Aykut, U.; Turhan, S. Influence of Wild Garlic on Color, Free Fatty Acids, and Chemical and Sensory Properties of Herby Pickled Cheese. *Int. J. Food Prop.* **2011**, *14*, 287–299. [CrossRef]
3. Gębczyński, P.; Bernaś, E.; Słupski, J. Usage of Wild-Growing Plants as Foodstuff. In *Cultural Heritage—Possibilities for Land-centered Societal Development*; Hernik, J., Walczycka, M., Sankowski, E., Harris, B.J., Eds.; Springer Nature: Cham, Switzerland, 2022; pp. 269–283. ISBN 9783030580919.
4. Sobolewska, D.; Podolak, I.; Makowska-Wąs, J. *Allium ursinum*: Botanical, Phytochemical and Pharmacological Overview. *Phytochem. Rev.* **2015**, *14*, 81–97. [CrossRef] [PubMed]
5. De Marchi, M.; Bittante, G.; Dal Zotto, R.; Dalvit, C.; Cassandro, M. Effect of Holstein Friesian and Brown Swiss Breeds on Quality of Milk and Cheese. *J. Dairy Sci.* **2008**, *91*, 4092–4102. [CrossRef]
6. Kalač, P. The Effects of Silage Feeding on Some Sensory and Health Attributes of Cow's Milk: A Review. *Food Chem.* **2011**, *125*, 307–317. [CrossRef]
7. Domagała, J.; Pluta-Kubica, A.; Sady, M.; Bonczar, G.; Duda, I.; Pustkowiak, H. Comparison of the Composition and Quality Properties of Fromage Frais-Type Cheese Manufactured from the Milk of Selected Cow Breeds. *Ann. Anim. Sci.* **2020**, *20*, 661–676. [CrossRef]
8. Hurtaud, C.; Peyraud, J.L.; Michel, G.; Berthelot, D.; Delaby, L. Winter Feeding Systems and Dairy Cow Breed Have an Impact on Milk Composition and Flavour of Two Protected Designation of Origin French Cheeses. *Animal* **2009**, *3*, 1327–1338. [CrossRef]
9. ISO 5534:2004; Cheese and Processed Cheese—Determination of the Total Solids Content. International Organization for Standardization: Geneva, Switzerland, 2004.
10. AOAC. *Official Methods of Analysis of AOAC International*, 18th ed.; Dairy Products: Arlington, VA, USA, 2007; Chapter 33.
11. ISO 3433:2008; Cheese–Determination of Fat Content–Van Gulik Method. International Organization for Standardization: Geneva, Switzerland, 2008; p. 3433.
12. ISO 18787:2017; Foodstuffs—Determination of Water Activity. International Organization for Standardization: Geneva, Switzerland, 2017.
13. Gawęcka, J.; Jędryka, T. Rozdział 5. Metody Punktowe. In *Analiza Sensoryczna: Wybrane Metody i Przykłady Zastosowań*; Wydawnictwo Akademii Ekonomicznej w Poznaniu: Poznań, Poland, 2001; pp. 57–76. ISBN 838876019X.
14. Baryłko-Pikielna, N.; Matuszewska, I. Rozdział 10. Metody Sensorycznej Analizy Opisowej. In *Sensoryczne Badania Żywności. Podstawy-Metody-Zastosowania*; Wydawnictwo Naukowe PTTŻ: Kraków, Poland, 2014; pp. 181–226. ISBN 978-83-935421-3-0.
15. Štefániková, J.; Ducková, V.; Miškeje, M.; Kačániová, M.; Canigová, M. The Impact of Different Factors on the Quality and Volatile Organic Compounds Profile in "Bryndza" Cheese. *Foods* **2020**, *9*, 1195. [CrossRef]
16. Štefániková, J.; Nagyová, V.; Hynšt, M.; Vietoris, V.; Martišová, P.; Nagyová, Ľ. Application of Electronic Nose for Determination of Slovak Cheese Authentication Based on Aroma Profile. *Potravin. Slovak J. Food Sci.* **2019**, *13*, 262–267. [CrossRef]
17. Sádecká, J.; Kolek, E.; Pangallo, D.; Valík, L.; Kuchta, T. Principal Volatile Odorants and Dynamics of Their Formation during the Production of May Bryndza Cheese. *Food Chem.* **2014**, *150*, 301–306. [CrossRef]
18. Najgebauer-Lejko, D.; Liszka, K.; Tabaszewska, M.; Domagała, J. Probiotic Yoghurts with Sea Buckthorn, Elderberry, and Sloe Fruit Purees. *Molecules* **2021**, *26*, 2345. [CrossRef] [PubMed]
19. Quintanilla, P.; Beltrán, M.C.; Molina, A.; Escriche, I.; Molina, M.P. Characteristics of Ripened Tronchón Cheese from Raw Goat Milk Containing Legally Admissible Amounts of Antibiotics. *J. Dairy Sci.* **2019**, *102*, 2941–2953. [CrossRef] [PubMed]
20. Filipczak-Fiutak, M.; Pluta-Kubica, A.; Domagała, J.; Duda, I.; Migdał, W. Nutritional Value and Organoleptic Assessment of Traditionally Smoked Cheeses Made from Goat, Sheep and Cow's Milk. *PLoS ONE* **2021**, *16*, e0254431. [CrossRef] [PubMed]
21. Gliguem, H.; Ben Hassine, D.; Ben Haj Said, L.; Ben Tekaya, I.; Rahmani, R.; Bellagha, S. Supplementation of Double Cream Cheese with Allium Roseum: Effects on Quality Improvement and Shelf-Life Extension. *Foods* **2021**, *10*, 1276. [CrossRef]
22. Güler, Z. Profiles of Organic Acid and Volatile Compounds in Acid-Type Cheeses Containing Herbs and Spices (Surk Cheese). *Int. J. Food Prop.* **2014**, *17*, 1379–1392. [CrossRef]
23. The Good Scents Company Information System [WWW Document], n.d. Available online: Http://Www.Thegoodscentscompany.Com/ (accessed on 14 March 2022).

24. Pluta-Kubica, A.; Domagała, J.; Gąsior, R.; Wojtycza, K.; Witczak, M. Characterisation of the Profile of Volatiles of Polish Emmental Cheese. *Int. Dairy J.* **2021**, *116*, 104954. [CrossRef]
25. Pillonel, L.; Ampuero, S.; Tabacchi, R.; Bosset, J.O. Analytical Methods for the Determination of the Geographic Origin of Emmental Cheese: Volatile Compounds by GC/MS-FID and Electronic Nose. *Eur. Food Res. Technol.* **2003**, *216*, 179–183. [CrossRef]
26. Frank, D.C.; Owen, C.M.; Patterson, J. Solid Phase Microextraction (SPME) Combined with Gas-Chromatography and Olfactometry-Mass Spectrometry for Characterization of Cheese Aroma Compounds. *LWT-Food Sci. Technol.* **2004**, *37*, 139–154. [CrossRef]
27. Marilley, L.; Casey, M.G. Flavours of Cheese Products: Metabolic Pathways, Analytical Tools and Identification of Producing Strains. *Int. J. Food Microbiol.* **2004**, *90*, 139–159. [CrossRef]
28. Endrizzi, I.; Fabris, A.; Biasioli, F.; Aprea, E.; Franciosi, E.; Poznanski, E.; Cavazza, A.; Gasperi, F. The Effect of Milk Collection and Storage Conditions on the Final Quality of Trentingrana Cheese: Sensory and Instrumental Evaluation. *Int. Dairy J.* **2012**, *23*, 105–114. [CrossRef]
29. Foroutan, A.; Guo, A.C.; Vazquez-Fresno, R.; Lipfert, M.; Zhang, L.; Zheng, J.; Badran, H.; Budinski, Z.; Mandal, R.; Ametaj, B.N.; et al. Chemical Composition of Commercial Cow's Milk. *J. Agric. Food Chem.* **2019**, *67*, 4897–4914. [CrossRef] [PubMed]
30. Mordenti, A.L.; Brogna, N.; Formigoni, A. REVIEW: The Link between Feeding Dairy Cows and Parmigiano-Reggiano Cheese Production Area. *Prof. Anim. Sci.* **2017**, *33*, 520–529. [CrossRef]
31. Condurso, C.; Verzera, A.; Romeo, V.; Ziino, M.; Conte, F. Solid-Phase Microextraction and Gas Chromatography Mass Spectrometry Analysis of Dairy Product Volatiles for the Determination of Shelf-Life. *Int. Dairy J.* **2008**, *18*, 819–825. [CrossRef]
32. Henneberry, S.; Wilkinson, M.G.; Kilcawley, K.N.; Kelly, P.M.; Guinee, T.P. Interactive Effects of Salt and Fat Reduction on Composition, Rheology and Functional Properties of Mozzarella-Style Cheese. *Dairy Sci. Technol.* **2015**, *95*, 613–638. [CrossRef]

MDPI
St. Alban-Anlage 66
4052 Basel
Switzerland
www.mdpi.com

Foods Editorial Office
E-mail: foods@mdpi.com
www.mdpi.com/journal/foods

Disclaimer/Publisher's Note: The statements, opinions and data contained in all publications are solely those of the individual author(s) and contributor(s) and not of MDPI and/or the editor(s). MDPI and/or the editor(s) disclaim responsibility for any injury to people or property resulting from any ideas, methods, instructions or products referred to in the content.

www.ingramcontent.com/pod-product-compliance
Lightning Source LLC
LaVergne TN
LVHW070654100526
838202LV00013B/957